Stochastic Programming

Wiley-Interscience Series In Systems And Optimization

GITTINS—Multi-armed Bandit Allocation Indices

KALL/WALLACE—Stochastic Programming

KAMP/HASLER—Recursive Neural Networks for Associative Memory

VAN DIJK—Queueing Networks and Product Forms: A Systems Approach

WHITTLE—Risk-sensitive Optimal Control

Stochastic Programming

Peter Kall
Institute for Operations Research and Mathematical Methods of Economics
University of Zurich
Switzerland

Stein W. Wallace
Department of Managerial Economics and Operations Research
Norwegian Institute of Technology
University of Trondheim
Norway

JOHN WILEY & SONS
Chichester · New York · Brisbane · Toronto · Singapore

Reprinted July 1995

Other Wiley Editorial Offices

John Wiley & Sons, Inc., 605 Third Avenue,
New York, NY 10158-0012, USA

Jacaranda Wiley Ltd, 33 Park Road, Milton,
Queensland 4064, Australia

John Wiley & Sons (Canada) Ltd, 22 Worcester Road,
Rexdale, Ontario M9W 1L1, Canada

John Wiley & Sons (SEA) Pte Ltd, 37 Jalan Pemimpin #05-04,
Block B, Union Industrial Building, Singapore 2057

British Library Cataloguing in Publication Data

A catalogue record for this book is available from the British Library

ISBN 0 471 95108 0; 0 471 95158 7 (pbk)

Produced from camera-ready copy supplied by the authors using LaTeX
Printed and bound in Great Britain by Biddles Ltd, Guildford and King's Lynn

Contents

Preface

Over the last few years, both of the authors, and also most others in the field of stochastic programming, have said that what we need more than anything just now is a basic textbook—a textbook that makes the area available not only to mathematicians, but also to students and other interested parties who cannot or will not try to approach the field via the journals. We also felt the need to provide an appropriate text for instructors who want to include the subject in their curriculum. It is probably not possible to write such a book without assuming some knowledge of mathematics, but it has been our clear goal to avoid writing a text readable only for mathematicians. We want the book to be accessible to any quantitatively minded student in business, economics, computer science and engineering, plus, of course, mathematics.

So what do we mean by a quantitatively minded student? We assume that the reader of this book has had a basic course in calculus, linear algebra and probability. Although most readers will have a background in linear programming (which replaces the need for a specific course in linear algebra), we provide an outline of all the theory we need from linear and nonlinear programming. We have chosen to put this material into Chapter 1, so that the reader who is familiar with the theory can drop it, and the reader who knows the material, but wonders about the exact definition of some term, or who is slightly unfamiliar with our terminology, can easily check how we see things. We hope that instructors will find enough material in Chapter 1 to cover specific topics that may have been omitted in the standard book on optimization used in their institution. By putting this material directly into the running text, we have made the book more readable for those with the minimal background. But, at the same time, we have found it best to separate what is new in this book—stochastic programming—from more standard material of linear and nonlinear programming.

Despite this clear goal concerning the level of mathematics, we must admit that when treating some of the subjects, like probabilistic constraints (Section 1.5 and Chapter 4), or particular solution methods for stochastic programs, like stochastic decomposition (Section 3.8) or quasi-gradient methods (Section 3.9), we have had to use a slightly more advanced language in probability. Although the actual information found in those parts of the book is made simple, some terminology may here and there not belong to the basic probability terminology. Hence, for these parts, the instructor must either provide some basic background in terminology, or the reader should at least consult carefully Section 1.3.1, where we have tried to put together those terms and concepts from probability theory used later in this text.

Within the mathematical programming community, it is common to split the field into topics such as linear programming, nonlinear programming, network flows, integer and combinatorial optimization, and, finally, stochastic programming. Convenient as that may be, it is conceptually inappropriate. It puts forward the idea that stochastic programming is distinct from integer programming the same way that linear programming is distinct from nonlinear programming. The counterpart of stochastic programming is, of course, deterministic programming. We have stochastic and deterministic linear programming, deterministic and stochastic network flow problems, and so on. Although this book mostly covers stochastic linear programming (since that is the best developed topic), we also discuss stochastic nonlinear programming, integer programming and network flows.

Since we have let subject areas guide the organization of the book, the chapters are of rather different lengths. Chapter 1 starts out with a simple example that introduces many of the concepts to be used later on. Tempting as it may be, we strongly discourage skipping these introductory parts. If these parts are skipped, stochastic programming will come forward as merely an algorithmic and mathematical subject, which will serve to limit the usefulness of the field. In addition to the algorithmic and mathematical facets of the field, stochastic programming also involves model creation and specification of solution characteristics. All instructors know that modelling is harder to teach than are methods. We are sorry to admit that this difficulty persists in this text as well. That is, we do not provide an in-depth discussion of modelling stochastic programs. The text is not free from discussions of models and modelling, however, and it is our strong belief that a course based on this text is better (and also easier to teach and motivate) when modelling issues are included in the course.

Chapter 1 contains a formal approach to stochastic programming, with a discussion of different problem classes and their characteristics. The chapter ends with linear and nonlinear programming theory that weighs heavily in stochastic programming. The reader will probably get the feeling that the parts concerned with chance-constrained programming are mathematically more complicated than some parts discussing recourse models. There is a good reason for that: whereas recourse models transform the randomness contained in a stochastic program into one special parameter of some random vector's distribution, namely its expectation, chance constrained models deal more explicitly with the distribution itself. Hence the latter models may be more difficult, but at the same time they also exhaust more of the information contained in the probability distribution. However, with respect to applications, there is no generally valid justification to state that any one of the two basic model types is "better" or "more relevant". As a matter of fact, we know of applications for which the recourse model is very appropriate and of others for which chance constraints have to be modelled, and even applications are known for which recourse terms for one part of the stochastic constraints and chance constraints for another part were designed. Hence, in a first reading or an introductory course, one or the other proof appearing too complicated can certainly be skipped without harm. However, to get a valid picture about stochastic programming, the statements about basic properties of both model types as well as the ideas underlying the various solution approaches should be noticed. Although the basic linear and nonlinear programming is put together in one specific part of the book, the instructor or the reader should pick up the subjects as they are needed for the understanding of the other chapters. That way, it will be easier to pick out exactly those parts of the theory that the students or readers do not know already.

Chapter 2 starts out with a discussion of the Bellman principle for solving dynamic problems, and then discusses decision trees and dynamic programming in both deterministic and stochastic settings. There then follows a discussion of the rather new approach of scenario aggregation. We conclude the chapter with a discussion of the value of using stochastic models.

Chapter 3 covers recourse problems. We first discuss some topics from Chapter 1 in more detail. Then we consider decomposition procedures especially designed for stochastic programs with recourse. We next turn to the questions of bounds and approximations, outlining some major ideas and indicating the direction for other approaches. The special case of simple recourse is then explained, before we show how decomposition procedures for stochastic programs fit into the framework of branch-and-cut procedures for integer programs. This makes it possible to develop an approach for stochastic integer programs. We conclude the chapter with a discussion of Monte-Carlo based methods, in particular stochastic decomposition and quasi-gradient methods.

Chapter 4 is devoted to probabilistic constraints. Based on convexity statements provided in Section 1.5, one particular solution method is described for the case of joint chance constraints with a multivariate normal distribution of the right-hand side. For separate probabilistic constraints with a joint normal distribution of the coefficients, we show how the problem can be transformed into a deterministic convex nonlinear program. Finally, we address a problem very relevant in dealing with chance constraints: the problem of how to construct efficiently lower and upper bounds for a multivariate distribution function, and give a first sketch of the ideas used in this area.

Preprocessing is the subject of Chapter 5. "Preprocessing" is any analysis that is carried out before the actual solution procedure is called. Preprocessing can be useful for simplifying calculations, but the main purpose is to facilitate a tool for model evaluation.

We conclude the book with a closer look at networks (Chapter 6). Since these are nothing else than specially structured linear programs, we can draw freely from the topics in Chapter 3. However, the added structure of networks allows many simplifications. We discuss feasibility, preprocessing and bounds. We conclude the chapter with a closer look at PERT networks.

Each chapter ends with a short discussion of where more literature can be found, some exercises, and, finally, a list of references.

Writing this book has been both interesting and difficult. Since it is the first basic textbook totally devoted to stochastic programming, we both enjoyed and suffered from the fact that there is, so far, no experience to suggest how such a book should be constructed. Are the chapters in the correct order? Is the level of difficulty even throughout the book? Have we really captured the basics of the field? In all cases the answer is probably NO. Therefore, dear reader, we appreciate all comments you may have, be they regarding misprints, plain errors, or simply good ideas about how this should have been done. And also, if you produce suitable exercises, we shall be very happy to receive them, and if this book ever gets revised, we shall certainly add them, and allude to the contributor.

About 50% of this text served as a basis for a course in stochastic programming at The Norwegian Institute of Technology in the fall of 1992. We wish to thank the students for putting up with a very preliminary text, and for finding such an astonishing number of errors and misprints. Last but not least, we owe sincere thanks to Julia Higle (University of Arizona, Tucson), Diethard Klatte (Univerity of Zurich), Janos Mayer (University of Zurich) and Pavel Popela (Technical University of Brno) who have read the manuscript[1] very carefully and fixed not only linguistic bugs but prevented us from quite a number of crucial mistakes. Finally we highly appreciate the good cooperation and very helpful comments provided by our publisher. The remaining errors are obviously the sole responsibility of the authors.

Zurich and Trondheim, February 1994 P. K. and S.W.W.

[1] Written in LaTeX

1

Basic Concepts

1.1 Preliminaries

Many practical decision problems—in particular, rather complex ones—can be modelled as *linear programs*

$$\left.\begin{array}{l} \min\{c_1x_1 + c_2x_2 + \cdots + c_nx_n\} \\ \text{subject to} \\ \begin{array}{rcl} a_{11}x_1 + \ a_{12}x_2 + \cdots + \ a_{1n}x_n & = & b_1 \\ a_{21}x_1 + \ a_{22}x_2 + \cdots + \ a_{2n}x_n & = & b_2 \\ \vdots & & \vdots \\ a_{m1}x_1 + a_{m2}x_2 + \cdots + a_{mn}x_n & = & b_m \\ x_1, x_2, \cdots, x_n & \geq & 0. \end{array} \end{array}\right\} \tag{1.1}$$

Using matrix–vector notation, the shorthand formulation of problem (1.1) would read as

$$\left.\begin{array}{r} \min c^{\mathrm{T}}x \\ \text{s.t. } Ax = b \\ x \geq 0. \end{array}\right\} \tag{1.2}$$

Typical applications may be found in the areas of industrial production, transportation, agriculture, energy, ecology, engineering, and many others. In problem (1.1) the coefficients c_j (e.g. factor prices), a_{ij} (e.g. productivities) and b_i (e.g. demands or capacities) are assumed to have fixed known real values and we are left with the task of finding an optimal combination of the values for the decision variables x_j (e.g. factor inputs, activity levels or energy flows) that have to satisfy the given constraints. Obviously, model (1.1) can only provide a reasonable representation of a real life problem when the functions involved (e.g. cost functions or production functions) are fairly linear in the decision variables. If this condition is substantially violated—for example, because of increasing marginal costs or decreasing marginal returns of production—we should use a more general form to model our problem:

$$\left.\begin{array}{l}\min g_0(x)\\ \text{s.t. } g_i(x) \le 0,\ i = 1,\cdots,m\\ \qquad x \in X \subset \mathbb{R}^n.\end{array}\right\} \tag{1.3}$$

The form presented in (1.3) is known as a *mathematical programming* problem. Here it is understood that the set X as well as the functions $g_i : \mathbb{R}^n \to \mathbb{R}, i = 0,\cdots,m$, are given by the modelling process.

Depending on the properties of the problem defining functions g_i and the set X, program (1.3) is called

(a) *linear*, if the set X is convex polyhedral and the functions $g_i,\ i = 0,\cdots,m$, are linear;

(b) *nonlinear*, if at least one of the functions $g_i,\ i = 0,\cdots,m$, is nonlinear or X is not a convex polyhedral set; among nonlinear programs, we denote a program as

   (b1) *convex*, if $X \cap \{x \mid g_i(x) \le 0,\ i = 1,\cdots,m\}$ is a convex set and g_0 is a convex function (in particular if the functions $g_i,\ i = 0,\cdots,m$ are convex and X is a convex set); and

   (b2) *nonconvex*, if either $X \cap \{x \mid g_i(x) \le 0,\ i = 1,\cdots,m\}$ is not a convex set or the objective function g_0 is not convex.

Case (b2) above is also referred to as *global optimization*. Another special class of problems, called *(mixed) integer programs*, arises if the set X requires (at least some of) the variables $x_j,\ j = 1,\cdots,n$, to take integer values only. We shall deal only briefly with discrete (i.e. mixed integer) problems, and there is a natural interest in avoiding nonconvex programs whenever possible for a very simple reason revealed by the following example from elementary calculus.

Example 1.1 Consider the optimization problem

$$\min_{x\in\mathbb{R}} \varphi(x), \tag{1.4}$$

where $\varphi(x) := \frac{1}{4}x^4 - 5x^3 + 27x^2 - 40x$. A necessary condition for solving problem (1.4) is

$$\varphi'(x) = x^3 - 15x^2 + 54x - 40 = 0.$$

Observing that

$$\varphi'(x) = (x-1)(x-4)(x-10),$$

we see that $x_1 = 1$, $x_2 = 4$ and $x_3 = 10$ are candidates to solve our problem. Moreover, evaluating the second derivative $\varphi''(x) = 3x^2 - 30x + 54$, we get

$$\begin{aligned}\varphi''(x_1) &= 27,\\ \varphi''(x_2) &= -18,\\ \varphi''(x_3) &= 54,\end{aligned}$$

indicating that x_1 and x_3 yield a *relative* minimum whereas in x_2 we find a *relative* maximum. However, evaluating the two relative minima yields $\varphi(x_1) = -17.75$ and $\varphi(x_3) = -200$. Hence, solving our little problem (1.4) with a numerical procedure that intends to satisfy the first- and second-order conditions for a minimum, we might (depending on the starting point of the procedure) end up with x_1 as a "solution" without realizing that there exists a (much) better possibility. □

As usual, a function ψ is said to attain a *relative* minimum—also called a *local* minimum—at some point $\hat{x}$ if there is a neighbourhood U of $\hat{x}$ (e.g. a ball with center $\hat{x}$ and radius $\varepsilon > 0$) such that $\psi(\hat{x}) \leq \psi(y)\ \forall y \in U$. A minimum $\psi(\bar{x})$ is called *global* if $\psi(\bar{x}) \leq \psi(z)\ \forall z$. As we just saw, a local minimum $\psi(\hat{x})$ need not be a global minimum.

A situation as in the above example cannot occur with convex programs because of the following.

Lemma 1.1 *If problem (1.3) is a convex program then any local (i.e. relative) minimum is a global minimum.*

Proof If $\bar{x}$ is a local minimum of problem (1.3) then $\bar{x}$ belongs to the feasible set $\mathcal{B} := X \cap \{x \mid g_i(x) \leq 0, i = 1, \cdots, m\}$. Further, there is an $\varepsilon_0 > 0$ such that for any ball $K_\varepsilon := \{x \mid \|x - \bar{x}\| \leq \varepsilon\}$, $0 < \varepsilon < \varepsilon_0$, we have that $g_0(\bar{x}) \leq g_0(x)\ \forall x \in K_\varepsilon \cap \mathcal{B}$. Choosing an arbitrary $y \in \mathcal{B}$, $y \neq \bar{x}$, we may choose an $\varepsilon > 0$ such that $\varepsilon < \|y - \bar{x}\|$ and $\varepsilon < \varepsilon_0$. Finally, since, from our assumption, $\mathcal{B}$ is a convex set and the objective g_0 is a convex function, the line segment $\overline{\bar{x}y}$ intersects the surface of the ball K_ε in a point $\hat{x}$ such that $\hat{x} = \alpha\bar{x} + (1 - \alpha)y$ for some $\alpha \in (0, 1)$, yielding $g_0(\bar{x}) \leq g_0(\hat{x}) \leq \alpha g_0(\bar{x}) + (1 - \alpha)g_0(y)$, which implies that $g_0(\bar{x}) \leq g_0(y)$. □

During the last four decades, progress in computational methods for solving mathematical programs has been impressive, and problems of considerable size may be solved efficiently, and with high reliability.

In many modelling situations it is unreasonable to assume that the coefficients c_j, a_{ij}, b_i or the functions g_i (and the set X) respectively in problems (1.1) and (1.3) are deterministically fixed. For instance, future productivities in a production problem, inflows into a reservoir connected to a hydro power station, demands at various nodes in a transportation network, and so on, are often appropriately modelled as uncertain parameters, which are at best characterized by probability distributions. The uncertainty about the realized values of those parameters cannot always be wiped out just by inserting their mean values or some other (fixed) estimates during the modelling process. That is, depending on the practical situation under consideration, problems (1.1) or (1.3) may not be the appropriate models for describing the problem we want to solve. In this chapter we emphasize—

and possibly clarify—the need to broaden the scope of modelling real life decision problems. Furthermore, we shall provide from *linear programming* and *nonlinear programming* the essential ingredients absolutely necessary for the understanding of the subsequent chapters. Obviously these latter sections may be skipped—or used as a quick revision—by readers who are already familiar with the related optimization courses.

Before coming to a more general setting we first derive some typical *stochastic programming* models, using a simplified production problem to illustrate the various model types.

1.2 An Illustrative Example

Let us consider the following problem, idealized for the purpose of easy presentation. From two raw materials, $raw1$ and $raw2$, we may simultaneously produce two different goods, $prod1$ and $prod2$ (as may happen for example in a refinery). The output of products per unit of the raw materials as well as the unit costs of the raw materials $c = (c_{raw1}, c_{raw2})^{\mathrm{T}}$ (yielding the production cost γ), the demands for the products $h = (h_{prod1}, h_{prod2})^{\mathrm{T}}$ and the production capacity $\hat{b}$, i.e. the maximal total amount of raw materials that can be processed, are given in Table 1.

According to this formulation of our production problem, we have to deal with the following linear program:

Table 1 Productivities $\pi(raw\,\mathrm{i}, prod\,\mathrm{j})$.

	Products			
Raws	$prod1$	$prod2$	c	$\hat{b}$
$raw1$	2	3	2	1
$raw2$	6	3	3	1
relation	$\geq$	$\geq$	$=$	$\leq$
h	180	162	γ	100

$$\left.\begin{array}{rl}\min(2x_{raw1} + 3x_{raw2}) & \\ \text{s.t. } x_{raw1} + x_{raw2} & \leq 100, \\ 2x_{raw1} + 6x_{raw2} & \geq 180, \\ 3x_{raw1} + 3x_{raw2} & \geq 162, \\ x_{raw1} \phantom{+ 3x_{raw2}} & \geq 0, \\ x_{raw2} & \geq 0.\end{array}\right\} \tag{2.1}$$

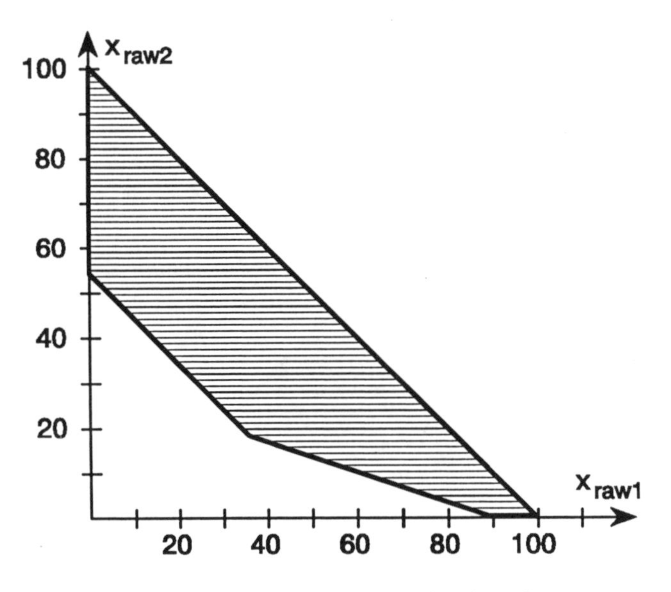

Figure 1 Deterministic LP: set of feasible production plans.

Due to the simplicity of the example problem, we can give a graphical representation of the set of feasible production plans (Figure 1).

Given the cost function $\gamma(x) = 2x_{raw1} + 3x_{raw2}$ we easily conclude (Figure 2) that

$$\hat{x}_{raw1} = 36, \hat{x}_{raw2} = 18, \gamma(\hat{x}) = 126 \tag{2.2}$$

is the unique optimal solution to our problem.

Our production problem is properly described by (2.1) and solved by (2.2) *provided* the productivities, the unit costs, the demands and the capacity (Table 1) are fixed data and known to us prior to making our decision on the production plan. However, this is obviously not always a realistic assumption. It may happen that at least some of the data—productivities and demands for instance—can vary within certain limits (for our discussion, randomly) and that we have to make our decision on the production plan before knowing the exact values of those data.

To be more specific, let us assume that

- our model describes the weekly production process of a refinery relying on two countries for the supply of crude oil (*raw*1 and *raw*2, respectively), supplying one big company with gasoline (*prod*1) for its distribution system of gas stations and another with fuel oil (*prod*2) for its heating and/or power plants;
- it is known that the productivities $\pi(raw1, prod1)$ and $\pi(raw2, prod2)$, i.e.

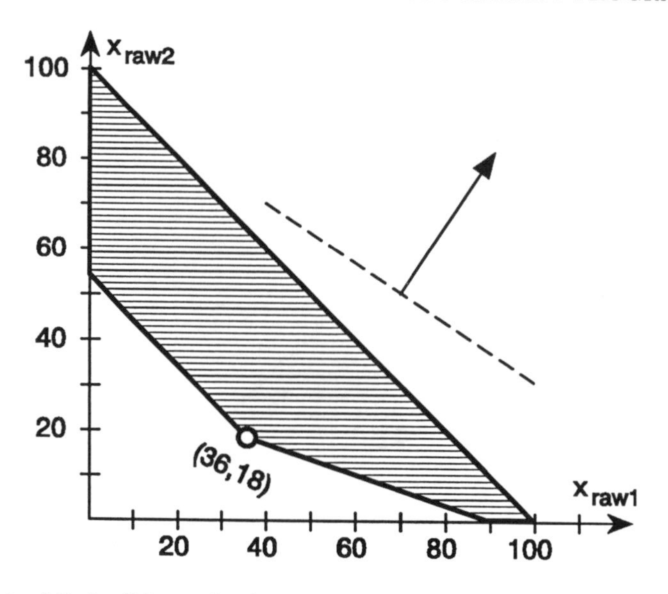

Figure 2 LP: feasible production plans and cost function for $\gamma = 290$.

the output of gas from $raw1$ and the output of fuel from $raw2$ may change randomly (whereas the other productivities are deterministic);

- simultaneously, the weekly demands of the clients, h_{prod1} for gas and h_{prod2} for fuel are varying randomly;
- the weekly production plan (x_{raw1}, x_{raw2}) has to be fixed in advance and cannot be changed during the week, whereas
- the actual productivities are only observed (measured) during the production process itself, and
- the clients expect their actual demand to be satisfied during the corresponding week.

Assume that, owing to statistics, we know that

$$\left.\begin{aligned} h_{prod1} &= 180 + \tilde{\zeta}_1, \\ h_{prod2} &= 162 + \tilde{\zeta}_2, \\ \pi(raw1, prod1) &= 2 + \tilde{\eta}_1, \\ \pi(raw2, prod2) &= 3.4 - \tilde{\eta}_2, \end{aligned}\right\} \tag{2.3}$$

where the random variables $\tilde{\zeta}_j$ are modelled using normal distributions, and $\tilde{\eta}_1$ and $\tilde{\eta}_2$ are distributed uniformly and exponentially respectively, with the following parameters:[1]

$$\left.\begin{array}{l} \text{distr } \tilde{\zeta}_1 \sim \mathcal{N}(0,12), \\ \text{distr } \tilde{\zeta}_2 \sim \mathcal{N}(0,9), \\ \text{distr } \tilde{\eta}_1 \sim \mathcal{U}[-0.8,0.8], \\ \text{distr } \tilde{\eta}_2 \sim \mathcal{EXP}(\lambda = 2.5). \end{array}\right\} \qquad (2.4)$$

For simplicity, we assume that these four random variables are mutually independent. Since the random variables $\tilde{\zeta}_1, \tilde{\zeta}_2$ and $\tilde{\eta}_2$ are unbounded, we restrict our considerations to their respective 99% confidence intervals (except for $\mathcal{U}$). So we have for the above random variables' realizations

$$\left.\begin{array}{l} \zeta_1 \in [-30.91, 30.91], \\ \zeta_2 \in [-23.18, 23.18], \\ \eta_1 \in [-0.8, 0.8], \\ \eta_2 \in [0.0, 1.84]. \end{array}\right\} \qquad (2.5)$$

Hence, instead of the linear program (2.1), we are dealing with the *stochastic linear program*

$$\left.\begin{array}{rrcl} \min(2x_{raw1} + 3x_{raw2}) & & & \\ \text{s.t.} \quad x_{raw1} + & x_{raw2} & \le & 100, \\ (2+\tilde{\eta}_1)x_{raw1} + & 6x_{raw2} & \ge & 180 + \tilde{\zeta}_1, \\ 3x_{raw1} + & (3.4 - \tilde{\eta}_2)x_{raw2} & \ge & 162 + \tilde{\zeta}_2, \\ x_{raw1} & & \ge & 0, \\ & x_{raw2} & \ge & 0. \end{array}\right\} \qquad (2.6)$$

This is not a well-defined decision problem, since it is not at all clear what the meaning of "min" can be before knowing a realization $(\zeta_1, \zeta_2, \eta_1, \eta_2)$ of $(\tilde{\zeta}_1, \tilde{\zeta}_2, \tilde{\eta}_1, \tilde{\eta}_2)$.

Geometrically, the consequence of our random parameter changes may be rather complex. The effect of only the right-hand sides ζ_i varying over the intervals given in (2.5) corresponds to parallel translations of the corresponding facets of the feasible set as shown in Figure 3.

We may instead consider the effect of only the η_i changing their values within the intervals mentioned in (2.5). That results in rotations of the related facets. Some possible situations are shown in Figure 4, where the centers of rotation are indicated by small circles.

Allowing for all the possible changes in the demands and in the productivities simultaneously yields a superposition of the two geometrical motions, i.e. the translations and the rotations. It is easily seen that the variation of the feasible set may be substantial, depending on the actual

[1] We use $\mathcal{N}(\mu, \sigma)$ to denote the normal distribution with mean μ and variance σ^2.

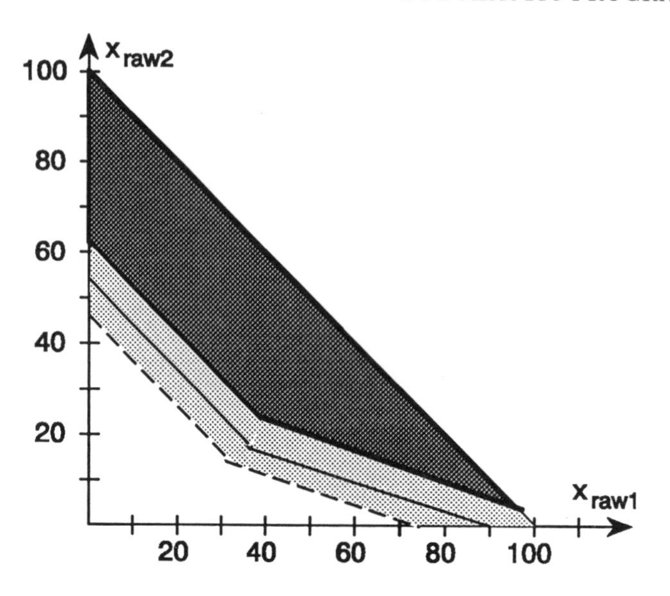

Figure 3 LP: feasible set varying with demands.

realizations of the random data. The same is also true for the so-called *wait-and-see* solutions, i.e. for those optimal solutions we should choose if we knew the realizations of the random parameters in advance. In Figure 5 a few possible situations are indicated. In addition to the deterministic solution

$$\hat{x} = (\hat{x}_{raw1}, \hat{x}_{raw2}) = (36, 18), \ \gamma = 126,$$

production plans such as

$$\left.\begin{array}{lll} \hat{y} = (\hat{y}_{raw1}, \hat{y}_{raw2}) = (20, 30), & \gamma = 130, \\ \hat{z} = (\hat{z}_{raw1}, \hat{z}_{raw2}) = (50, 22), & \gamma = 166, \\ \hat{v} = (\hat{v}_{raw1}, \hat{v}_{raw2}) = (58, 6), & \gamma = 134 \end{array}\right\} \tag{2.7}$$

may be wait-and-see solutions.

Unfortunately, wait-and-see solutions are not what we need. We have to decide production plans *under uncertainty*, since we only have statistical information about the distributions of the random demands and productivities.

A first possibility would consist in looking for a "safe" production program: one that will be feasible for all possible realizations of the productivities and demands. A production program like this is called a *fat solution* and reflects total risk aversion of the decision maker. Not surprisingly, fat solutions are usually rather expensive. In our example we can conclude from Figure 5 that

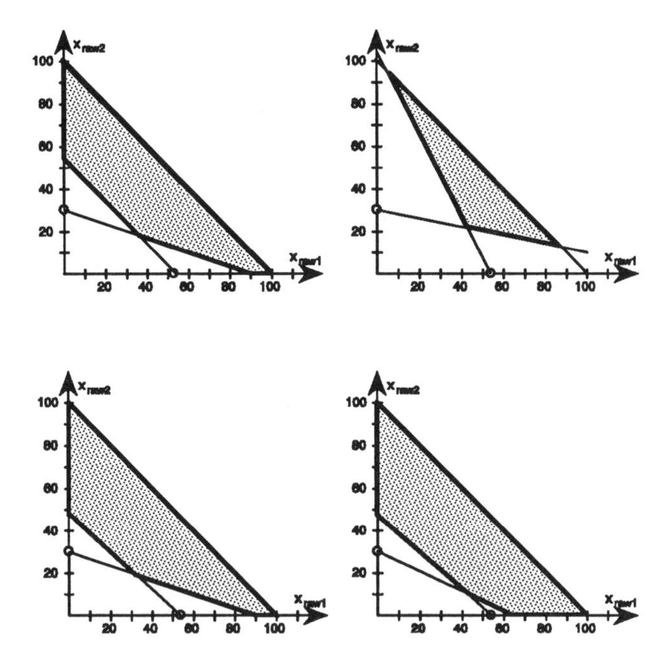

Figure 4 LP: feasible set varying with productivities.

a fat solution exists at the intersection of the two rightmost constraints for *prod*1 and *prod*2, which is easily computed as

$$x^* = (x^*_{raw1}, x^*_{raw2}) = (48.018, 25.548),\ \gamma^* = 172.681. \tag{2.8}$$

To introduce another possibility, let us assume that the refinery has made the following arrangement with its clients. In principle, the clients expect the refinery to satisfy their weekly demands. However, very likely—according to the production plan and the unforeseen events determining the clients' demands and/or the refinery's productivity—the demands cannot be covered by the production, which will cause "penalty" costs to the refinery. The amount of shortage has to be bought from the market. These penalties are supposed to be proportional to the respective shortage in products, and we assume that per unit of undeliverable products they amount to

$$q_{prod1} = 7, q_{prod2} = 12. \tag{2.9}$$

The costs due to shortage of production—or in general due to the amount of violation in the constraints—are actually determined after the observation of the random data and are denoted as *recourse* costs. In a case (like ours) of repeated execution of the production program it makes sense—according to what we have learned from statistics—to apply an *expected value criterion.*

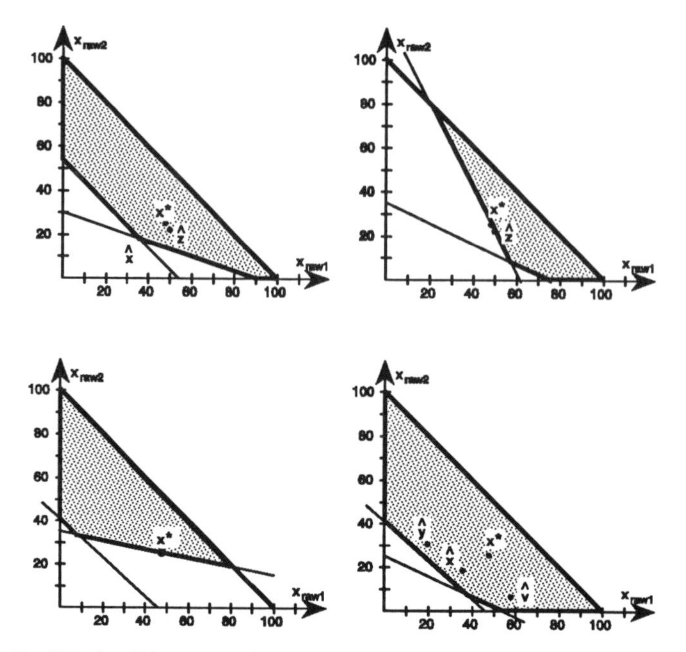

Figure 5 LP: feasible set varying with productivities and demands; some wait-and-see solutions.

More precisely, we may want to find a production plan that minimizes the sum of our original *first-stage* (i.e. production) costs and the *expected recourse* costs. To formalize this approach, we abbreviate our notation. Instead of the four single random variables $\tilde{\zeta}_1, \tilde{\zeta}_2, \tilde{\eta}_1$ and $\tilde{\eta}_2$, it seems convenient to use the random vector $\tilde{\xi} = (\tilde{\zeta}_1, \tilde{\zeta}_2, \tilde{\eta}_1, \tilde{\eta}_2)^{\mathrm{T}}$. Further, we introduce for each of the two stochastic constraints in (2.6) a *recourse variable* $y_i(\tilde{\xi}),\ i = 1, 2$, which simply measures the corresponding shortage in production if there is any; since shortage depends on the realizations of our random vector $\tilde{\xi}$, so does the corresponding recourse variable, i.e. the $y_i(\tilde{\xi})$ are themselves random variables. Following the approach sketched so far, we now replace the vague stochastic program (2.6) by the well defined *stochastic program with recourse*, using

$$h_1(\tilde{\xi}) := h_{prod1} = 180 + \tilde{\zeta}_1,\ h_2(\tilde{\xi}) := h_{prod2} = 162 + \tilde{\zeta}_2,$$

$$\alpha(\tilde{\xi}) := \pi(raw1, prod1) = 2 + \tilde{\eta}_1,\ \beta(\tilde{\xi}) := \pi(raw2, prod2) = 3.4 + \tilde{\eta}_2:$$

$$
\left.\begin{array}{llllll}
\min\{2x_{raw1} + 3x_{raw2} + E_{\tilde{\xi}}[7y_1(\tilde{\xi}) + 12y_2(\tilde{\xi})]\} \\
\text{s.t.} \quad x_{raw1} + x_{raw2} & \leq 100, \\
\alpha(\tilde{\xi})x_{raw1} + 6x_{raw2} + y_1(\tilde{\xi}) & \geq h_1(\tilde{\xi}), \\
3x_{raw1} + \beta(\tilde{\xi})x_{raw2} + y_2(\tilde{\xi}) & \geq h_2(\tilde{\xi}), \\
x_{raw1} & \geq 0, \\
x_{raw2} & \geq 0, \\
y_1(\tilde{\xi}) & \geq 0, \\
y_2(\tilde{\xi}) & \geq 0.
\end{array}\right\} \tag{2.10}
$$

In (2.10) $E_{\tilde{\xi}}$ stands for the *expected value* with respect to the distribution of $\tilde{\xi}$, and in general, it is understood that the stochastic constraints have to hold *almost surely (a.s.)* (i.e., they are to be satisfied with probability 1). Note that if $\tilde{\xi}$ has a finite discrete distribution $\{(\xi^i, p_i),\ i = 1, \cdots, r\}$ ($p_i > 0\ \forall i$) then (2.10) is just an ordinary linear program with a so-called *dual decomposition structure*:

$$
\left.\begin{array}{ll}
\min\{2x_{raw1} + 3x_{raw2} + \sum_{i=1}^{r} p_i[7y_1(\xi^i) + 12y_2(\xi^i)]\} \\
\text{s.t.} \quad x_{raw1} + x_{raw2} & \leq 100, \\
\alpha(\xi^i)x_{raw1} + 6x_{raw2} + y_1(\xi^i) & \geq h_1(\xi^i)\ \forall i, \\
3x_{raw1} + \beta(\xi^i)x_{raw2} + y_2(\xi^i) & \geq h_2(\xi^i)\ \forall i, \\
x_{raw1} & \geq 0, \\
x_{raw2} & \geq 0, \\
y_1(\xi^i) & \geq 0\ \forall i, \\
y_2(\xi^i) & \geq 0\ \forall i.
\end{array}\right\} \tag{2.11}
$$

Depending on the number of realizations of $\tilde{\xi}$, r, this linear program may become (very) large in scale, but its particular block structure is amenable to specially designed algorithms. Linear programs with dual decomposition structure will be introduced in general in Section 1.4 on page 36. A basic solution method for these problems will be described in Section 1.6.4 (page 70).

To further analyse our refinery problem, let us first assume that only the demands, $h_i(\tilde{\xi}), i = 1, 2$, are changing their values randomly, whereas the productivities are fixed. In this case we are in the situation illustrated in Figure 3. Even this small idealized problem can present numerical difficulties if solved as a nonlinear program. The reason for this lies in the fact that the evaluation of the expected value which appears in the objective function requires

- multivariate numerical integration;
- implicit definition of the functions $\hat{y}_i(\xi)$ (these functions yielding for a fixed x the optimal solutions of (2.10) for every possible realization ξ of $\tilde{\xi}$),

both of which are rather cumbersome tasks. To avoid these difficulties, we shall try to approximate the normal distributions by discrete ones. For this purpose, we

- generate large samples $\zeta_i^\mu, \mu = 1, 2, \cdots, K,\ i = 1, 2$, restricted to the 99% intervals of (2.5), sample size K =10 000;
- choose equidistant partitions of the 99% intervals into $r_i, i = 1, 2$, subintervals (e.g. $r_1 = r_2 = 15$);
- calculate for every subinterval $I_{i\nu}, \nu = 1, \cdots, r_i,\ i = 1, 2$, the arithmetic mean $\bar{\zeta}_i^\nu$ of sample values $\zeta_i^\nu \in I_{i\nu}$, yielding an estimate for the conditional expectation of $\tilde{\zeta}_i$ given $I_{i\nu}$;
- calculate for every subinterval $I_{i\nu}$ the relative frequency $p_{i\nu}$ for $\zeta_i^\mu \in I_{i\nu}$ (i.e. $p_{i\nu} = k_{i\nu}/K$, where $k_{i\nu}$ is the number of sample values ζ_i^μ contained in $I_{i\nu}$). This yields an estimate for the probability of $\{\zeta_i \in I_{i\nu}\}$.

The discrete distributions $\{(\bar{\zeta}_i^\nu, p_{i\nu}), \nu = 1, \cdots, r_i\},\ i = 1, 2$, are then used as approximations for the given normal distributions. Figure 6 shows these discrete distributions for $\mathcal{N}(0, 12)$ and $\mathcal{N}(0, 9)$, with 15 realizations each.

Obviously, these discrete distributions with 15 realizations each can only be rough approximations of the corresponding normal distributions. Therefore approximating probabilities of particular events using these discrete distributions can be expected to cause remarkable discretization errors. This will become evident in the following numerical examples.

Using these latter distributions, with 15 realizations each, we get $15^2 = 225$ realizations for the joint distribution, and hence 225 blocks in our decomposition problem. This yields as an optimal solution for the linear program (2.11) (with $\gamma(\cdot)$ the total objective of (2.11) and $\gamma_I(x) = 2x_{raw1} + 3x_{raw2}$)

$$\tilde{x} = (\tilde{x}_1, \tilde{x}_2) = (38.539, 20.539),\ \gamma(\tilde{x}) = 140.747, \tag{2.12}$$

with corresponding *first-stage costs* of

$$\gamma_I(\tilde{x}) = 138.694.$$

Defining $\rho(x)$ as the empirical *reliability* (i.e. the probability to be feasible) for any production plan x, we find—with respect to the approximating discrete distribution—for our solution $\tilde{x}$ that

$$\rho(\tilde{x}) = 0.9541,$$

whereas using our original linear program's solution $\hat{x} = (36, 18)$ would yield the total expected cost

$$\gamma(\hat{x}) = 199.390$$

and an empirical reliability of

$$\rho(\hat{x}) = 0.3188,$$

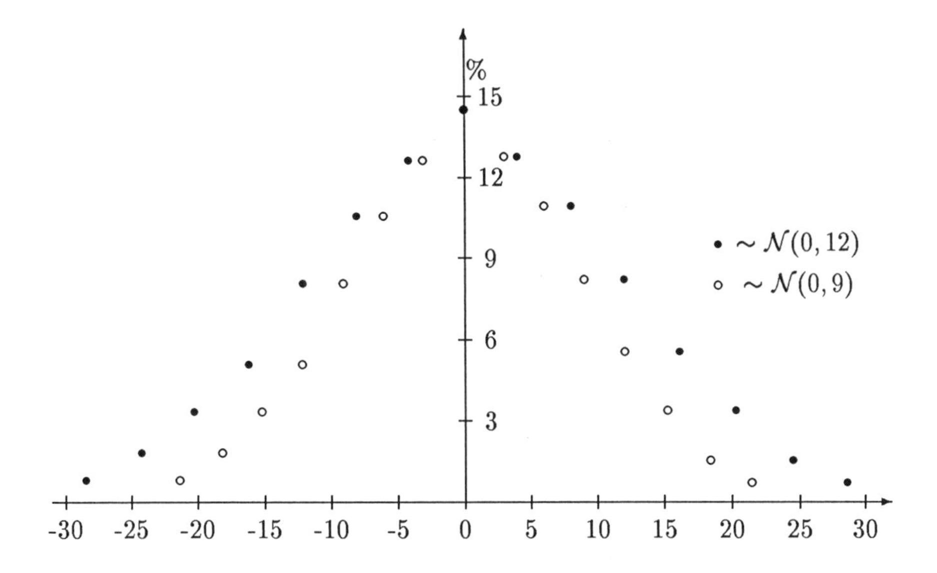

Figure 6 Discrete distribution generated from $\mathcal{N}(0,12), \mathcal{N}(0,9)$; $(r_1, r_2) = (15, 15)$.

which is clearly overestimated (compared with its theoretical value of 0.25), which indicates that the crude method of discretization we use here just for demonstration has to be refined, either by choosing a finer discretization or preferably—in view of the numerical workload drastically increasing with the size of the discrete distributions support—by finding a more appropriate strategy for determining the subintervals of the partition.

Let us now consider the effect of randomness in the productivities. To this end, we assume that $h_i(\tilde{\xi})$, $i = 1, 2$, are fixed at their expected values and the two productivities $\alpha(\tilde{\xi})$ and $\beta(\tilde{\xi})$ behave according to their distributions known from (2.3) and (2.4). Again we discretize the given distributions confining ourselves to 15 and 18 subintervals for the uniform and the exponential distributions respectively, yielding $15 \times 18 = 270$ blocks in (2.11). Solving the resulting stochastic program with recourse (2.11) as an ordinary linear program, we get as solution $\bar{x}$

$$\bar{x} = (37.566, 22.141),\ \gamma(\bar{x}) = 144.179,\ \gamma_I(\bar{x}) = 141.556,$$

whereas the solution of our original LP (2.1) would yield as total expected costs

$$\gamma(\hat{x}) = 204.561.$$

For the reliability, we now get

$$\rho(\bar{x}) = 0.9497,$$

in contrast to

$$\rho(\hat{x}) = 0.2983$$

for the LP solution $\hat{x}$.

Finally we consider the most general case of $\alpha(\tilde{\xi}), \beta(\tilde{\xi}), h_1(\tilde{\xi})$ and $h_2(\tilde{\xi})$ varying randomly where the distributions are discretely approximated by 5-, 9-, 7- and 11-point distributions respectively, in an analogous manner to the above. This yields a joint discrete distribution of $5 \times 9 \times 7 \times 11 = 3465$ realizations and hence equally many blocks in the recourse problem (2.11); in other words, we have to solve a linear program with $2 \times 3465 + 1 = 6931$ constraints! The solution $\breve{x}$ amounts to

$$\breve{x} = (37.754, 23.629),\ \gamma(\breve{x}) = 150.446,\ \gamma_I(\breve{x}) = 146.396,$$

with a reliability of

$$\rho(\breve{x}) = 0.9452,$$

whereas the LP solution $\hat{x} = (36, 18)$ would yield

$$\gamma(\hat{x}) = 232.492,\ \rho(\hat{x}) = 0.2499.$$

So far we have focused on the case where decisions, turning out *post festum* to be the wrong ones, imply penalty costs that depend on the magnitude of constraint violations. Afterwards, we were able to determine the *reliability* of the resulting decisions, which represents a measure of feasibility. Note that the reliability provides no indication of the size of possible constraint violations and corresponding penalty costs. Nevertheless, there are many real life decision situations where reliability is considered to be the most important issue—either because it seems impossible to quantify a penalty or because of questions of image or ethics. Examples may be found in various areas such as medical problems as well as technical applications.

For instance, suppose once again that only the demands are random. Suppose further that the management of our refinery is convinced that it is absolutely necessary—in order to maintain a client base—to maintain a reliability of 95% with respect to satisfying their demands. In this case we may formulate the following stochastic program with *joint probabilistic constraints*:

$$\begin{aligned}
&\min(2x_{raw1} + 3x_{raw2}) \\
&\text{s.t.} \quad \begin{array}{rrl} x_{raw1} + & x_{raw2} & \le 100, \\ x_{raw1} & & \ge 0, \\ & x_{raw2} & \ge 0, \end{array} \\
&P\begin{pmatrix} 2x_{raw1} + 6x_{raw2} \ge h_1(\tilde{\xi}) \\ 3x_{raw1} + 3x_{raw2} \ge h_2(\tilde{\xi}) \end{pmatrix} \ge 0.95.
\end{aligned}$$

This problem can be solved with appropriate methods, one of which will be presented later in this text. It seems worth mentioning that in this case using the normal distributions instead of their discrete approximations is appropriate owing to theoretical properties of probabilistic constraints to be discussed later on. The solution of the probabilistically constrained program is

$$z = (37.758, 21.698),\ \gamma_I(z) = 140.612.$$

So the costs—i.e. the first-stage costs—are only slightly increased compared with the LP solution if we observe the drastic increase of reliability. There seems to be a contradiction on comparing this last result with the solution (2.12) in that $\gamma_I(\tilde{x}) < \gamma_I(z)$ and $\rho(\tilde{x}) > 0.95$; however, this discrepancy is due to the discretization error made by replacing the true normal distribution of $(\tilde{\xi}_1, \tilde{\xi}_2)$ by the 15×15 discrete distribution used for the computation of the solution (2.12). Using the correct normal distribution would obviously yield $\gamma_I(\tilde{x}) = 138.694$ (as in (2.12)), but only $\rho(\tilde{x}) = 0.9115$!

1.3 Stochastic Programs: General Formulation

In the same way as random parameters in (2.1) led us to the stochastic (linear) program (2.6), random parameters in (1.3) may lead to the problem

$$\left.\begin{array}{l} \text{“min”} g_0(x, \tilde{\xi}) \\ \text{s.t. } g_i(x, \tilde{\xi}) \le 0,\ i = 1, \cdots, m, \\ \qquad\quad x \in X \subset \mathbb{R}^n, \end{array}\right\} \tag{3.1}$$

where $\tilde{\xi}$ is a random vector varying over a set $\Xi \subset \mathbb{R}^k$. More precisely, we assume throughout that a family $\mathcal{F}$ of “events”, i.e. subsets of Ξ, and the probability distribution P on $\mathcal{F}$ are given. Hence for every subset $A \subset \Xi$ that is an event, i.e. $A \in \mathcal{F}$, the probability $P(A)$ is known. Furthermore, we assume that the functions $g_i(x, \cdot) : \Xi \longrightarrow \mathbb{R}\ \forall x, i$ are random variables themselves, and that the probability distribution P is independent of x.

However, problem (3.1) is not well defined since the meanings of “min” as well as of the constraints are not clear at all, if we think of taking a decision on x before knowing the realization of $\tilde{\xi}$. Therefore a revision of the modelling process is necessary, leading to so-called *deterministic equivalents* for (3.1), which can be introduced in various ways, some of which we have seen for our example in the previous section. Before discussing them, we review some basic concepts in probability theory, and fix the terminology and notation used throughout this text.

1.3.1 Measures and Integrals

In $\mathbb{R}^k$ we denote sets of the type

$$I_{[a,b)} = \{x \in \mathbb{R}^k \mid a_i \le x_i < b_i,\ i = 1, \cdots, k\}$$

as (half-open) intervals. In geometric terms, depending on the dimension k of $\mathbb{R}^k$, $I_{[a,b)}$ is

- an interval if $k = 1$,
- a rectangle if $k = 2$,
- a cube if $k = 3$,

while for $k > 3$ there is no common language term for these objects since geometric imagination obviously ends there.

Sometimes we want to know something about the "size" of a set in $\mathbb{R}^k$, e.g. the length of a beam, the area of a piece of land or the volume of a building; in other words, we want to *measure* it. One possibility to do this is to fix first how we determine the measure of intervals, and a "natural" choice of a measure μ would be

- in $\mathbb{R}^1$: $\mu(I_{[a,b)}) = \begin{cases} b - a & \text{if } a \le b, \\ 0 & \text{otherwise}, \end{cases}$
- in $\mathbb{R}^2$: $\mu(I_{[a,b)}) = \begin{cases} (b_1 - a_1)(b_2 - a_2) & \text{if } a \le b, \\ 0 & \text{otherwise}, \end{cases}$
- in $\mathbb{R}^3$: $\mu(I_{[a,b)}) = \begin{cases} (b_1 - a_1)(b_2 - a_2)(b_3 - a_3) & \text{if } a \le b, \\ 0 & \text{otherwise}. \end{cases}$

Analogously, in general for $I_{[a,b)} \subset \mathbb{R}^k$ with arbitrary k, we have

$$\mu(I_{[a,b)}) = \begin{cases} \prod_{i=1}^{k} (b_i - a_i) & \text{if } a \le b \\ 0 & \text{else.} \end{cases} \tag{3.2}$$

Obviously for a set A that is the disjoint finite union of intervals, i.e. $A = \cup_{n=1}^{M} I^{(n)}$, $I^{(n)}$ being intervals such that $I^{(n)} \cap I^{(m)} = \emptyset$ for $n \ne m$, we define its measure as $\mu(A) = \sum_{n=1}^{M} \mu(I^{(n)})$. In order to measure a set A that is not just an interval or a finite union of disjoint intervals, we may proceed as follows.

Any finite collection of pairwise-disjoint intervals contained in A forms a *packing* C of A, C being the union of those intervals, with a well-defined measure $\mu(C)$ as mentioned above. Analogously, any finite collection of pairwise disjoint intervals, with their union containing A, forms a *covering* D of A with a well-defined measure $\mu(D)$.

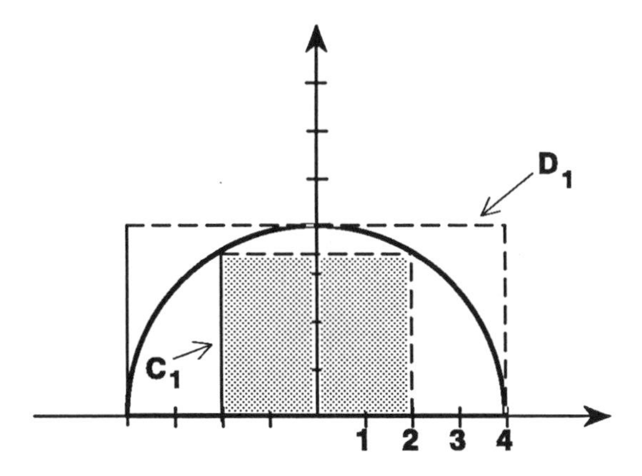

Figure 7 Measure of a half-circle: first approximation.

Take for example in $\mathbb{R}^2$ the set

$$A_{circ} = \{(x, y) \mid x^2 + y^2 \leq 16,\ y \geq 0\},$$

i.e. the half-circle illustrated in Figure 7, which also shows a first possible packing C_1 and covering D_1. Obviously we learned in high school that the area of A_{circ} is computed as $\mu(A_{circ}) = \frac{1}{2} \times \pi \times (radius)^2 = 25.1327$, whereas we easily compute $\mu(C_1) = 13.8564$ and $\mu(D_1) = 32$. If we forgot all our wisdom from high school, we would only be able to conclude that the measure of the half-circle A_{circ} is between 13.8564 and 32. To obtain a more precise estimate, we can try to improve the packing and the covering in such a way that the new packing C_2 exhausts more of the set A_{circ} and the new covering D_2 becomes a tighter outer approximation of A_{circ}. This is shown in Figure 8, for which we get $\mu(C_2) = 19.9657$ and $\mu(D_2) = 27.9658$.

Hence the measure of A_{circ} is between 19.9657 and 27.9658. If this is still not precise enough, we may further improve the packing and covering. For the half-cirle A_{circ}, it is easily seen that we may determine its measure in this way with any arbitrary accuracy.

In general, for any closed bounded set $A \subset \mathbb{R}^k$, we may try a similar procedure to measure A. Denote by $\mathcal{C}_A$ the set of all packings for A and by

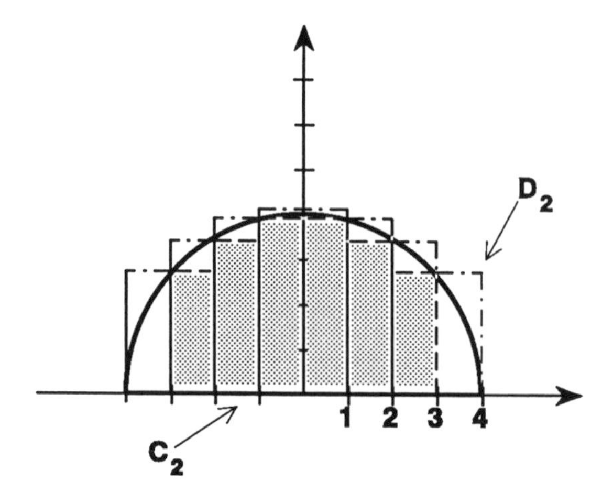

Figure 8 Improved approximate measure of a half-circle.

$\mathcal{D}_A$ the set of all coverings of A. Then we make the following definition.

The closed bounded set A is *measurable* if

$$\sup\{\mu(C) \mid C \in \mathcal{C}_A\} = \inf\{\mu(D) \mid D \in \mathcal{D}_A\},$$

with the measure $\mu(A) = \sup\{\mu(C) \mid C \in \mathcal{C}_A\}$.

To get rid of the boundedness restriction, we may extend this definition immediately by saying:

An arbitrary closed set $A \subset \mathbb{R}^k$ is measurable iff[2] for every interval $I_{[a,b)} \subset \mathbb{R}^k$ the set $A \cap I_{[a,b)}$ is measurable (in the sense defined before).

This implies that $\mathbb{R}^k$ itself is measurable. Observing that there always exist collections of countably many pairwise-disjoint intervals $I_{[a^\nu,b^\nu)}$, $\nu = 1, 2, \cdots$, covering $\mathbb{R}^k$, i.e. $\bigcup_{\nu=1}^{\infty} I_{[a^\nu,b^\nu)} = \mathbb{R}^k$ (e.g. take intervals with all edges having length 1), we get $\mu(A) = \sum_{\nu=1}^{\infty} \mu(A \cap I_{[a^\nu,b^\nu)})$ as the measure of A. Obviously $\mu(A) = \infty$ may happen, as it does for instance with $A = \mathbb{R}^2_+$ (i.e. the positive orthant of $\mathbb{R}^2$) or with $A = \{(x, y) \in \mathbb{R}^2 \mid x \geq 1,\ 0 \leq y \leq \frac{1}{x}\}$. But we also may find unbounded sets with finite measure as e.g. $A = \{(x, y) \in \mathbb{R}^2 \mid x \geq 0,\ 0 \leq y \leq e^{-x}\}$ (see the exercises at the end of this chapter).

[2] "iff" stands for "if and only if"

The measure introduced this way for closed sets and based on the elementary measure for intervals as defined in (3.2) may be extended as a "natural" measure for the class $\mathcal{A}$ of measurable sets in $\mathbb{R}^k$, and will be denoted throughout by μ. We just add that $\mathcal{A}$ is characterized by the following properties:

$$\text{if } A \in \mathcal{A} \text{ then also } \mathbb{R}^k - A \in \mathcal{A}; \tag{3.3 i}$$

$$\text{if } A_i \in \mathcal{A},\ i = 1, 2, \cdots, \text{ then also } \bigcup_{i=1}^{\infty} A_i \in \mathcal{A}. \tag{3.3 ii}$$

This implies that with $A_i \in \mathcal{A},\ i = 1, 2, \cdots$, also $\bigcap_{i=1}^{\infty} A_i \in \mathcal{A}$.

As a consequence of the above construction, we have, for the natural measure μ defined in $\mathbb{R}^k$, that

$$\mu(A) \geq 0\ \forall A \in \mathcal{A} \text{ and } \mu(\emptyset) = 0; \tag{3.4 i}$$

$$\begin{array}{l}\text{if } A_i \in \mathcal{A},\ i = 1, 2, \cdots, \text{ and } A_i \cap A_j = \emptyset \text{ for } i \neq j, \\ \text{then } \mu(\bigcup_{i=1}^{\infty} A_i) = \sum_{i=1}^{\infty} \mu(A_i).\end{array} \tag{3.4 ii}$$

In other words, the measure of a countable disjoint union of measurable sets equals the countable sum of the measures of these sets.

These properties are also familiar from probability theory: there we have some space Ω of *outcomes* ω (e.g. the results of random experiments), a collection $\mathcal{F}$ of subsets $F \subset \Omega$ called *events*, and a probability measure (or probability distribution) P assigning to each $F \in \mathcal{F}$ the probability with which it occurs. To set up probability theory, it is then required that

(i) Ω is an event, i.e. $\Omega \in \mathcal{F}$, and, with $F \in \mathcal{F}$, it holds that also $\Omega - F \in \mathcal{F}$, i.e. if F is an event then so also is its complement (or not$F$);

(ii) the countable union of events is an event.

Observe that these formally coincide with (3.3) except that Ω can be any space of objects and need not be $\mathbb{R}^k$.

For the probability measure, it is required that

(i) $P(F) \geq 0\ \forall F \in \mathcal{F}$ and $P(\Omega) = 1$;

(ii) if $F_i \in \mathcal{F},\ i = 1, 2, \cdots$, and $F_i \cap F_j = \emptyset$ for $i \neq j$, then $P(\bigcup_{i=1}^{\infty} F_i) = \sum_{i=1}^{\infty} P(F_i)$.

The only difference with (3.4) is that P is bounded to $P(F) \leq 1\ \forall F \in \mathcal{F}$, whereas μ is unbounded on $\mathbb{R}^k$. The triple $(\Omega, \mathcal{F}, P)$ with the above properties is called a *probability space*.

In addition, in probability theory we find *random variables* and *random vectors* $\tilde{\xi}$. With $\mathcal{A}$ the collection of naturally measurable sets in $\mathbb{R}^k$, a random vector is a function (i.e. a single-valued mapping)

$$\tilde{\xi} : \Omega \longrightarrow \mathbb{R}^k \text{such that, for all } A \in \mathcal{A},\ \tilde{\xi}^{-1}[A] := \{\omega \mid \tilde{\xi}(\omega) \in A\} \in \mathcal{F}. \tag{3.5}$$

This requires the "inverse" (with respect to the function $\tilde{\xi}$) of any measurable set in $\mathbb{R}^k$ to be an event in Ω.

Observe that a random vector $\tilde{\xi} : \Omega \longrightarrow \mathbb{R}^k$ induces a probability measure $P_{\tilde{\xi}}$ on $\mathcal{A}$ according to

$$P_{\tilde{\xi}}(A) = P(\{\omega \mid \tilde{\xi}(\omega) \in A\}) \; \forall A \in \mathcal{A}.$$

Example 1.2 At a market hall for the fruit trade you find a particular species of apples. These apples are traded in certain lots (e.g. of 1000 lb). Buying a lot involves some risk with respect to the quality of apples contained in it. What does "quality" mean in this context? Obviously quality is a conglomerate of criteria described in terms like size, ripeness, flavour, colour and appearance. Some of the criteria can be expressed through quantitative measurement, while others cannot (they have to be judged upon by experts). Hence the set Ω of all possible "qualities" cannot as such be represented as a subset of some $\mathbb{R}^k$.

Having bought a lot, the trader has to sort his apples according to their "outcomes" (i.e. qualities), which could fall into "events" like "unusable" (e.g. rotten or too unripe), "cooking apples"and "low (medium, high) quality eatable apples". Having sorted out the "unusable" and the "cooking apples", for the remaining apples experts could be asked to judge on ripeness, flavour, colour and appearance, by assigning real values between 0 and 1 to parameters r, f, c and a respectively, corresponding to the "degree (or percentage) of achieving" the particular criterion.

Now we can construct a scalar value for any particular outcome (quality) ω, for instance as

$$\tilde{v}(\omega) := \begin{cases} 0 & \text{if } \omega \in \text{"unusable"}, \\ \frac{1}{2} & \text{if } \omega \in \text{"cooking apples"}, \\ (1+r)(1+f)(1+c)(1+a) & \text{otherwise.} \end{cases}$$

Obviously $\tilde{v}$ has the range $\tilde{v}[\Omega] = \{0\} \cup \{\frac{1}{2}\} \cup \{[1, 16]\}$. Denoting the events "unusable" by U and "cooking apples" by C, we may define the collection $\mathcal{F}$ of events as follows. With $\mathcal{G}$ denoting the family of all subsets of $\Omega - (U \cup C)$ let $\mathcal{F}$ contain all unions of $U, C, \emptyset$ or Ω with any element of $\mathcal{G}$. Assume that after long series of observations we have a good estimate for the probabilities $P(A), A \in \mathcal{F}$.

According to our scale, we could classify the apples as

- eatable and
  - 1st class for $\tilde{v}(\omega) \in [12, 16]$ (high selling price),
  - 2nd class for $\tilde{v}(\omega) \in [8, 12)$ (medium price),
  - 3rd class for $\tilde{v}(\omega) \in [1, 8)$ (low price);
- good for cooking for $\tilde{v}(\omega) = \frac{1}{2}$ (cheap);

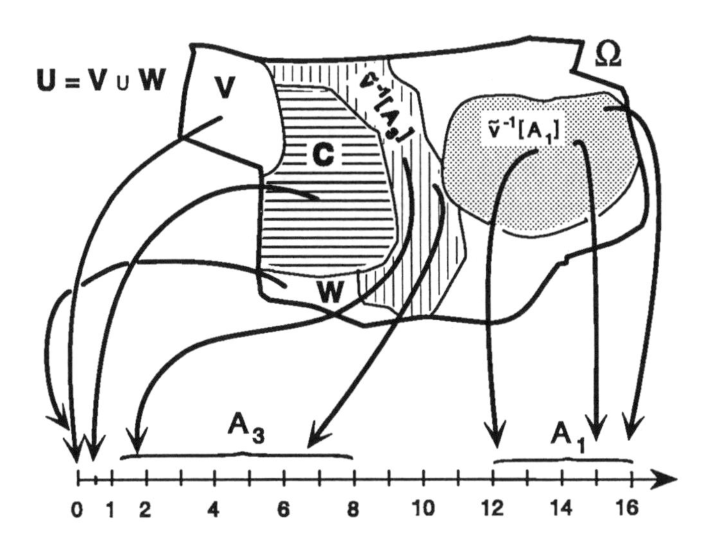

Figure 9 Classification of apples by quality.

- waste for $\tilde{v}(\omega) = 0$.

Obviously the probabilities to have 1st-class apples in our lot is $P_{\tilde{v}}(\{[12,16]\}) = P(\tilde{v}^{-1}[\{[12,16]\}])$, whereas the probability for having 3rd-class or cooking apples amounts to

$$\begin{aligned} P_{\tilde{v}}(\{[1,8)\} \cup \{\tfrac{1}{2}\}) &= P(\tilde{v}^{-1}[\{[1,8)\} \cup \{\tfrac{1}{2}\}]) \\ &= P(\tilde{v}^{-1}[\{[1,8)\}]) + P(C), \end{aligned}$$

using the fact that $\tilde{v}$ is single-valued and $\{[1,8)\}$, $\{\frac{1}{2}\}$ and hence $\tilde{v}^{-1}[\{[1,8)\}]$, $\tilde{v}^{-1}[\{\frac{1}{2}\}] = C$ are disjoint. For an illustration, see Figure 9. □

If it happens that $\Omega \subset \mathbb{R}^k$ and $\mathcal{F} \subset \mathcal{A}$ (i.e. every event is a "naturally" measurable set) then we may replace ω trivially by $\tilde{\xi}(\omega)$ by just applying the identity mapping $\tilde{\xi}(\omega) \equiv \omega$, which preserves the probability measure $P_{\tilde{\xi}}$ on $\mathcal{F}$, i.e.

$$P_{\tilde{\xi}}(A) = P(A) \text{ for } A \in \mathcal{F}$$

since obviously $\{\omega \mid \tilde{\xi}(\omega) \in A\} = A$ if $A \in \mathcal{F}$.

In any case, given a random vector $\tilde{\xi}$ with $\Xi \in \mathcal{A}$ such that $\{\omega \mid \tilde{\xi}(\omega) \in \Xi\} = \Omega$ (observe that $\Xi = \mathbb{R}^k$ always satisfies this, but there may be smaller sets in $\mathcal{A}$ that do so), with $\hat{\mathcal{F}} = \{B \mid B = A \cap \Xi,\ A \in \mathcal{A}\}$, instead of the abstract probability space $(\Omega, \mathcal{F}, P)$ we may equivalently consider the induced probability space $(\Xi, \hat{\mathcal{F}}, P_{\tilde{\xi}})$, which we shall use henceforth and therefore denote as $(\Xi, \mathcal{F}, P)$. We shall use $\tilde{\xi}$ for the random vector and ξ for the elements of Ξ (i.e. for the possible realizations of $\tilde{\xi}$).

Sometimes we like to assert a special property (like continuity or differentiability of some function $f : \mathbb{R}^k \longrightarrow \mathbb{R}$) everywhere in $\mathbb{R}^k$. But it may happen that this property almost always holds except on some particular points of $\mathbb{R}^k$ like $N_1 = \{\text{finitely many isolated points}\}$ or (for $k \geq 2$) $N_2 = \{\text{finitely many segments of straight lines}\}$, the examples mentioned being ('naturally') measurable and having the natural measure $\mu(N_1) = \mu(N_2) = 0$. In a situation like this, more precisely if there is a set $N_\delta \in \mathcal{A}$ with $\mu(N_\delta) = 0$, and if our property holds for all $x \in \mathbb{R}^k - N_\delta$, we say that it holds *almost everywhere (a.e.)*. In the context of a probability space $(\Xi, \mathcal{F}, P)$, if there is an event $N_\delta \in \mathcal{F}$ with $P(N_\delta) = 0$ such that a property holds on $\Xi - N_\delta$, owing to the practical interpretation of probabilities, we say that the property holds *almost surely (a.s.)*.

Next let us briefly review integrals. Consider first $\mathbb{R}^k$ with $\mathcal{A}$, its measurable sets, and the natural measure μ, and choose some bounded measurable set $B \in \mathcal{A}$. Further, let $\{A_1, \cdots, A_r\}$ be a *partition* of B into measurable sets, i.e. $A_i \in \mathcal{A}$, $A_i \cap A_j = \emptyset$ for $i \neq j$, and $\bigcup_{i=1}^r A_i = B$. Given the *indicator functions* $\chi_{A_i} : B \longrightarrow \mathbb{R}$ defined by

$$\chi_{A_i}(x) = \begin{cases} 1 & \text{if } x \in A_i, \\ 0 & \text{otherwise,} \end{cases}$$

we may introduce a so-called *simple function* $\varphi : B \longrightarrow \mathbb{R}$ given with some constants c_i by

$$\begin{aligned} \varphi(x) &= \sum_{i=1}^r c_i \chi_{A_i}(x) \\ &= c_i \quad \text{for } x \in A_i. \end{aligned}$$

Then the *integral* $\int_B \varphi(x) d\mu$ is defined as

$$\int_B \varphi(x) d\mu = \sum_{i=1}^r c_i \mu(A_i). \tag{3.6}$$

In Figure 10 the integral would result by accumulating the shaded areas with their respective signs as indicated.

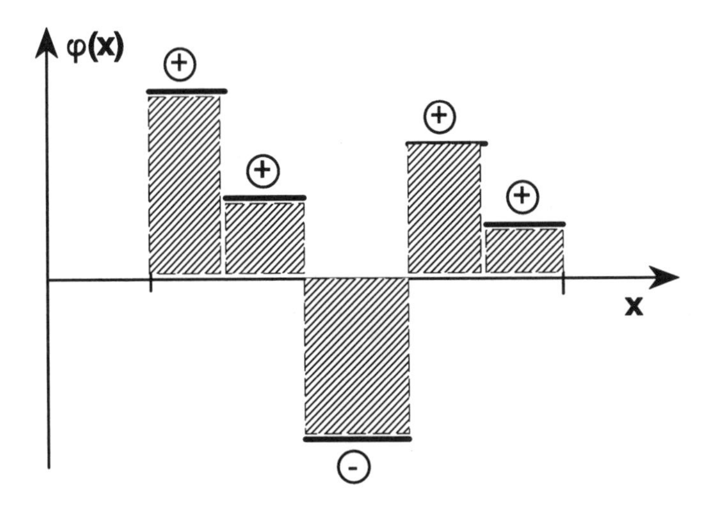

Figure 10 Integrating a simple function.

Observe that the sum (or difference) of simple functions φ_1 and φ_2 is again a simple function and that

$$\int_B [\varphi_1(x) + \varphi_2(x)]d\mu = \int_B \varphi_1(x)d\mu + \int_B \varphi_2(x)d\mu$$

$$\left|\int_B \varphi(x)d\mu\right| \leq \int_B |\varphi(x)|d\mu$$

from the elementary properties of finite sums. Furthermore, it is easy to see that for disjoint measurable sets (i.e. $B_j \in \mathcal{A}$, $j = 1, \cdots, s$, and $B_j \cap B_l = \emptyset$ for $j \neq l$) such that $\bigcup_{j=1}^{s} B_j = B$, it follows that

$$\int_B \varphi(x)d\mu = \sum_{j=1}^{s} \int_{B_j} \varphi(x)d\mu.$$

To integrate any other function $\psi : B \longrightarrow \mathbb{R}$ that is not a simple function, we use simple functions to approximate ψ (see Figure 11), and whose integrals converge.

Any sequence $\{\varphi_n\}$ of simple functions on B satisfying

$$\int_B |\varphi_n(x) - \varphi_m(x)|d\mu \longrightarrow 0 \text{ for } n, m \longrightarrow \infty$$

is called *mean fundamental.* If there exists a sequence $\{\varphi_n\}$ such that

$$\varphi_n(x) \longrightarrow \psi(x) \text{ a.e.}^3 \text{ and } \{\varphi_n\} \text{ is mean fundamental}$$

[3] The convergence a.e. can be replaced by another type of convergence, which we omit here.

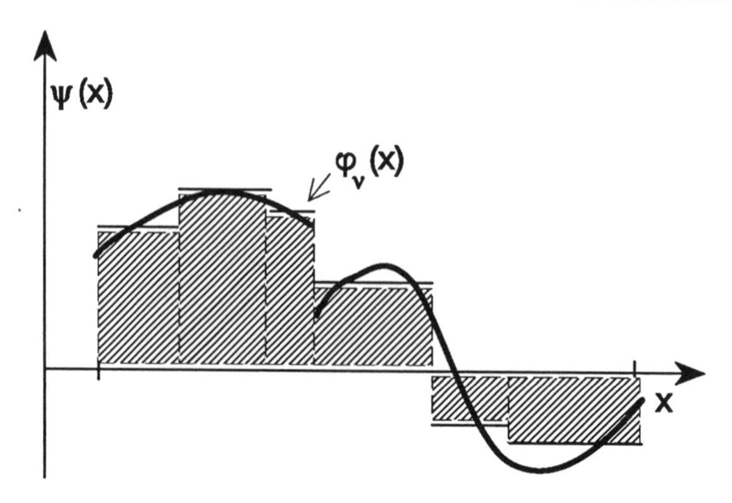

Figure 11 Integrating an arbitrary function.

then the *integral* $\int_B \psi(x)d\mu$ is defined by

$$\int_B \psi(x)d\mu = \lim_{n\to\infty} \int_B \varphi_n(x)d\mu$$

and ψ is called *integrable.*

Observe that

$$\left|\int_B \varphi_n(x)d\mu - \int_B \varphi_m(x)d\mu\right| \leq \int_B |\varphi_n(x) - \varphi_m(x)|d\mu,$$

such that $\{\int_B \varphi_n(x)d\mu\}$ is a Cauchy sequence. Therefore $\lim_{n\to\infty} \int_B \varphi_n(x)d\mu$ exists. It can be shown that this definition yields a uniquely determined value for the integral, i.e. it cannot happen that a choice of another mean fundamental sequence of simple functions converging a.e. to ψ yields a different value for the integral.

The boundedness of B is not absolutely essential here; with a slight modification of the assumption "$\varphi_n(x) \longrightarrow \psi(x)$ a.e." the integrability of ψ may be defined analogously.

Now it should be obvious that, given a probability space $(\Xi, \mathcal{F}, P)$—assumed to be introduced by a random vector $\tilde{\xi}$ in $\mathbb{R}^k$—and a function $\psi : \Xi \longrightarrow \mathbb{R}$, the integral with respect to the probability measure P, denoted by

$$E_{\tilde{\xi}}\psi(\tilde{\xi}) = \int_\Xi \psi(\xi)dP,$$

can be derived exactly as above if we simply replace the measure μ by the probability measure P. Here E refers to *expectation* and $\tilde{\xi}$ indicates that we are integrating with respect to the probability measure P induced by the random vector $\tilde{\xi}$.

Finally, we recall that in probability theory the probability measure P of a probability space $(\Xi, \mathcal{F}, P)$ in $\mathbb{R}^k$ is equivalently described by the *distribution function* $F_{\tilde{\xi}}$ defined by

$$F_{\tilde{\xi}}(x) = P(\{\xi \mid \xi \le x\}), \; x \in \mathbb{R}^k.$$

If there exists a function $f_{\tilde{\xi}} : \Xi \longrightarrow \mathbb{R}$ such that the distribution function can be represented by an integral with respect to the natural measure μ as

$$F_{\tilde{\xi}}(\hat{x}) = \int_{x \le \hat{x}} f_{\tilde{\xi}}(x) d\mu, \; \hat{x} \in \mathbb{R}^k,$$

then $f_{\tilde{\xi}}$ is called the *density function* of P. In this case the distribution function is called *of continuous type*. It follows that for any event $A \in \mathcal{F}$ we have $P(A) = \int_A f_{\tilde{\xi}}(x) d\mu$. This implies in particular that for any $A \in \mathcal{F}$ such that $\mu(A) = 0$ also $P(A) = 0$ has to hold. This fact is referred to by saying that the probability measure P is *absolutely continuous* with respect to the natural measure μ. It can be shown that the reverse statement is also true: given a probability space $(\Xi, \mathcal{F}, P)$ in $\mathbb{R}^k$ with P absolutely continuous with respect to μ (i.e. every event $A \in \mathcal{F}$ with the natural measure $\mu(A) = 0$ has also a probability of zero), there exists a density function $f_{\tilde{\xi}}$ for P.

1.3.2 Deterministic Equivalents

Let us now come back to deterministic equivalents for (3.1). For instance, in analogy to the particular *stochastic linear program with recourse* (2.10), for problem (3.1) we may proceed as follows. With

$$g_i^+(x,\xi) = \begin{cases} 0 & \text{if } g_i(x,\xi) \le 0, \\ g_i(x,\xi) & \text{otherwise,} \end{cases}$$

the ith constraint of (3.1) is violated if and only if $g_i^+(x,\xi) > 0$ for a given decision x and realization ξ of $\tilde{\xi}$. Hence we could provide for each constraint a *recourse* or *second-stage activity* $y_i(\xi)$ that, after observing the realization ξ, is chosen such as to compensate its constraint's violation—if there is one—by satisfying $g_i(x,\xi) - y_i(\xi) \le 0$. This extra effort is assumed to cause an extra cost or penalty of q_i per unit, i.e. our additional costs (called the *recourse function*) amount to

$$Q(x,\xi) = \min_y \left\{ \sum_{i=1}^m q_i y_i(\xi) \,\middle|\, y_i(\xi) \ge g_i^+(x,\xi), \; i = 1, \cdots, m \right\}, \qquad (3.7)$$

yielding a total cost—*first-stage* and *recourse* cost—of

$$f_0(x,\xi) = g_0(x,\xi) + Q(x,\xi). \tag{3.8}$$

Instead of (3.7), we might think of a more general linear recourse program with a recourse vector $y(\xi) \in Y \subset \mathbb{R}^{\overline{n}}$ (Y is some given polyhedral set, such as $\{y \mid y \geq 0\}$), an arbitrary fixed $m \times \overline{n}$ matrix W (the *recourse matrix*) and a corresponding unit cost vector $q \in \mathbb{R}^{\overline{n}}$, yielding for (3.8) the recourse function

$$Q(x,\xi) = \min_y \{q^{\mathrm{T}} y \mid Wy \geq g^+(x,\xi), y \in Y\}, \tag{3.9}$$

where $g^+(x,\xi) = (g_1^+(x,\xi), \cdots, g_m^+(x,\xi))^{\mathrm{T}}$.

If we think of a factory producing m products, $g_i(x,\xi)$ could be understood as the difference $\{demand\} - \{output\}$ of a product i. Then $g_i^+(x,\xi) > 0$ means that there is a shortage in product i, relative to the demand. Assuming that the factory is committed to cover the demands, problem (3.7) could for instance be interpreted as buying the shortage of products at the market. Problem (3.9) instead could result from a *second-stage* or *emergency* production program, carried through with the factor input y and a technology represented by the matrix W. Choosing $W = I$, the $m \times m$ identity matrix, (3.7) turns out to be a special case of (3.9).

Finally we also could think of a nonlinear recourse program to define the recourse function for (3.8); for instance, $Q(x,\xi)$ could be chosen as

$$Q(x,\xi) = \min\{q(y) \mid H_i(y) \geq g_i^+(x,\xi), i = 1,\cdots,m; y \in Y \subset \mathbb{R}^{\overline{n}}\}, \tag{3.10}$$

where $q : \mathbb{R}^{\overline{n}} \to \mathbb{R}$ and $H_i : \mathbb{R}^{\overline{n}} \to \mathbb{R}$ are supposed to be given.

In any case, if it is meaningful and acceptable to the decision maker to minimize the expected value of the total costs (i.e. first-stage and recourse costs), instead of problem (3.1) we could consider its deterministic equivalent, the (two-stage) *stochastic program with recourse*

$$\min_{x \in X} E_{\tilde{\xi}} f_0(x,\tilde{\xi}) = \min_{x \in X} E_{\tilde{\xi}} \{g_0(x,\tilde{\xi}) + Q(x,\tilde{\xi})\}. \tag{3.11}$$

The above two-stage problem is immediately extended to the *multistage recourse program* as follows: instead of the two decisions x and y, to be taken at stages 1 and 2, we are now faced with $K+1$ sequential decisions $x_0, x_1, \cdots, x_K$ ($x_\tau \in \mathbb{R}^{n_\tau}$), to be taken at the subsequent stages $\tau = 0, 1, \cdots, K$. The term "stages" can, but need not, be interpreted as "time periods".

Assume for simplicity that the objective of (3.1) is deterministic, i.e. $g_0(x,\xi) \equiv g_0(x_0)$. At stage τ ($\tau \geq 1$) we know the realizations $\xi_1, \cdots, \xi_\tau$ of the random vectors $\tilde{\xi}_1, \cdots, \tilde{\xi}_\tau$ as well as the previous decisions $x_0, \cdots, x_{\tau-1}$, and we have to decide on x_τ such that the constraint(s) (with vector valued constraint functions g_τ)

$$g_\tau(x_0,\cdots,x_\tau,\xi_1,\cdots,\xi_\tau)\le 0$$

are satisfied, which—as stated—at this stage can only be achieved by the proper choice of x_τ, based on the knowledge of the previous decisions and realizations. Hence, assuming a cost function $q_\tau(x_\tau)$, at stage $\tau\ge 1$ we have a recourse function

$$Q_\tau(x_0,x_1,\cdots,x_{\tau-1},\xi_1,\cdots,\xi_\tau)=\min_{x_\tau}\{q_\tau(x_\tau)\mid g_\tau(x_0,\cdots,x_\tau,\xi_1,\cdots,\xi_\tau)\le 0\}$$

indicating that the optimal recourse action $\hat{x}_\tau$ at time τ depends on the previous decisions and the realizations observed until stage τ, i.e.

$$\hat{x}_\tau=\hat{x}_\tau(x_0,\cdots,x_{\tau-1},\xi_1,\cdots,\xi_\tau),\ \tau\ge 1.$$

Hence, taking into account the multiple stages, we get as total costs for the multistage problem

$$f_0(x_0,\xi_1,\cdots,\xi_K)=g_0(x_0)+\sum_{\tau=1}^{K}Q_\tau(x_0,\hat{x}_1,\cdots,\hat{x}_{\tau-1},\xi_1,\cdots,\xi_\tau) \tag{3.12}$$

yielding the deterministic equivalent for the described dynamic decision problem, the *multistage stochastic program with recourse*

$$\min_{x_0\in X}[g_0(x_0)+\sum_{\tau=1}^{K}E_{\tilde{\xi}_1,\cdots,\tilde{\xi}_\tau}Q_\tau(x_0,\hat{x}_1,\cdots,\hat{x}_{\tau-1},\tilde{\xi}_1,\cdots,\tilde{\xi}_\tau)], \tag{3.13}$$

obviously a straight generalization of our former (two-stage) stochastic program with recourse (3.11).

For the two-stage case, in view of their practical relevance it is worthwile to describe briefly some variants of recourse problems in the *stochastic linear programming setting*. Assume that we are given the following stochastic linear program

$$\left.\begin{aligned}\text{“min”}\,&c^{\mathrm{T}}x\\ \text{s.t.}\quad Ax&=b,\\ T(\tilde{\xi})x&=h(\tilde{\xi}),\\ x&\ge 0.\end{aligned}\right\} \tag{3.14}$$

Comparing this with the general stochastic program (3.1), we see that the set $X\subset\mathbb{R}^n$ is specified as

$$X=\{x\in\mathbb{R}^n\mid Ax=b,\ x\ge 0\},$$

where the $m_0 \times n$ matrix A and the vector b are assumed to be deterministic. In contrast, the $m_1 \times n$ matrix $T(\cdot)$ and the vector $h(\cdot)$ are allowed to depend on the random vector $\tilde{\xi}$, and therefore to have random entries themselves. In general, we assume that this dependence on $\xi \in \Xi \subset \mathbb{R}^k$ is given as

$$\left.\begin{array}{l} T(\xi) = \hat{T}^0 + \xi_1 \hat{T}^1 + \cdots + \xi_k \hat{T}^k, \\ h(\xi) = \hat{h}^0 + \xi_1 \hat{h}^1 + \cdots + \xi_k \hat{h}^k, \end{array}\right\} \tag{3.15}$$

with deterministic matrices $\hat{T}^0, \cdots, \hat{T}^k$ and vectors $\hat{h}^0, \cdots, \hat{h}^k$. Observing that the stochastic constraints in (3.14) are equalities (instead of inequalities, as in the general problem formulation (3.1)), it seems meaningful to equate their deficiencies, which, using linear recourse and assuming that $Y = \{y \in \mathbb{R}^{\overline{n}} \mid y \geq 0\}$, according to (3.9) yields the *stochastic linear program with fixed recourse*

$$\left.\begin{array}{l} \min_x E_{\tilde{\xi}}\{c^T x + Q(x, \tilde{\xi})\} \\ \quad \text{s.t. } Ax = b \\ \qquad\quad x \geq 0, \\ \text{where} \\ Q(x, \xi) = \min\{q^T y \mid Wy = h(\xi) - T(\xi)x, \; y \geq 0\} \; . \end{array}\right\} \tag{3.16}$$

In particular, we speak of *complete fixed recourse* if the *fixed* $m_1 \times \overline{n}$ recourse matrix W satisfies

$$\{z \mid z = Wy, \; y \geq 0\} = \mathbb{R}^{m_1}. \tag{3.17}$$

This implies that, whatever the first-stage decision x and the realization ξ of $\tilde{\xi}$ turn out to be, the second-stage program

$$Q(x, \xi) = \min\{q^T y \mid Wy = h(\xi) - T(\xi)x, \; y \geq 0\}$$

will always be feasible. A special case of complete fixed recourse is *simple recourse,* where with the identity matrix I of order m_1:

$$W = (I, -I). \tag{3.18}$$

Then the second-stage program reads as

$$Q(x, \xi) = \min\{(q^+)^T y^+ + (q^-)^T y^- \mid y^+ - y^- = h(\xi) - T(\xi)x, \; y^+ \geq 0, y^- \geq 0\},$$

i.e., for $q^+ + q^- \geq 0$, the recourse variables y^+ and y^- can be chosen to measure (positively) the absolute deficiencies in the stochastic constraints.

Generally, we may put all the above problems into the following form:

$$\left.\begin{array}{l} \min E_{\tilde{\xi}} f_0(x, \tilde{\xi}) \\ \text{s.t. } E_{\tilde{\xi}} f_i(x, \tilde{\xi}) \leq 0, \; i = 1, \cdots, s, \\ \qquad E_{\tilde{\xi}} f_i(x, \tilde{\xi}) = 0, \; i = s+1, \cdots, \overline{m}, \\ \qquad\qquad x \in X \subset \mathbb{R}^n, \end{array}\right\} \tag{3.19}$$

where the f_i are constructed from the objective and the constraints in (3.1) or (3.14) respectively. So far, f_0 represented the total costs (see (3.8) or (3.12)) and $f_1, \cdots, f_{\bar{m}}$ could be used to describe the first-stage feasible set X. However, depending on the way the functions f_i are derived from the problem functions g_j in (3.1), this general formulation also includes other types of deterministic equivalents for the stochastic program (3.1).

To give just two examples showing how other deterministic equivalent problems for (3.1) may be generated, let us choose first $\alpha \in [0, 1]$ and define a "payoff" function for all constraints as

$$\varphi(x, \xi) := \begin{cases} 1 - \alpha & \text{if } g_i(x, \xi) \leq 0, \; i = 1, \cdots, m, \\ -\alpha & \text{otherwise.} \end{cases}$$

Consequently, for x infeasible at ξ we have an absolute loss of α, whereas for x feasible at ξ we have a return of $1 - \alpha$. It seems natural to aim for decisions on x that, at least in the mean (i.e. on average), avoid an absolute loss. This is equivalent to the requirement

$$E_{\tilde{\xi}} \varphi(x, \tilde{\xi}) = \int_{\Xi} \varphi(x, \xi) dP \geq 0.$$

Defining $f_0(x, \xi) = g_0(x, \xi)$ and $f_1(x, \xi) := -\varphi(x, \xi)$, we get

$$\left. \begin{aligned} & f_0(x, \xi) = g_0(x, \xi), \\ & f_1(x, \xi) = \begin{cases} \alpha - 1 & \text{if } g_i(x, \xi) \leq 0, \; i = 1, \cdots, m, \\ \alpha & \text{otherwise,} \end{cases} \end{aligned} \right\} \tag{3.20}$$

implying

$$E_{\tilde{\xi}} f_1(x, \tilde{\xi}) = -E_{\tilde{\xi}} \varphi(x, \tilde{\xi}) \leq 0,$$

where, with the vector-valued function $g(x, \xi) = (g_1(x, \xi), \cdots, g_m(x, \xi))^{\mathrm{T}}$,

$$\begin{aligned} E_{\tilde{\xi}} f_1(x, \tilde{\xi}) &= \int_{\Xi} f_1(x, \xi) dP \\ &= \int_{\{g(x,\xi) \leq 0\}} (\alpha - 1) dP + \int_{\{g(x,\xi) \not\leq 0\}} \alpha dP \\ &= (\alpha - 1) P(\{\xi \mid g(x, \xi) \leq 0\}) + \alpha P(\{\xi \mid g(x, \xi) \not\leq 0\}) \\ &= \alpha \underbrace{[P(\{\xi \mid g(x, \xi) \leq 0\}) + P(\{\xi \mid g(x, \xi) \not\leq 0\})]}_{=1} \\ &\quad - P(\{\xi \mid g(x, \xi) \leq 0\}). \end{aligned}$$

Therefore the constraint $E_{\tilde{\xi}} f_1(x, \tilde{\xi}) \leq 0$ is equivalent to $P(\{\xi \mid g(x, \xi) \leq 0\}) \geq \alpha$. Hence, under these assumptions, (3.19) reads as

$$\left. \begin{aligned} & \min_{x \in X} E_{\tilde{\xi}} g_0(x, \tilde{\xi}) \\ & \text{s.t. } P(\{\xi \mid g_i(x, \xi) \leq 0, \; i = 1, \cdots, m\}) \geq \alpha. \end{aligned} \right\} \tag{3.21}$$

Problem (3.21) is called a *probabilistically constrained* or *chance constrained program* (or a problem with *joint probabilistic constraints*).

If instead of (3.20) we define $\alpha_i \in [0,1], i = 1,\cdots,m$, and analogous "payoffs" for every single constraint, resulting in

$$f_0(x,\xi) = g_0(x,\xi)$$

$$f_i(x,\xi) = \begin{cases} \alpha_i - 1 & \text{if } g_i(x,\xi) \le 0, \\ \alpha_i & \text{otherwise,} \end{cases} \qquad i = 1,\cdots,m,$$

then we get from (3.19) the problem with *single (or separate) probabilistic constraints*:

$$\left.\begin{array}{l} \min_{x\in X} E_{\tilde{\xi}} g_0(x,\tilde{\xi}) \\ \text{s.t. } P(\{\xi \mid g_i(x,\xi) \le 0\}) \ge \alpha_i, \; i = 1,\cdots,m. \end{array}\right\} \tag{3.22}$$

If, in particular, we have that the functions $g_i(x,\xi)$ are linear in x, and if furthermore the set X is convex polyhedral, i.e. we have the stochastic linear program

$$\left.\begin{array}{rl} \text{"min"} & c^{\mathrm{T}}(\tilde{\xi})x \\ \text{s.t.} & Ax = b, \\ & T(\tilde{\xi})x \ge h(\tilde{\xi}), \\ & x \ge 0, \end{array}\right\}$$

then problems (3.21) and (3.22) become

$$\left.\begin{array}{l} \min_{x\in X} E_{\tilde{\xi}} c^{\mathrm{T}}(\tilde{\xi})x \\ \text{s.t. } P(\{\xi \mid T(\xi)x \ge h(\xi)\}) \ge \alpha, \end{array}\right\} \tag{3.23}$$

and, with $T_i(\cdot)$ and $h_i(\cdot)$ denoting the ith row and ith component of $T(\cdot)$ and $h(\cdot)$ respectively,

$$\left.\begin{array}{l} \min_{x\in X} E_{\tilde{\xi}} c^{T}(\tilde{\xi})x \\ \text{s.t. } P(\{\xi \mid T_i(\xi)x \ge h_i(\xi)\}) \ge \alpha_i, \; i = 1,\cdots,m, \end{array}\right\} \tag{3.24}$$

the stochastic *linear* programs with joint and with single chance constraints respectively.

Obviously there are many other possibilities to generate types of deterministic equivalents for (3.1) by constructing the f_i in different ways out of the objective and the constraints of (3.1).

Formally, all problems derived, i.e. all the above deterministic equivalents, are mathematical programs. The first question is, whether or under which assumptions do they have properties like convexity and smoothness such that we have any reasonable chance to deal with them computationally using the toolkit of mathematical programming methods.

1.4 Properties of Recourse Problems

Convexity may be shown easily for the recourse problem (3.11) under rather mild assumptions (given the integrability of $g_0 + Q$).

Proposition 1.1 *If $g_0(\cdot,\xi)$ and $Q(\cdot,\xi)$ are convex in x $\forall \xi \in \Xi$, and if X is a convex set, then (3.11) is a convex program.*

Proof For $\hat{x}, \bar{x} \in X, \lambda \in (0,1)$ and $\check{x} := \lambda\hat{x} + (1-\lambda)\bar{x}$ we have

$$g_0(\check{x},\xi) + Q(\check{x},\xi) \le \lambda[g_0(\hat{x},\xi) + Q(\hat{x},\xi)] + (1-\lambda)[g_0(\bar{x},\xi) + Q(\bar{x},\xi)] \;\forall \xi \in \Xi$$

implying

$$E_{\tilde{\xi}}\{g_0(\check{x},\tilde{\xi})+Q(\check{x},\tilde{\xi})\} \le \lambda E_{\tilde{\xi}}\{g_0(\hat{x},\tilde{\xi})+Q(\hat{x},\tilde{\xi})\}+(1-\lambda)E_{\tilde{\xi}}\{g_0(\bar{x},\tilde{\xi})+Q(\bar{x},\tilde{\xi})\}.$$

□

Remark 1.1 Observe that for $Y = \mathbb{R}_+^{\bar{n}}$ the convexity of $Q(\cdot,\xi)$ can immediately be asserted for the linear case (3.16) and that it also holds for the nonlinear case (3.10) if the functions $q(\cdot)$ and $g_i(\cdot,\xi)$ are convex and the $H_i(\cdot)$ are concave. Just to sketch the argument, assume that $\bar{y}$ and $\check{y}$ solve (3.10) for $\bar{x}$ and $\check{x}$ respectively, at some realization $\xi \in \Xi$. Then, by the convexity of g_i and the concavity of H_i, $i = 1,\cdots,m$, we have, for any $\lambda \in (0,1)$,

$$\begin{aligned} g_i(\lambda\bar{x} + (1-\lambda)\check{x},\xi) &\le \lambda g_i(\bar{x},\xi) + (1-\lambda)g_i(\check{x},\xi) \\ &\le \lambda H_i(\bar{y}) + (1-\lambda)H_i(\check{y}) \\ &\le H_i(\lambda\bar{y} + (1-\lambda)\check{y}). \end{aligned}$$

Hence $\hat{y} = \lambda\bar{y} + (1-\lambda)\check{y}$ is feasible in (3.10) for $\hat{x} = \lambda\bar{x} + (1-\lambda)\check{x}$, and therefore, by the convexity of q,

$$\begin{aligned} Q(\hat{x},\xi) &\le q(\hat{y}) \\ &\le \lambda q(\bar{y}) + (1-\lambda)q(\check{y}) \\ &= \lambda Q(\bar{x},\xi) + (1-\lambda)Q(\check{x},\xi). \end{aligned}$$

□

Smoothness (i.e. partial differentiability of $\mathcal{Q}(x) = \int_\Xi Q(x,\xi)\,dP$) of recourse problems may also be asserted under fairly general conditions. For example, suppose that $\varphi : \mathbb{R}^2 \longrightarrow \mathbb{R}$, so that $\varphi(x,y) \in \mathbb{R}$. Recalling that φ is partially differentiable at some point $(\hat{x},\hat{y})$ with respect to x, this means that there exists a function, called the partial derivative and denoted by $\dfrac{\partial\varphi(x,y)}{\partial x}$, such that

$$\frac{\varphi(\hat{x}+h,\hat{y}) - \varphi(\hat{x},\hat{y})}{h} = \frac{\partial\varphi(\hat{x},\hat{y})}{\partial x} + \frac{r(\hat{x},\hat{y};h)}{h},$$

where the "residuum" r satisfies

$$\frac{r(\hat{x},\hat{y};h)}{h} \overset{h\to 0}{\longrightarrow} 0.$$

The recourse function is partially differentiable with respect to x_j in $(\hat{x},\hat{\xi})$ if there is a function $\frac{\partial Q(x,\xi)}{\partial x_j}$ such that

$$\frac{Q(\hat{x}+he^j,\hat{\xi})-Q(\hat{x},\hat{\xi})}{h} = \frac{\partial Q(\hat{x},\hat{\xi})}{\partial x_j} + \frac{\rho_j(\hat{x},\hat{\xi};h)}{h}$$

with

$$\frac{\rho_j(\hat{x},\hat{\xi};h)}{h} \overset{h\to 0}{\longrightarrow} 0,$$

where e^j is the jth unit vector. The vector $(\frac{\partial Q(x,\xi)}{\partial x_1},\cdots,\frac{\partial Q(x,\xi)}{\partial x_n})^{\mathrm{T}}$ is called the *gradient of* $Q(x,\xi)$ *with respect to* x and is denoted by $\nabla_x Q(x,\xi)$. Now we are not only interested in the partial differentiability of the recourse function $Q(x,\xi)$ but also in that of the expected recourse function $\mathcal{Q}(x)$. Provided that $Q(x,\xi)$ is partially differentiable at $\hat{x}$ a.s., we get

$$\begin{aligned}
\frac{\mathcal{Q}(\hat{x}+he^j)-\mathcal{Q}(\hat{x})}{h} &= \int_\Xi \frac{Q(\hat{x}+he^j,\xi)-Q(\hat{x},\xi)}{h}dP \\
&= \int_{\Xi-N_\delta}\left[\frac{\partial Q(\hat{x},\xi)}{\partial x_j}+\frac{\rho_j(\hat{x},\xi;h)}{h}\right]dP \\
&= \int_{\Xi-N_\delta}\frac{\partial Q(\hat{x},\xi)}{\partial x_j}dP + \int_{\Xi-N_\delta}\frac{\rho_j(\hat{x},\xi;h)}{h}dP,
\end{aligned}$$

where $N_\delta \in \mathcal{F}$ and $P(N_\delta)=0$. Hence, under these assumptions, $\mathcal{Q}$ is partially differentiable if

$$\int_{\Xi-N_\delta}\frac{\partial Q(\hat{x},\xi)}{\partial x_j}dP \text{ exists and } \frac{1}{h}\int_{\Xi-N_\delta}\rho_j(\hat{x},\xi;h)dP \overset{h\to 0}{\longrightarrow} 0.$$

This yields the following.

Proposition 1.2 *If $Q(x,\tilde{\xi})$ is partially differentiable with respect to x_j at some $\hat{x}$ a.s. (i.e. for all ξ except maybe those belonging to an event with probability zero), if its partial derivative $\frac{\partial Q(\hat{x},\xi)}{\partial x_j}$ is integrable and if the residuum satisfies $(1/h)\int_\Xi \rho_j(\hat{x},\xi;h)dP \overset{h\to 0}{\longrightarrow} 0$ then $\frac{\partial \mathcal{Q}(\hat{x})}{\partial x_j}$ exists as well and*

$$\frac{\partial \mathcal{Q}(\hat{x})}{\partial x_j} = \int_\Xi \frac{\partial Q(\hat{x},\xi)}{\partial x_j}dP.$$

Questions arise as a result of the general formulation of the assumptions of this proposition. It is often possible to decide that the recourse function is partially differentiable a.s. and the partial derivative is integrable. However, the requirement that the residuum is integrable and—roughly speaking—its integral converges to zero faster than h can be difficult to check. Hence we leave the general case and focus on stochastic linear programs with complete fixed recourse (3.16) in the following remark.

Remark 1.2 In the linear case (3.16) with complete fixed recourse it is known from linear programming (see Section 1.6) that the optimal value function $Q(x,\xi)$ is continuous and piecewise linear in $h(\xi) - T(\xi)x$. In other words, there exist finitely many convex polyhedral cones $B_l \subset \mathbb{R}^{m_1}$ with nonempty interiors such that any two of them have at most boundary points in common and $\cup_l B_l = \mathbb{R}^{m_1}$, and $Q(x,\xi)$ is given as $Q(x,\xi) = d^{l\mathrm{T}}(h(\xi) - T(\xi)x) + \delta_l$ for $h(\xi) - T(\xi)x \in B_l$. Then, for $h(\xi) - T(\xi)x \in \mathrm{int}B_l$ (i.e. for $h(\xi) - T(\xi)x$ an interior point of B_l), the function $Q(x,\xi)$ is partially differentiable with respect to any component of x. Hence for the gradient with respect to x we get from the chain rule that $\nabla_x Q(x,\xi) = -T^{\mathrm{T}}(\xi)d^l$ for $h(\xi) - T(\xi)x \in \mathrm{int}B_l$.

Assume for simplicity that Ξ is a bounded interval in $\mathbb{R}^k$ and keep x fixed. Then, by (3.15), we have a linear affine mapping $\psi(\cdot) := h(\cdot) - T(\cdot)x : \Xi \longrightarrow \mathbb{R}^{m_1}$. Therefore the sets

$$\hat{D}_l(x) = \psi^{-1}[B_l] := \{\xi \in \Xi \mid \psi(\xi) \in B_l\}$$

are convex polyhedra (see Figure 12) satisfying $\bigcup_l \hat{D}_l(x) = \Xi$.

Define $D_l(x) := \mathrm{int}\hat{D}_l(x)$. To get the intended differentiability result, the following assumption is crucial:

$$\xi \in D_l(x) \Longrightarrow \psi(\xi) = h(\xi) + T(\xi)x \in \mathrm{int}B_l \ \forall l. \tag{4.1}$$

By this assumption, we enforce the event $\{\xi \in \Xi \mid \psi(\xi) \in B_l - \mathrm{int}B_l\}$ to have the natural measure $\mu(\{\xi \in \Xi \mid \psi(\xi) \in B_l - \mathrm{int}B_l\}) = 0$, which need not be true in general, as illustrated in Figure 13.

Since the B_l are convex polyhedral cones in $\mathbb{R}^{m_1}$ (see Section 1.6) with nonempty interiors, they may be represented by inequality systems

$$C^l z \le 0,$$

where $C^l \neq 0$ is an appropriate matrix with no row equal to zero. Fix l and let $\xi \in D_l(x)$ such that, by (4.1), $h(\xi) - T(\xi)x \in \mathrm{int}B_l$. Then

$$C^l[h(\xi) - T(\xi)x] < 0,$$

i.e. for any fixed j there exists a $\hat{\tau}_{lj} > 0$ such that

$$C^l[h(\xi) - T(\xi)(x \pm \tau_{lj} e^j)] \le 0$$

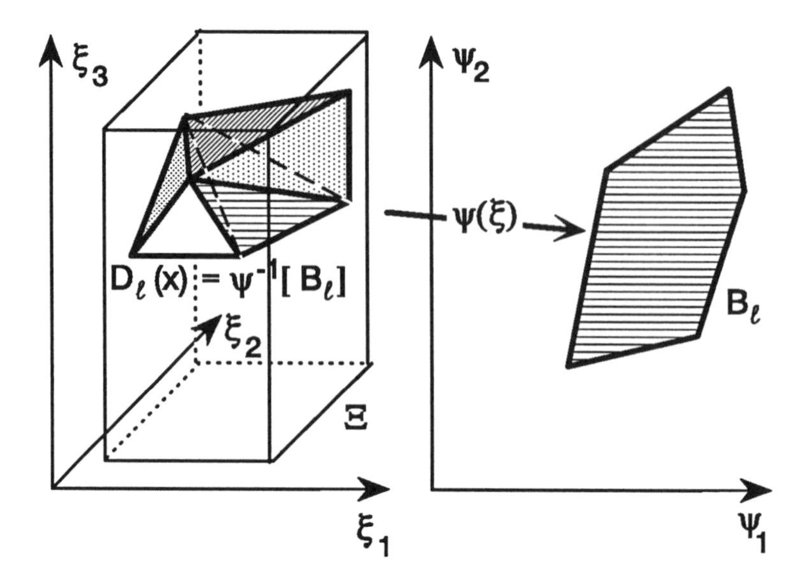

Figure 12 Linear affine mapping of a polyhedron.

or, equivalently,

$$C^l[h(\xi) - T(\xi)x] \leq \mp\tau_{lj} C^l T(\xi)e^j \ \forall\tau_{lj} \in [0, \hat{\tau}_{lj}].$$

Hence for $\gamma(\xi) = \max_i \left|(C^l T(\xi)e^j)_i\right|$ there is a

$$t_l > 0 : \ C^l[h(\xi) - T(\xi)x] \leq -|t|\gamma(\xi)e \ \forall|t| < t_l,$$

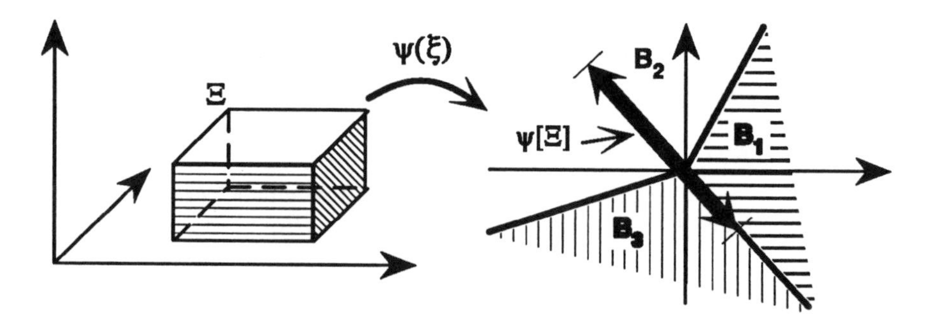

Figure 13 Linear mapping violating assumption (4.1).

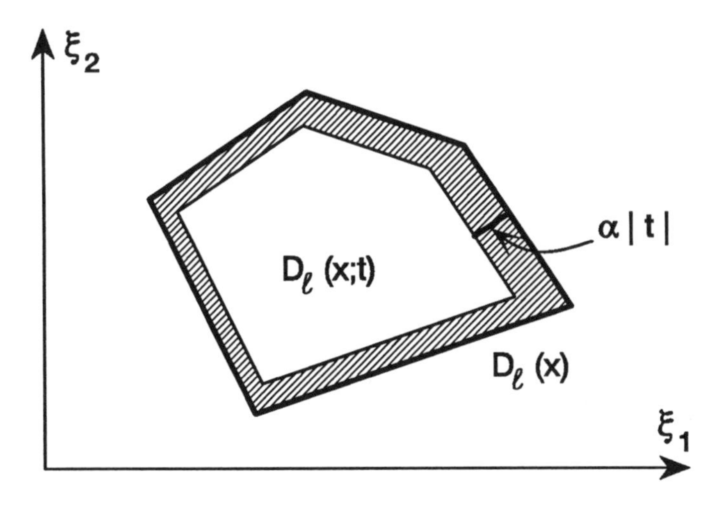

Figure 14 Difference set $D_l(x) - D_l(x;t)$.

$e = (1, \cdots, 1)^{\mathrm{T}}$. This implies that for $\gamma := \max_{\xi\in\Xi} \gamma(\xi)$ there exists a $t_0 > 0$ such that

$$C^l[h(\xi) - T(\xi)x] \leq -|t|\gamma e \ \forall |t| < t_0$$

(choose, for example, $t_0 = t_l/\gamma$). In other words, there exists a $t_0 > 0$ such that

$$D_l(x;t) := \{\xi \mid C^l[h(\xi) - T(\xi)x] \leq -|t|\gamma e\} \neq \emptyset \ \forall |t| < t_0,$$

and obviously $D_l(x;t) \subset D_l(x)$. Furthermore, by elementary geometry, the natural measure μ satisfies

$$\mu(D_l(x) - D_l(x;t)) \leq |t|v$$

with some constant v (see Figure 14).

For $\xi \in D_l(x;t)$ it follows that

$$\begin{aligned} C^l[h(\xi) - T(\xi)(x + te^j)] &= C^l[h(\xi) - T(\xi)x] - tC^lT(\xi)e^j \\ &\leq -|t|\gamma e - tC^lT(\xi)e^j \\ &\leq 0, \end{aligned}$$

owing to the fact that each component of $C^lT(\xi)e^j$ is absolutely bounded by γ. Hence in this case we have $h(\xi) - T(\xi)(x + te^j) \in B_l$, and so

$$\frac{Q(x + te^j, \xi) - Q(x, \xi)}{t} = -d^{l\mathrm{T}}T(\xi)e^j = \frac{\partial Q(x, \xi)}{\partial x_j} \quad \forall |t| < t_0,$$

i.e. in this case we have the residuum $\rho_j(x, \xi; t) \equiv 0$.

For $\xi \in D_l(x) - D_l(x;t)$ we have, considering that $h(\xi) - T(\xi)(x + te^j)$ could possibly belong to some other $B_{\bar{l}}$, at least the estimate

$$\frac{|Q(x+te^j,\xi) - Q(x,\xi)|}{|t|} \leq \max\{|d^{\bar{l}\mathrm{T}}T(\xi)e^j| \mid \xi \in \Xi,\ \forall \bar{l}\} =: \beta.$$

Assuming now that we have a continuous density $\varphi(\xi)$ for P, we know already from (4.1) that $\mu(\{\xi \in \Xi \mid \psi(\xi) \in B_l - \mathrm{int}B_l\}) = 0$. Hence it follows that

$$\begin{aligned} E_{\tilde{\xi}}Q(x,\tilde{\xi}) &= \sum_l \int_{D_l(x)} Q(x,\xi)\varphi(\xi)d\xi \\ &= \sum_l \int_{D_l(x)} \{d^{l\mathrm{T}}[h(\xi) - T(\xi)x] - \delta_l\}\varphi(\xi)d\xi, \end{aligned}$$

and, since

$$\left| \int_{D_l(x)-D_l(x;t)} \frac{Q(x+te^j,\xi) - Q(x,\xi)}{t}\varphi(\xi)d\mu \right| \leq \beta \max_{\xi\in\Xi}\varphi(\xi)|t|v \ \stackrel{t\to 0}{\longrightarrow} 0,$$

$$\begin{aligned} \nabla E_{\tilde{\xi}}Q(x,\tilde{\xi}) &= \sum_l \int_{D_l(x)} \nabla_x Q(x,\xi)\varphi(\xi)d\xi \\ &= \sum_l \int_{D_l(x)} T^{\mathrm{T}}(\xi)d^l\varphi(\xi)d\xi. \end{aligned}$$

Hence for the linear case—observing (3.15)—we get the differentiability statement of Proposition 1.2 provided that (4.1) is satisfied and P has a continuous density on Ξ. □

Summarizing the statements given so far, we see that *stochastic programs with recourse* are likely to have such properties as convexity (Proposition 1.1) and, given continuous-type distributions, differentiability (Proposition 1.2), which—from the viewpoint of mathematical programming—are appreciated. On the other hand, if we have a joint finite discrete probability distribution $\{(\xi^k, p_k), k = 1, \cdots, r\}$ of the random data then, for example, problem (3.16) becomes—similarly to the special example (2.11)—a linear program

$$\left.\begin{aligned} &\min_{x\in X}\left\{c^{\mathrm{T}}x + \sum_{k=1}^{r} p_k q^{\mathrm{T}} y^k\right\} \\ &\text{s.t. } T(\xi^k)x + Wy^k = h(\xi^k),\ k = 1,\cdots,r, \\ &\qquad y^k \geq 0 \end{aligned}\right\} \quad (4.2)$$

having the so-called *dual decomposition structure*, as mentioned already for our special example (2.11) and demonstrated in Figure 15 (see also Section 1.6.4).

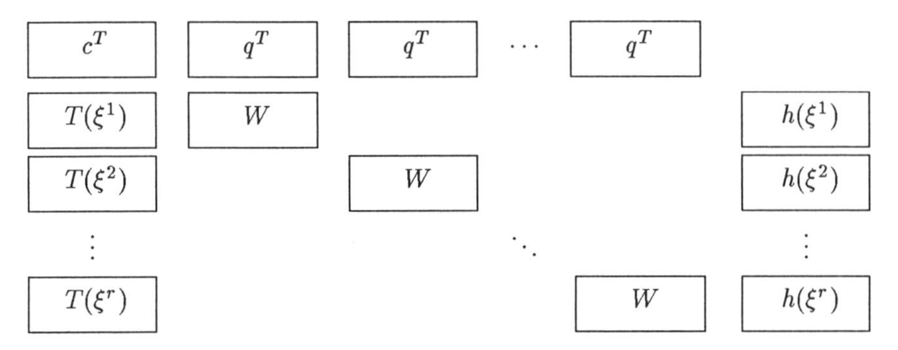

Figure 15 Dual decomposition data structure.

However—for finite discrete as well as for continuous distributions—we are faced with a further problem, which we might discuss for the linear case (i.e. for stochastic linear programs with fixed recourse (3.16)). By $\operatorname{supp} P$ we denote the *support of the probability measure* P, i.e. the smallest closed set $\Xi \subset \mathbb{R}^k$ such that $P_{\tilde{\xi}}(\Xi) = 1$. With the practical interpretation of the second-stage problem as given, for example, in Section 1.2, and assuming that $\Xi = \operatorname{supp} P_{\tilde{\xi}}$, we should expect that for any first-stage decision $x \in X$ the compensation of deficiencies in the stochastic constraints is possible whatever $\xi \in \Xi$ will be realized for $\tilde{\xi}$. In other words, we expect the program

$$\left.\begin{aligned} Q(x, \xi) = \min\ & q^{\mathrm{T}} y \\ \text{s.t. } & Wy = h(\xi) - T(\xi)x, \\ & y \geq 0 \end{aligned}\right\} \tag{4.3}$$

to be feasible $\forall \xi \in \Xi$. Depending on the defined recourse matrix W and the given support Ξ, this need not be true for all first-stage decisions $x \in X$. Hence it may become necessary to impose—in addition to $x \in X$—further restrictions on our first-stage decisions called *induced constraints*. To be more specific, let us assume that Ξ is a (bounded) convex polyhedron, i.e. the convex hull of finitely many points $\xi^j \in \Xi \subset \mathbb{R}^k$:

$$\begin{aligned} \Xi &= \operatorname{conv}\{\xi^1, \cdots, \xi^r\} \\ &= \left\{ \xi \,\middle|\, \xi = \sum_{j=1}^{r} \lambda_j \xi^j,\ \sum_{j=1}^{r} \lambda_j = 1,\ \lambda_j \geq 0\ \forall j \right\}. \end{aligned}$$

From the definition of a support, it follows that $x \in \mathbb{R}^n$ allows for a feasible solution of the second-stage program for all $\xi \in \Xi$ if and only if this is true for all ξ^j, $j = 1, \cdots, r$. In other words, the *induced first-stage feasibility set* K is given as

$$K = \{x \mid T(\xi^j)x + Wy^j = h(\xi^j),\ y^j \geq 0,\ j = 1, \cdots, r\}.$$

From this formulation of K (which obviously also holds if $\tilde{\xi}$ has a finite discrete distribution, i.e. $\Xi = \{\xi^1, \cdots, \xi^r\}$), we evidently get the following.

Proposition 1.3 *If the support Ξ of the distribution of $\tilde{\xi}$ is either a finite set or a (bounded) convex polyhedron then the induced first-stage feasibility set K is a convex polyhedral set. The first-stage decisions are restricted to $x \in X \bigcap K$.*

Example 1.3 Consider the following first-stage feasible set:

$$X = \{x \in \mathbb{R}^2_+ \mid x_1 - 2x_2 \geq -4, x_1 + 2x_2 \leq 8, 2x_1 - x_2 \leq 6\}.$$

For the second-stage constraints choose

$$W = \begin{pmatrix} -1 & 3 & 5 \\ 2 & 2 & 2 \end{pmatrix}, \quad T(\xi) \equiv T = \begin{pmatrix} 2 & 3 \\ 3 & 1 \end{pmatrix}$$

and a random vector $\tilde{\xi}$ with the support $\Xi = [4, 19] \times [13, 21]$. Then the constraints to be satisfied for all $\xi \in \Xi$ are

$$Wy = \xi - Tx, \ y \geq 0.$$

Observing that the second column W_2 of W is a positive linear combination of W_1 and W_3, namely $W_2 = \frac{1}{3}W_1 + \frac{2}{3}W_3$, the above second-stage constraints reduce to the requirement that for all $\xi \in \Xi$ the right-hand side $\xi - Tx$ can be written as

$$\xi - Tx = \lambda W_1 + \mu W_3, \ \lambda, \mu \geq 0,$$

or in detail as

$$\begin{aligned} \xi_1 - 2x_1 - 3x_2 &= -\lambda + 5\mu, \\ \xi_2 - 3x_1 - x_2 &= 2\lambda + 2\mu, \\ \lambda, \mu &\geq 0. \end{aligned}$$

Multiplying this system of equations with the regular matrix $S = \begin{pmatrix} 2 & 1 \\ -2 & 5 \end{pmatrix}$, which corresponds to adding 2 times the first equation to the second and adding -2 times the first to 5 times the second, respectively, we get the equivalent system

$$\begin{aligned} 2\xi_1 + \xi_2 - 7x_1 - 7x_2 &= 12\mu \geq 0, \\ -2\xi_1 + 5\xi_2 - 11x_1 + x_2 &= 12\lambda \geq 0. \end{aligned}$$

Because of the required nonnegativity of λ and μ, this is equivalent to the system of inequalities

$$\begin{aligned} 7x_1 + 7x_2 &\leq 2\xi_1 + \xi_2 \ (\geq 21 \ \forall \xi \in \Xi), \\ 11x_1 - x_2 &\leq -2\xi_1 + 5\xi_2 \ (\geq 27 \ \forall \xi \in \Xi). \end{aligned}$$

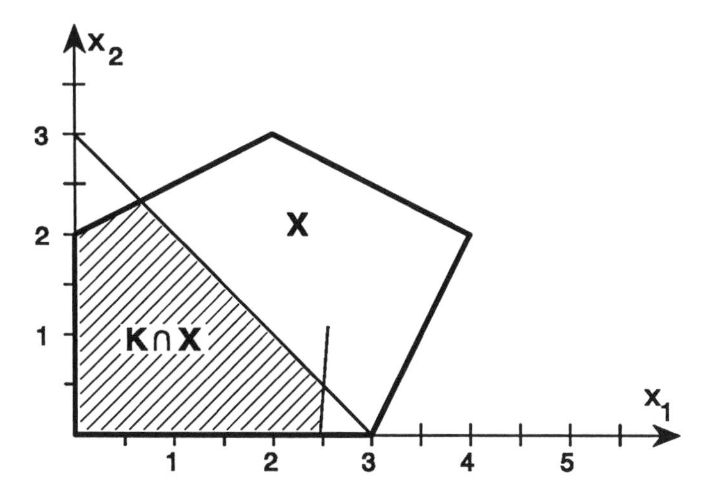

Figure 16 Induced constraints K.

Since these inequalities have to be satisfied for all $\xi \in \Xi$, choosing the minimal right-hand sides (for $\xi \in \Xi$) yields the induced constraints as

$$K = \{x \mid 7x_1 + 7x_2 \leq 21, \; 11x_1 - x_2 \leq 27\}.$$

The first-stage feasible set X together with the induced feasible set are illustrated in Figure 16. □

It might happen that $X \bigcap K = \emptyset$; then we should check our model very carefully to figure out whether we really modelled what we had in mind or whether we can find further possibilities for compensation that are not yet contained in our model. On the other hand, we have already mentioned the case of a complete fixed recourse matrix (see (3.17) on page 28), for which $K = \mathbb{R}^n$ and therefore the problem of induced constraints does not exist. Hence it seems interesting to recognize complete recourse matrices.

Proposition 1.4 *An* $m_1 \times \overline{n}$ *matrix* W *is a complete recourse matrix iff*[4]

- *it has rank* $\mathrm{rk}(W) = m_1$ *and,*

[4] We use *"iff"* as short-hand for *"if and only if"*.

- *assuming without loss of generality that its first m_1 columns $W_1, W_2, \cdots, W_{m_1}$ are linearly independent, the linear constraints*

$$\left.\begin{aligned} Wy &= 0 \\ y_i &\geq 1, \ i = 1, \cdots, m_1, \\ y &\geq 0 \end{aligned}\right\} \tag{4.4}$$

have a feasible solution.

Proof W is a complete recourse matrix iff

$$\{z \mid z = Wy, \ y \geq 0\} = \mathbb{R}^{m_1}.$$

From this condition, it follows immediately that $\operatorname{rk}(W) = m_1$ necessarily has to hold. In addition, for $\hat{z} = -\sum_{i=1}^{m_1} W_i \in \mathbb{R}^{m_1}$ the second-stage constraints $Wy = \hat{z}, y \geq 0$ have a feasible solution $\check{y}$ such that

$$\begin{aligned} \sum_{i=1}^{m_1} W_i \check{y}_i + \sum_{i=m_1+1}^{\overline{n}} W_i \check{y}_i &= \hat{z} \\ &= -\sum_{i=1}^{m_1} W_i, \\ \check{y}_i \geq 0, \quad i &= 1, \cdots, \overline{n}. \end{aligned}$$

With

$$y_i = \begin{cases} \check{y}_i + 1, & i = 1, \cdots, m_1, \\ \check{y}_i, & i > m_1, \end{cases}$$

this implies that the constraints (4.4) are necessarily feasible.

To show that the above conditions are also sufficient for complete recourse let us choose an arbitrary $\bar{z} \in \mathbb{R}^{m_1}$. Since the columns $W_1, \cdots, W_{m_1}$ are linearly independent, the system of linear equations

$$\sum_{i=1}^{m_1} W_i y_i = \bar{z}$$

has a unique solution $\bar{y}_1, \cdots, \bar{y}_{m_1}$. If $\bar{y}_i \geq 0, \ i = 1, \cdots, m_1$, we are finished; otherwise, we define $\gamma := \min\{\bar{y}_1, \cdots, \bar{y}_{m_1}\}$. By assumption, the constraints (4.4) have a feasible solution $\check{y}$. Now it is immediate that $\hat{y}$ defined by

$$\hat{y}_i = \begin{cases} \bar{y}_i - \gamma \check{y}_i, & i = 1, \cdots, m_1, \\ -\gamma \check{y}_i, & i = m_1 + 1, \cdots, \overline{n}, \end{cases}$$

solves $Wy = \bar{z}, \ y \geq 0$. □

Finally, if (4.3) is feasible for all $\xi \in \Xi$ and at least for all $x \in X = \{x \mid Ax = b, x \geq 0\}$ then (3.16) is said to be of *relatively complete recourse.*

1.5 Properties of Probabilistic Constraints

For chance constrained problems, the situation becomes more difficult, in general. Consider the constraint of (3.21),

$$P(\{\xi \mid g(x,\xi) \leq 0\}) \geq \alpha,$$

where the g_i were replaced by the vector-valued function g defined by $g(x,\xi) := \left(g_1(x,\xi), \cdots, g_m(x,\xi)\right)^{\mathrm{T}}$: a point $\hat{x}$ is feasible iff the set

$$S(\hat{x}) = \{\xi \mid g(\hat{x},\xi) \leq 0\} \tag{5.1}$$

has a probability measure $P(S(\hat{x}))$ of at least α. In other words, if $\mathcal{G} \subset \mathcal{F}$ is the collection of all events of $\mathcal{F}$ such that $P(G) \geq \alpha \; \forall G \in \mathcal{G}$ then $\hat{x}$ is feasible iff we find at least one event $\tilde{G} \in \mathcal{G}$ such that for all $\xi \in \tilde{G}$, $g(\hat{x},\xi) \leq 0$. Formally, $\hat{x}$ is feasible iff $\exists G \in \mathcal{G}$:

$$\hat{x} \in \bigcap_{\xi \in G} \{x \mid g(x,\xi) \leq 0\}. \tag{5.2}$$

Hence the feasible set

$$\mathcal{B}(\alpha) = \{x \mid P(\{\xi \mid g(x,\xi) \leq 0\}) \geq \alpha\}$$

is the union of all those vectors x feasible according to (5.2), and consequently may be rewritten as

$$\mathcal{B}(\alpha) = \bigcup_{G \in \mathcal{G}} \bigcap_{\xi \in G} \{x \mid g(x,\xi) \leq 0\}. \tag{5.3}$$

Since a union of convex sets need not be convex, this presentation demonstrates that in general we may not expect $\mathcal{B}(\alpha)$ to be convex, even if $\{x \mid g(x,\xi) \leq 0\}$ are convex $\forall \xi \in \Xi$. Indeed, there are simple examples for nonconvex feasible sets.

Example 1.4 Assume that in our refinery problem (2.1) the demands are random with the following discrete joint distribution:

$$P\begin{pmatrix} h_1(\xi^1) = 160 \\ h_2(\xi^1) = 135 \end{pmatrix} = 0.85,$$

$$P\begin{pmatrix} h_1(\xi^2) = 150 \\ h_2(\xi^2) = 195 \end{pmatrix} = 0.08,$$

$$P\begin{pmatrix} h_1(\xi^3) = 200 \\ h_2(\xi^3) = 120 \end{pmatrix} = 0.07.$$

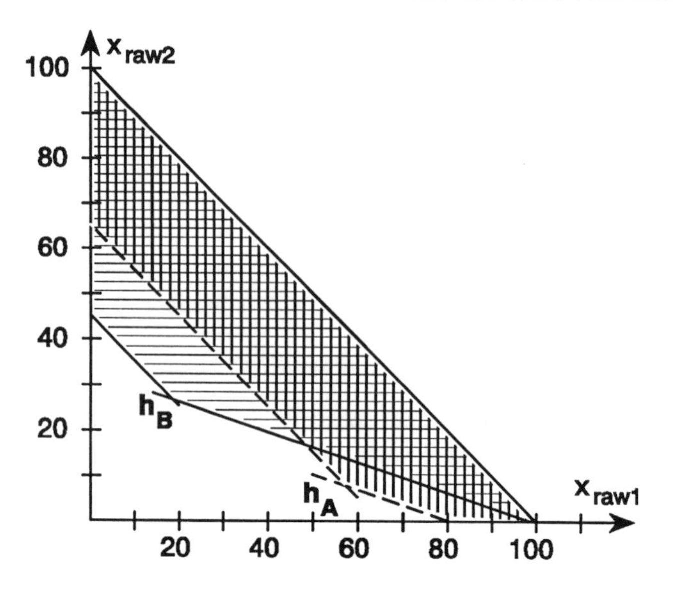

Figure 17 Chance constraints: nonconvex feasible set.

Then the constraints

$$\begin{aligned} x_{raw1} + x_{raw2} &\le 100 \\ x_{raw1} \qquad &\ge 0 \\ x_{raw2} &\ge 0 \end{aligned}$$

$$P\begin{pmatrix} 2x_{raw1} + 6x_{raw2} \ge h_1(\tilde{\xi}) \\ 3x_{raw1} + 3x_{raw2} \ge h_2(\tilde{\xi}) \end{pmatrix} \ge \alpha$$

for any $\alpha \in (0.85, 0.92]$ require that we

- <u>either</u> satisfy the demands $h_i(\xi^1)$ <u>and</u> $h_i(\xi^2)$, $i = 1, 2$ (enforcing a reliability of 93%) and hence choose a production program to cover a demand $h_A = \begin{pmatrix} 160 \\ 195 \end{pmatrix}$
- <u>or</u> satisfy the demands $h_i(\xi^1)$ <u>and</u> $h_i(\xi^3)$, $i = 1, 2$ (enforcing a reliability of 92%) such that our production plan is designed to cope with the demand $h_B = \begin{pmatrix} 200 \\ 135 \end{pmatrix}$.

It follows that the feasible set for the above constraints is nonconvex, as shown in Figure 17. □

As above, define $S(x) := \{\xi \mid g(x,\xi) \le 0\}$. If $g(\cdot,\cdot)$ is jointly convex in (x,ξ) then, with $x^i \in \mathcal{B}(\alpha), i = 1,2,\ \xi^i \in S(x^i)$ and $\lambda \in [0,1]$, for $(\bar{x},\bar{\xi}) = \lambda(x^1,\xi^1) + (1-\lambda)(x^2,\xi^2)$ it follows that

$$g(\bar{x},\bar{\xi}) \le \lambda g(x^1,\xi^1) + (1-\lambda)g(x^2,\xi^2) \le 0,$$

i.e. $\bar{\xi} = \lambda\xi^1 + (1-\lambda)\xi^2 \in S(\bar{x})$, and hence[5]

$$S(\bar{x}) \supset [\lambda S(x^1) + (1-\lambda)S(x^2)]$$

implying

$$P(S(\bar{x})) \ge P(\lambda S(x^1) + (1-\lambda)S(x^2)).$$

By our assumption on g (joint convexity), any set $S(x)$ is convex. Now we conclude immediately that $\mathcal{B}(\alpha)$ is convex $\forall\alpha \in [0,1]$, if

$$P(\lambda S_1 + (1-\lambda)S_2) \ge \min[P(S_1), P(S_2)] \quad \forall\lambda \in [0,1]$$

for all convex sets $S_i \in \mathcal{F},\ i = 1,2$, i.e. if P is *quasi-concave*. Hence we have proved the following

Proposition 1.5 *If $g(\cdot,\cdot)$ is jointly convex in (x,ξ) and P is quasi-concave, then the feasible set $\mathcal{B}(\alpha) = \{x \mid P(\{\xi \mid g(x,\xi) \le 0\}) \ge \alpha\}$ is convex $\forall\alpha \in [0,1]$.*

Remark 1.3 The assumption of joint convexity of $g(\cdot,\cdot)$ is so strong that it is even not satisfied in the linear case (3.23), in general. However, if in (3.23) $T(\xi) \equiv T$ (constant) and $h(\xi) \equiv \xi$ then it is satisfied and the constraints of (3.23), $F_{\tilde{\xi}}$ being the distribution function of $\tilde{\xi}$, read as

$$P(\{\xi \mid Tx \ge \xi\}) = F_{\tilde{\xi}}(Tx) \ge \alpha.$$

Therefore $\mathcal{B}(\alpha)$ is convex $\forall\alpha \in [0,1]$ in this particular case if $F_{\tilde{\xi}}$ is a quasi-concave function, i.e. if $F_{\tilde{\xi}}(\lambda\xi^1 + (1-\lambda)\xi^2) \ge \min[F_{\tilde{\xi}}(\xi^1), F_{\tilde{\xi}}(\xi^2)]$ for any two $\xi^1,\xi^2 \in \Xi$ and $\forall\lambda \in [0,1]$. □

It seems worthwile to mention the following facts. If the probability measure P is quasi-concave then the corresponding distribution function $F_{\tilde{\xi}}$ is quasi-concave. This follows from observing that by the definition of distribution functions $F_{\tilde{\xi}}(\xi^i) = P(S_i)$ with $S_i = \{\xi \mid \xi \le \xi^i\},\ i = 1,2$, and that for $\hat{\xi} = \lambda\xi^1 + (1-\lambda)\xi^2,\ \lambda \in [0,1]$, we have $\hat{S} = \{\xi \mid \xi \le \hat{\xi}\} = \lambda S_1 + (1-\lambda)S_2$ (see Figure 18). With P being quasi-concave, this yields

$$F_{\tilde{\xi}}(\hat{\xi}) = P(\hat{S}) \ge \min[P(S_1), P(S_2)] = \min[F_{\tilde{\xi}}(\xi^1), F_{\tilde{\xi}}(\xi^2)].$$

[5] The algebraic sum of sets $\rho S_1 + \sigma S_2 := \{\xi := \rho\xi^1 + \sigma\xi^2 \mid \xi^1 \in S_1,\ \xi^2 \in S_2\}$.

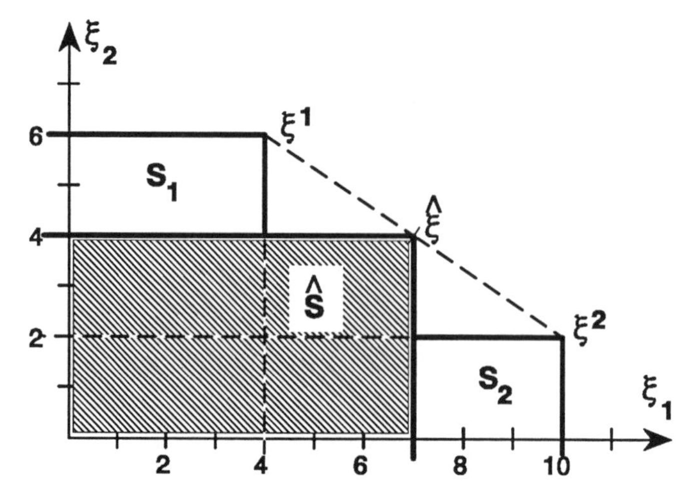

Figure 18 Convex combination of events involved by distribution functions, $\lambda = \frac{1}{2}$.

On the other hand, $F_{\tilde{\xi}}$ being a quasi-concave function does not imply in general that the corresponding probability measure P is quasi-concave. For instance, in $\mathbb{R}^1$ every monotone function is easily seen to be quasi-concave, such that every distribution function of a random variable (always being monotonically increasing) is quasi-concave.But not every probability measure P on $\mathbb{R}$ is quasi-concave (see Figure 19 for a counterexample).

Hence we stay with the question of when a probability measure—or its distribution function—is quasi-concave. This question was answered first for the subclass of *log-concave probability measures*, i.e. measures satisfying

$$P(\lambda S_1 + (1-\lambda)S_2) \geq P^{\lambda}(S_1)\, P^{1-\lambda}(S_2)$$

for all convex $S_i \in \mathcal{F}$ and $\lambda \in [0,1]$. That the class of log-concave measures is really a subclass of the class of quasi-concave measures is easily seen.

Lemma 1.2 *If P is a log-concave measure on $\mathcal{F}$ then P is quasi-concave.*

Proof Let $S_i \in \mathcal{F}$, $i = 1,2$, be convex sets such that $P(S_i) > 0$, $i = 1,2$ (otherwise there is nothing to prove, since $P(S) \geq 0\ \forall S \in \mathcal{F}$). By assumption, for any $\lambda \in (0,1)$ we have

$$P(\lambda S_1 + (1-\lambda)S_2) \geq P^{\lambda}(S_1)\, P^{1-\lambda}(S_2).$$

By the monotonicity of the logarithm, it follows that

$$\begin{aligned}\ln[P(\lambda S_1 + (1-\lambda)S_2)] &\geq \lambda \ln[P(S_1)] + (1-\lambda)\ln[P(S_2)] \\ &\geq \min\{\ln[P(S_1)], \ln[P(S_2)]\},\end{aligned}$$

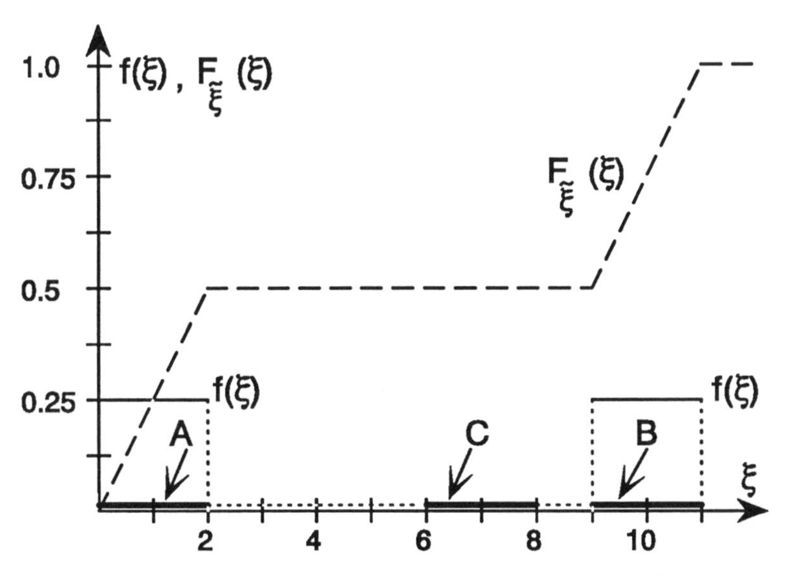

Figure 19 P here is not quasi-concave: $P(C) = P(\frac{1}{3}A + \frac{2}{3}B) = 0$, but $P(A) = P(B) = \frac{1}{2}$.

and hence

$$P(\lambda S_1 + (1-\lambda)S_2) \geq \min[P(S_1), P(S_2)].$$

□

As mentioned above, for the log-concave case necessary and sufficient conditions were derived first, and later corresponding conditions for quasi-concave measures were found.

Proposition 1.6 *Let P on $\Xi = \mathbb{R}^k$ be of the continuous type, i.e. have a density f. Then the following statements hold:*

- *P is log-concave iff f is log-concave (i.e. if the logarithm of f is a concave function);*
- *P is quasi-concave iff $f^{-1/k}$ is convex.*

The proof has to be omitted here, since it would require a rather advanced knowledge of measure theory.

Remark 1.4 Consider

(a) the k-dimensional uniform distribution on a convex body $S \subset \mathbb{R}^k$ (with positive natural measure μ) given by the density

$$\varphi_{\mathcal{U}}(x) := \begin{cases} 1/\mu(S) & \text{if } x \in S, \\ 0 & \text{otherwise} \end{cases}$$

(μ is the natural measure in $\mathbb{R}^k$, see Section 1.3.1);

(b) the exponential distribution with density

$$\varphi_{\mathcal{EXP}}(x) := \begin{cases} 0 & \text{if } x < 0, \\ \lambda e^{-\lambda x} & \text{if } x \geq 0 \end{cases}$$

($\lambda > 0$ is constant);

(c) the multivariate normal distribution in $\mathbb{R}^k$ described by the density

$$\varphi_{\mathcal{N}}(x) := \gamma e^{-\frac{1}{2}(x-m)^{\mathrm{T}}\Sigma^{-1}(x-m)}$$

($\gamma > 0$ is constant, m is the vector of expected values and Σ is the covariance matrix).

Then we get immediately

(a) $$\varphi_{\mathcal{U}}^{-\frac{1}{k}}(x) = \begin{cases} \sqrt[k]{\mu(S)} & \text{if } x \in S, \\ \infty & \text{otherwise,} \end{cases}$$

implying by Proposition 1.6 that the corresponding propability measure $P_{\mathcal{U}}$ is quasi-concave.

(b) Since

$$\ln[\varphi_{\mathcal{EXP}}(x)] = \begin{cases} -\infty & \text{if } x < 0, \\ \ln\lambda - \lambda x & \text{if } x \geq 0, \end{cases}$$

the density of the exponential distribution is obviously log-concave, implying by Proposition 1.6 that the corresponding measure $P_{\mathcal{EXP}}$ is log-concave and hence, by Lemma 1.2, also quasi-concave.

(c) Taking the logarithm

$$\ln[\varphi_{\mathcal{N}}(x)] = \ln\gamma - \tfrac{1}{2}(x-m)^{\mathrm{T}}\Sigma^{-1}(x-m)$$

and observing that the covariance matrix Σ and hence its inverse Σ^{-1} are positive definite, we see that this density is log-concave, and therefore the corresponding measure $P_{\mathcal{N}}$ is log-concave (by Proposition 1.6) as well as quasiconcave (by Lemma 1.2).

There are many other classes of widely used continuous type probability measures, which—according to Proposition 1.6—are either log-concave or at least quasi-concave. □

In addition to Proposition 1.5, we have the following statement, which is of interest because, for mathematical programs in general, we cannot assert the existence of solutions if the feasible sets are not known to be closed.

Proposition 1.7 *If $g : \mathbb{R}^n \times \Xi \longrightarrow \mathbb{R}^m$ is continuous then the feasible set $\mathcal{B}(\alpha)$ is closed.*

Proof Consider any sequence $\{x^\nu\}$ such that $x^\nu \longrightarrow \hat{x}$ and $x^\nu \in \mathcal{B}(\alpha)\ \forall \nu$. To prove the assertion, we have to show that $\hat{x} \in \mathcal{B}(\alpha)$. Define $A(x) := \{\xi \mid g(x, \xi) \leq 0\}$. Let V_k be the open ball with center $\hat{x}$ and radius $1/k$. Then we show first that

$$A(\hat{x}) = \bigcap_{k=1}^{\infty} \operatorname{cl} \bigcup_{x \in V_k} A(x). \tag{5.4}$$

Here the inclusion "$\subset$" is obvious since $\hat{x} \in V_k\ \forall k$, so we have only to show that

$$A(\hat{x}) \supset \bigcap_{k=1}^{\infty} \operatorname{cl} \bigcup_{x \in V_k} A(x).$$

Assume that $\hat{\xi} \in \bigcap_{k=1}^{\infty} \operatorname{cl} \bigcup_{x \in V_k} A(x)$. This means that for every k we have $\hat{\xi} \in \operatorname{cl} \bigcup_{x \in V_k} A(x)$; in other words, for every k there exists a $\xi^k \in \bigcup_{x \in V_k} A(x)$ and hence some $x^k \in V_k$ with $\xi^k \in A(x^k)$ such that $\|\xi^k - \hat{\xi}\| \leq 1/k$ (and obviously $\|x^k - \hat{x}\| \leq 1/k$ since $x^k \in V_k$). Hence $(x^k, \xi^k) \longrightarrow (\hat{x}, \hat{\xi})$. Since $\xi^k \in A(x^k)$, $g(x^k, \xi^k) \leq 0\ \forall k$ and therefore, by the continuity of $g(\cdot, \cdot)$, $\hat{\xi} \in A(\hat{x})$, which proves (5.4) to be true.

The sequence of sets

$$B_K := \bigcap_{k=1}^{K} \operatorname{cl} \bigcup_{x \in V_k} A(x)$$

is monotonically decreasing to the set $A(\hat{x})$. Since $x^\nu \longrightarrow \hat{x}$, for every K there exists a ν_K such that $x^{\nu_K} \in V_K \subset V_{K-1} \subset \cdots \subset V_1$, implying that $A(x^{\nu_K}) \subset B_K$ and hence $P(B_K) \geq P(A(x^{\nu_K})) \geq \alpha\ \forall K$. Hence, by the well-known continuity of probability measures on monotonic sequences, we have $P(A(\hat{x})) \geq \alpha$, i.e. $\hat{x} \in \mathcal{B}(\alpha)$. □

For *stochastic programs with joint chance constraints* the situation appears to be more difficult than for *stochastic programs with recourse*. But, at least under certain additional assumptions, we may assert convexity and closedness of the feasible sets as well (Proposition 1.5, Remark 1.3 and Proposition 1.7).

For stochastic linear programs with single chance constraints, convexity statements have been derived without the joint convexity assumption on $g_i(x, \xi) := h_i(\xi) - T_i(\xi)x$, for special distributions and special intervals for the values of α_i. In particular, if $T_i(\xi) \equiv T_i$ (constant), the situation becomes rather convenient: with F_i the distribution function of $h_i(\tilde{\xi})$, we have

$$P(\{\xi \mid T_i x \geq h_i(\xi)\}) = F_i(T_i x) \geq \alpha_i,$$

or equivalently

$$T_i x \geq F_i^{-1}(\alpha_i),$$

where $F_i^{-1}(\alpha_i)$ is assumed to be the smallest real value η such that $F_i(\eta) \geq \alpha_i$. Hence in this special case any single chance constraint turns out to be just a linear constraint, and the only additional work to do is to compute $F_i^{-1}(\alpha_i)$.

1.6 Linear Programming

Throughout this section we shall discuss linear programs in the following standard form

$$\left.\begin{aligned} &\min c^{\mathrm{T}} x \\ &\text{s.t. } Ax = b, \\ &\qquad\ x \geq 0, \end{aligned}\right\} \tag{6.1}$$

where the vectors $c \in \mathbb{R}^n$, $b \in \mathbb{R}^m$ and the $m \times n$ matrix A are given and $x \in \mathbb{R}^n$ is to be determined. Any other LP[6] formulation can easily be transformed to assume the form (6.1). If, for instance, we have the problem

$$\begin{aligned} &\min c^{\mathrm{T}} x \\ &\text{s.t. } Ax \geq b \\ &\qquad\ x \geq 0, \end{aligned}$$

then, by introducing a vector $y \in \mathbb{R}^m_+$ of *slack variables*, we get the problem

$$\begin{aligned} &\min c^{\mathrm{T}} x \\ &\text{s.t. } Ax - y = b \\ &\qquad\qquad x \geq 0 \\ &\qquad\qquad y \geq 0, \end{aligned}$$

which is of the form (6.1). This LP is equivalent to (6.1) in the sense that the x part of its solution set and the solution set of (6.1) as well as the two optimal values obviously coincide. Instead, we may have the problem

$$\begin{aligned} &\min c^{\mathrm{T}} x \\ &\text{s.t. } Ax \geq b, \end{aligned}$$

where the decision variables are not required to be nonnegative—so-called *free variables*. In this case we may introduce a vector $y \in \mathbb{R}^m_+$ of slack variables and—observing that any real number may be presented as the difference of two nonnegative numbers—replace the original decision vector x by the difference

[6] We use occasionally "LP" as abbreviation for "linear program(ming)".

$z^+ - z^-$ of the new decision vectors $z^+, z^- \in \mathbb{R}^n_+$ yielding the problem

$$\begin{aligned} \min\{c^{\mathrm{T}} z^+ - c^{\mathrm{T}} z^-\} \\ \text{s.t. } Az^+ - Az^- - y = b, \\ z^+ \geq 0, \\ z^- \geq 0, \\ y \geq 0, \end{aligned}$$

which is again of the form (6.1). Furthermore, it is easily seen that this transformed LP and its original formulation are equivalent in the sense that

- given any solution $(\hat{z}^+, \hat{z}^-, \hat{y})$ of the transformed LP, $\hat{x} := \hat{z}^+ - \hat{z}^-$ is a solution of the original version,
- given any solution $\check{x}$ of the original LP, the vectors $\check{y} := A\check{x} - b$ and $\check{z}^+, \check{z}^- \in \mathbb{R}^n_+$, chosen such that $\check{z}^+ - \check{z}^- = \check{x}$, solve the transformed version,

and the optimal values of both versions of the LP coincide.

1.6.1 The Feasible Set and Solvability

From linear algebra, we know that the system $Ax = b$ of linear equations in (6.1) is solvable if and only if the rank condition

$$\text{rk}(A, b) = \text{rk}(A) \tag{6.2}$$

is satisfied. Given this condition, it may happen that $\text{rk}(A) < m$, but then we may drop one or more equations from the system without changing its solution set. Therefore we assume throughout this section that

$$\text{rk}(A) = m, \tag{6.3}$$

which obviously implies that $m \leq n$.

Let us now investigate the *feasible set*

$$\mathcal{B} := \{x \mid Ax = b,\ x \geq 0\}$$

of (6.1). A central concept in linear programming is that of a *feasible basic solution* defined as follows: $\hat{x} \in \mathcal{B}$ is a feasible basic solution if with $I(\hat{x}) := \{i \mid \hat{x}_i > 0\}$ the set $\{A_i \mid i \in I(\hat{x})\}$ of columns of A is linearly independent.[7] Hence the components $\hat{x}_i$, $i \in I(\hat{x})$, are the unique solution of the system of linear equations

$$\sum_{i \in I(\hat{x})} A_i x_i = b.$$

[7] According to this definition, for $I(\hat{x}) = \emptyset$, i.e. $\hat{x} = 0$ and hence $b = 0$, it follows that $\hat{x}$ is a feasible basic solution as well.

In general, the set $I(\hat{x})$ and hence also the column set $\{A_i \mid i \in I(\hat{x})\}$ may have less than m elements, which can cause some inconvenience—at least in formulating the statements we want to present.

Proposition 1.8 *Given assumption (6.3), for any basic solution $\hat{x}$ of $\mathcal{B}$ there exists at least one index set $I_B(\hat{x}) \supset I(\hat{x})$ such that the corresponding column set $\{A_i \mid i \in I_B(\hat{x})\}$ is a basis of $\mathbb{R}^m$. The components $\hat{x}_i$, $i \in I_B(\hat{x})$, of $\hat{x}$ uniquely solve the linear system $\sum_{i \in I_B(\hat{x})} A_i x_i = b$ with the nonsingular matrix $(A_i \mid i \in I_B(\hat{x}))$.*

Proof Assume that $\check{x} \in \mathcal{B}$ is a basic solution and that $\{A_i \mid i \in I(\check{x})\}$ contains k columns, $k < m$, of A. By (6.3), there exists at least one index set $J_m \subset \{1, \cdots, n\}$ with m elements such that the columns $\{A_i \mid i \in J_m\}$ are linearly independent and hence form a basis of $\mathbb{R}^m$. A standard result in linear algebra asserts that, given a basis of an m-dimensional vector space and a linear independent subset of $k < m$ vectors, it is possible, by adding $m - k$ properly chosen vectors from the basis, to complement the subset to become a basis itself. Hence in our case it is possible to choose $m - k$ indices from J_m and to add them to $I(\check{x})$, yielding $I_B(\check{x})$ such that $\{A_i \mid i \in I_B(\check{x})\}$ is a basis of $\mathbb{R}^m$. □

Given a basic solution $\hat{x} \in \mathcal{B}$, by this proposition the matrix A can be partitioned into two parts (corresponding to $\hat{x}$): a *basic* part

$$B = (A_i \mid i \in I_B(\hat{x}))$$

and a *nonbasic* part

$$N = (A_i \mid i \in \{1, \cdots, n\} - I_B(\hat{x})).$$

Introducing the vectors $x^{\{B\}} \in \mathbb{R}^m$—the vector of *basic variables*—and $x^{\{NB\}} \in \mathbb{R}^{n-m}$—the vector of *nonbasic variables*—and assigning

$$\begin{aligned} x_k^{\{B\}} &= x_i,\ i \text{ the } k\text{th element of } I_B(\hat{x}),\ k = 1, \cdots, m, \\ x_l^{\{NB\}} &= x_i,\ i \text{ the } l\text{th element of } \{1, \cdots, n\} - I_B(\hat{x}),\ l = 1, \cdots, n - m, \end{aligned} \tag{6.4}$$

the linear system $Ax = b$ of (6.1) may be rewritten as

$$Bx^{\{B\}} + Nx^{\{NB\}} = b$$

or equivalently as

$$x^{\{B\}} = B^{-1}b - B^{-1}Nx^{\{NB\}}, \tag{6.5}$$

which—using the assignment (6.4) – yields for any choice of the nonbasic variables $x^{\{NB\}}$ a solution of our system $Ax = b$, and in particular for $x^{\{NB\}} = 0$ reproduces our feasible basic solution $\hat{x}$.

Proposition 1.9 *If $\mathcal{B} \neq \emptyset$ then there exists at least one feasible basic solution.*

Proof Assume that for $\hat{x}$

$$A\hat{x} = b, \ \hat{x} \geq 0.$$

If for $I(\hat{x}) = \{i \mid \hat{x}_i > 0\}$ the column set $\{A_i \mid i \in I(\hat{x})\}$ is linearly dependent, then the linear homogeneous system of equations

$$\begin{aligned} \textstyle\sum_{i \in I(\hat{x})} A_i y_i &= 0, \\ y_i &= 0, \ i \notin I(\hat{x}), \end{aligned}$$

has a solution $\check{y} \neq 0$ with $\check{y}_i < 0$ for at least one $i \in I(\hat{x})$—if this does not hold for $\check{y}$, we could take $-\check{y}$, which solves the above homogeneous system as well. Hence for

$$\bar{\lambda} := \max\{\lambda \mid \hat{x} + \lambda\check{y} \geq 0\}$$

we have $0 < \bar{\lambda} < \infty$. Since $A\check{y} = 0$ obviously holds for $\check{y}$, it follows—observing the definition of $\bar{\lambda}$—that for $z := \hat{x} + \bar{\lambda}\check{y}$

$$\begin{aligned} Az &= A\hat{x} + \bar{\lambda}A\check{y} \\ &= b, \\ z &\geq 0, \end{aligned}$$

i.e. $z \in \mathcal{B}$, and $I(z) \subset I(\hat{x})$, $I(z) \neq I(\hat{x})$, such that we have "reduced" our original feasible solution $\hat{x}$ to another one with fewer positive components. Now either z is a basic solution or we repeat the above "reduction" with $\hat{x} := z$. Obviously there are only finitely many reductions of the number of positive components in feasible solutions possible. Hence we have to end up—after finitely many of these steps—with a feasible basic solution. □

With an elementary exercise, we see that the feasible set $\mathcal{B} = \{x \mid Ax = b, \ x \geq 0\}$ of our linear program (6.1) is convex. We want now to point out that feasible basic solutions play a dominant role in describing feasible sets of linear programs.

Proposition 1.10 *If $\mathcal{B}$ is a bounded set and $\mathcal{B} \neq \emptyset$ then $\mathcal{B}$ is the convex hull (i.e. the set of all convex linear combinations) of the set of its feasible basic solutions.*

Proof To avoid trivialities or statements on empty sets, we assume that the right-hand side $b \neq 0$. For any feasible solution $x \in \mathcal{B}$ we again have the index set $I(x) := \{i \mid x_i > 0\}$, and we denote by $|I(x)|$ the number of elements of $I(x)$. Obviously we have—recalling our assumption that $b \neq 0$—that for any feasible solution $1 \leq |I(x)| \leq n$. We may prove the proposition by induction on $|I(x)|$, the number of positive components of any feasible solution x. To begin with, we define $k_0 := \min_{x \in \mathcal{B}} |I(x)| \geq 1$. For a feasible x with $|I(x)| = k_0$ it follows that x is a basic solution—otherwise, by the proof of Proposition 1.9, there would exist a feasible basic solution with less than k_0 positive components—and we have $x = 1 \cdot x$, i.e. a convex linear combination of itself and hence of the set of feasible basic solutions. Let us now assume that for some $k \geq k_0$ and for all feasible solutions x such that $|I(x)| \leq k$ the hypothesis is true. Then, given a feasible solution $\hat{x}$ with $|I(\hat{x})| = k + 1$, for $\hat{x}$ a basic solution we again have $\hat{x} = 1 \cdot \hat{x}$ and thus the hypothesis holds. Otherwise, i.e. if $\hat{x}$ is not a basic solution, the homogeneous system

$$\begin{aligned} \textstyle\sum_{i \in I(\hat{x})} A_i y_i &= 0 \\ y_i &= 0, \quad i \notin I(\hat{x}) \end{aligned}$$

has a solution $\tilde{y} \neq 0$, for which at least one component is strictly negative and another is strictly positive, since otherwise we could assume $\tilde{y} \geq 0$, $\tilde{y} \neq 0$, to solve the homogeneous system $Ay = 0$, implying that $\hat{x} + \lambda\tilde{y} \in \mathcal{B}\ \forall \lambda \geq 0$, which, according to the inequality $\|\hat{x} + \lambda\tilde{y}\| \geq \lambda\|\tilde{y}\| - \|\hat{x}\|$, contradicts the assumed boundedness of $\mathcal{B}$. Hence we find for

$$\alpha := \max\{\lambda \mid \hat{x} + \lambda\tilde{y} \geq 0\},$$

$$\beta := \min\{\lambda \mid \hat{x} + \lambda\tilde{y} \geq 0\}$$

that $0 < \alpha < \infty$ and $0 > \beta > -\infty$. Defining $v := \hat{x} + \alpha\tilde{y}$ and $w := \hat{x} + \beta\tilde{y}$, we have $v, w \in \mathcal{B}$ and—by the definitions of α and β—that $|I(v)| \leq k$ and $|I(w)| \leq k$ such that, according to our induction assumption, with $\{x^{\{i\}},\ i = 1, \cdots, r\}$ the set of all feasible basic solutions, $v = \sum_{i=1}^{r} \lambda_i x^{\{i\}}$, where $\sum_{i=1}^{r} \lambda_i = 1$, $\lambda_i \geq 0\ \forall i$, and $w = \sum_{i=1}^{r} \mu_i x^{\{i\}}$, where $\sum_{i=1}^{r} \mu_i = 1$, $\mu_i \geq 0\ \forall i$. As is easily checked, we have $\hat{x} = \rho v + (1 - \rho)w$ with $\rho = -\beta/(\alpha - \beta) \in (0, 1)$. This implies immediately that $\hat{x}$ is a convex linear combination of $\{x^{\{i\}},\ i = 1, \cdots, r\}$. □

The convex hull of finitely many points $\{x^{\{1\}}, \cdots, x^{\{r\}}\}$, formally denoted by $\operatorname{conv}\{x^{\{1\}}, \cdots, x^{\{r\}}\}$, is called a *convex polyhedron* or a *bounded convex polyhedral set* (see Figure 20). Take for instance in $\mathbb{R}^2$ the points $z^1 = (2, 2), z^2 = (8, 1), z^3 = (4, 3), z^4 = (7, 7)$ and $z^5 = (1, 6)$. In Figure 21 we have $\tilde{\mathcal{P}} = \operatorname{conv}\{z^1, \cdots, z^5\}$, and it is obvious that z^3 is not necessary to generate $\tilde{\mathcal{P}}$; in other words, $\tilde{\mathcal{P}} = \operatorname{conv}\{z^1, z^2, z^3, z^4, z^5\} = \operatorname{conv}\{z^1, z^2, z^4, z^5\}$. Hence we may drop z^3 without any effect on the polyhedron $\tilde{\mathcal{P}}$, whereas omitting any other of the five points would essentially change the shape of the polyhedron. The points that really count in the definition of a convex polyhedron are its vertices (z^1, z^2, z^4 and z^5 in the example). Whereas in two- or three-dimensional spaces, we know by intuition what we mean by a vertex, we need a formal definition for higher-dimensional cases: A *vertex of a convex*

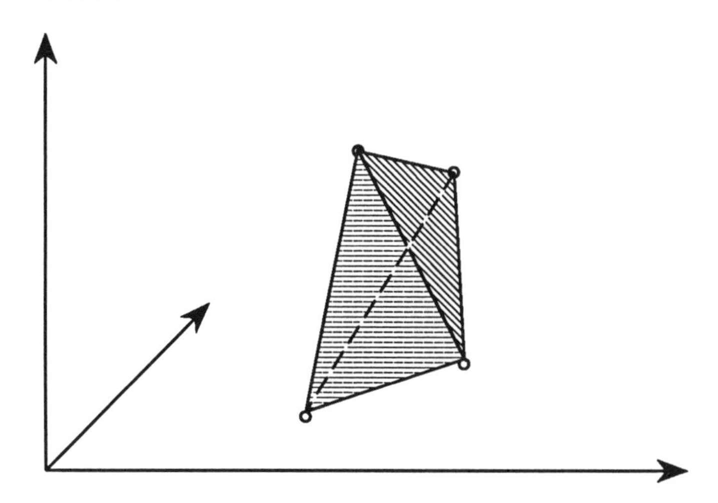

Figure 20 LP: bounded feasible set.

polyhedron $\mathcal{P}$ is a point $\hat{x} \in \mathcal{P}$ such that the line segment connecting any two points in $\mathcal{P}$, both different from $\hat{x}$, does not contain $\hat{x}$. Formally,

$$\nexists y, z \in \mathcal{P}, \ y \neq \hat{x} \neq z, \ \lambda \in (0,1), \text{ such that } \hat{x} = \lambda y + (1-\lambda)z.$$

It may be easily shown that for an LP with a bounded feasible set $\mathcal{B}$ the feasible basic solutions $x^{\{i\}}$, $i = 1, \cdots, r$, coincide with the vertices of $\mathcal{B}$.

By Proposition 1.10, the feasible set of a linear program is a convex polyhedron provided that $\mathcal{B}$ is bounded. Hence we have to find out under what conditions $\mathcal{B}$ is bounded or unbounded respectively. For $\mathcal{B} \neq \emptyset$ we have seen already in the proof of Proposition 1.10 that the existence of a $\tilde{y} \neq 0$ such that $A\tilde{y} = 0$, $\tilde{y} \geq 0$, would imply that $\mathcal{B}$ is unbounded. Therefore, for $\mathcal{B}$ to be bounded, the condition $\{y \mid Ay = 0, \ y \geq 0\} = \{0\}$ is necessary. Moreover, we have the following.

Proposition 1.11 *The feasible set $\mathcal{B} \neq \emptyset$ is bounded iff*

$$\{y \mid Ay = 0, \ y \geq 0\} = \{0\}.$$

Proof Given the above observations, it is only left to show that the condition $\{y \mid Ay = 0, \ y \geq 0\} = \{0\}$ is sufficient for the boundedness of $\mathcal{B}$. Assume in contrast that $\mathcal{B}$ is unbounded. This means that we have feasible solutions arbitrarily large in norm. Hence for any natural number K there exists an $x^K \in \mathcal{B}$ such that $\|x^K\| \geq K$. Defining

$$z^K := \frac{x^K}{\|x^K\|} \ \forall K,$$

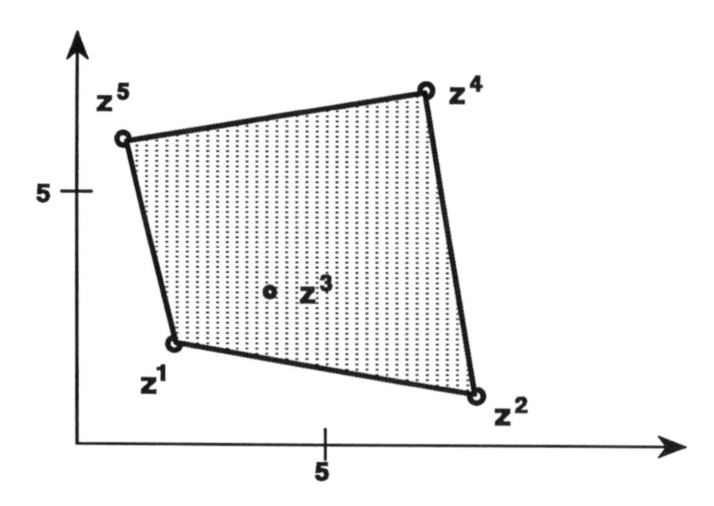

Figure 21 Polyhedron generated by its vertices.

we have

$$\left.\begin{aligned} z^K &\geq 0, \\ \|z^K\| &= 1, \\ Az^K &= b/\|x^K\|, \\ \text{and hence} \\ \|Az^K\| &\leq \|b\|/K \end{aligned}\right\} \forall K. \tag{6.6}$$

Therefore the sequence $\{z^K,\ K = 1,2,\cdots\}$ has an accumulation point $\hat{z}$, for which, according to (6.6), $\hat{z} \geq 0,\ \|\hat{z}\| = 1$ and $\|A\hat{z}\| = 0$, and hence $A\hat{z} = 0,\ \hat{z} \geq 0,\ \hat{z} \neq 0$. □

According to Proposition 1.11, the set $\mathcal{C} := \{y \mid Ay = 0,\ y \geq 0\}$ plays a decisive role for the boundedness or unboundedness of the feasible set $\mathcal{B}$. We see immediately that $\mathcal{C}$ is a *convex cone*, which means that for any two elements $y, z \in \mathcal{C}$ it follows that $\lambda y + \mu z \in \mathcal{C}\ \forall \lambda, \mu \geq 0$. In addition, we may show that $\mathcal{C}$ is a *convex polyhedral cone*, i.e. there exist finitely many $y^{\{i\}} \in \mathcal{C},\ i = 1, \cdots, s$, such that any $y \in \mathcal{C}$ may be represented as $y = \sum_{i=1}^{s} \alpha_i y^{\{i\}},\ \alpha_i \geq 0\ \forall i$. Formally, we also may speak of the *positive hull* denoted by $\mathrm{pos}\{y^{\{1\}}, \cdots, y^{\{s\}}\} := \{y \mid y = \sum_{i=1}^{s} \alpha_i y^{\{i\}},\ \alpha_i \geq 0\ \forall i\}$.

Proposition 1.12 *The set* $\mathcal{C} = \{y \mid Ay = 0,\ y \geq 0\}$ *is a convex polyhedral cone.*

Proof Since for $\mathcal{C} = \{0\}$ the statement is trivial, we assume that $\mathcal{C} \neq \{0\}$. For any arbitrary $\hat{y} \in \mathcal{C}$ such that $\hat{y} \neq 0$ and hence $\sum_{i=1}^{n} \hat{y}_i > 0$ we have, with $\mu := 1/\sum_{i=1}^{n} \hat{y}_i$ for $\tilde{y} := \mu\hat{y}$, that $\tilde{y} \in \overline{\mathcal{C}} := \{y \mid Ay = 0,\ \sum_{i=1}^{n} y_i = 1,\ y \geq 0\}$. Obviously $\overline{\mathcal{C}} \subset \mathcal{C}$ and, owing to the constraints $\sum_{i=1}^{n} y_i = 1,\ y \geq 0$, the set $\overline{\mathcal{C}}$

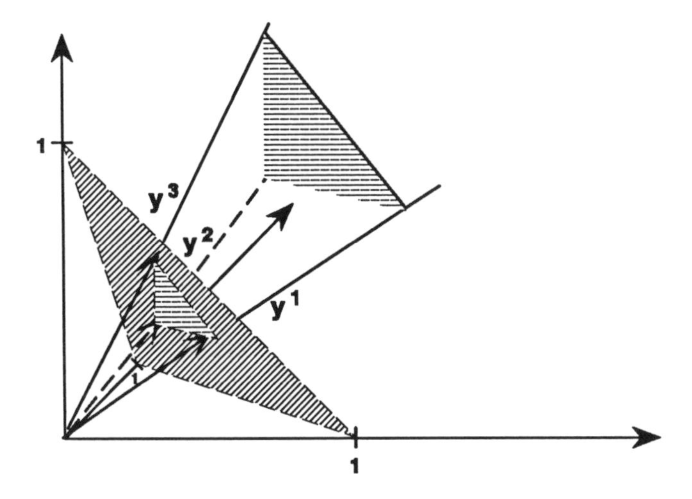

Figure 22 Polyhedral cone intersecting the hyperplane $H = \{y \mid e^{\mathrm{T}}y = 1\}$.

is bounded. Hence, by Proposition 1.10, $\overline{\mathcal{C}}$ is a convex polyhedron generated by its feasible basic solutions $\{y^{\{1\}}, \cdots, y^{\{s\}}\}$ such that $\tilde{y}$ has a representation $\tilde{y} = \sum_{i=1}^{s} \lambda_i y^{\{i\}}$ with $\sum_{i=1}^{s} \lambda_i = 1,\ \lambda_i \geq 0\ \forall i$, implying that $\hat{y} = (1/\mu)\tilde{y} = \sum_{i=1}^{s}(\lambda_i/\mu)y^{\{i\}}$. This shows that $\mathcal{C} = \{y \mid y = \sum_{i=1}^{s} \alpha_i y^{\{i\}},\ \alpha_i \geq 0\ \forall i\}$. □

In Figure 22 we see a convex polyhedral cone $\mathcal{C}$ and its intersection $\overline{\mathcal{C}}$ with the hyperplane $H = \{y \mid e^{\mathrm{T}}y = 1\}$ $(e = (1, \cdots, 1)^{\mathrm{T}})$. The vectors $y^{\{1\}}, y^{\{2\}}$ and $y^{\{3\}}$ are the generating elements (feasible basic solutions) of $\overline{\mathcal{C}}$, as discussed in the proof of Proposition 1.12, and therefore they are also the generating elements of the cone $\mathcal{C}$.

Now we are ready to describe the feasible set $\mathcal{B}$ of the linear program (6.1) in general. Given the convex polyhedron $\mathcal{P} := \mathrm{conv}\{x^{\{1\}}, \cdots, x^{\{r\}}\}$ generated by the feasible basic solutions $\{x^{\{1\}}, \cdots, x^{\{r\}}\} \subset \mathcal{B}$ and the convex polyhedral cone $\mathcal{C} = \{y \mid Ay = 0,\ y \geq 0\}$—given by its generating elements as $\mathrm{pos}\{y^{\{1\}}, \cdots, y^{\{s\}}\}$ as discussed in Proposition 1.12—we get the following.

Proposition 1.13 *$\mathcal{B}$ is the algebraic sum of $\mathcal{P}$ and $\mathcal{C}$, formally $\mathcal{B} = \mathcal{P} + \mathcal{C}$, meaning that every $\tilde{x} \in \mathcal{B}$ may be represented as $\tilde{x} = \tilde{z} + \tilde{y}$, where $\tilde{z} \in \mathcal{P}$ and $\tilde{y} \in \mathcal{C}$.*

Proof Choose an arbitrary $\tilde{x} \in \mathcal{B}$. Since $\{y \mid Ay = 0,\ 0 \leq y \leq \tilde{x}\}$ is compact, the continuous function $\varphi(y) := e^{\mathrm{T}}y$, where $e = (1, \cdots, 1)^{\mathrm{T}}$, attains its maximum on this set. Hence there exists a $\tilde{y}$ such that

$$\left.\begin{aligned} A\tilde{y} &= 0, \\ \tilde{y} &\le \tilde{x}, \\ \tilde{y} &\ge 0, \\ e^{\mathrm{T}}\tilde{y} &= \max\{e^{\mathrm{T}}y \mid Ay = 0,\ 0 \le y \le \tilde{x}\}. \end{aligned}\right\} \tag{6.7}$$

Let $\hat{x} := \tilde{x} - \tilde{y}$. Then $\hat{x} \in \mathcal{B}$ and $\{y \mid Ay = 0,\ 0 \le y \le \hat{x}\} = \{0\}$, since otherwise we should have a contradiction to (6.7). Hence for $I(\hat{x}) = \{i \mid \hat{x}_i > 0\}$ we have

$$\{y \mid Ay = 0,\ y_i = 0, i \notin I(\hat{x}),\ y \ge 0\} = \{0\}$$

and therefore, by Proposition 1.11, the feasible set

$$\mathcal{B}_1 := \{x \mid Ax = b,\ x_i = 0, i \notin I(\hat{x}),\ x \ge 0\}$$

is bounded and, observing that $\hat{x} \in \mathcal{B}_1$, nonempty. From Proposition 1.10, it follows that $\hat{x}$ is a convex linear combination of the feasible basic solutions of

$$\begin{aligned} Ax &= b \\ x_i &= 0,\ i \notin I(\hat{x}) \\ x &\ge 0 \end{aligned}$$

which are obviously feasible basic solutions of our original constraints

$$\begin{aligned} Ax &= b \\ x &\ge 0 \end{aligned}$$

as well. It follows that $\hat{x} \in \mathcal{P}$, and, by the above construction, we have $\tilde{y} \in \mathcal{C}$ and $\tilde{x} = \hat{x} + \tilde{y}$. □

According to this proposition, the feasible set of any LP is constructed as follows. First we determine the convex hull $\mathcal{P}$ of all feasible basic solutions, which might look like that in Figure 20, for example; then we add (algebraically) the convex polyhedral cone $\mathcal{C}$ (owing to Proposition 1.10 associated with the constraints of the LP) to $\mathcal{P}$, which is indicated in Figure 23.

The result of this operation—for an unbounded feasible set—is shown in Figure 24; in the bounded case $\mathcal{P}$ would remain unchanged (as, for example, in Figure 20), since then, according to Proposition 1.11, we have $\mathcal{C} = \{0\}$.

A set given as algebraic sum of a convex polyhedron and a convex polyhedral cone is called a *convex polyhedral set*. Observe that this definition contains the convex polyhedron as well as the convex polyhedral cone as special cases. We shall see later in this text that it is sometimes of interest to identify so-called facets of convex polyhedral sets. Consider for instance a pyramid (in $\mathbb{R}^3$). You will certainly agree that this is a three-dimensional convex

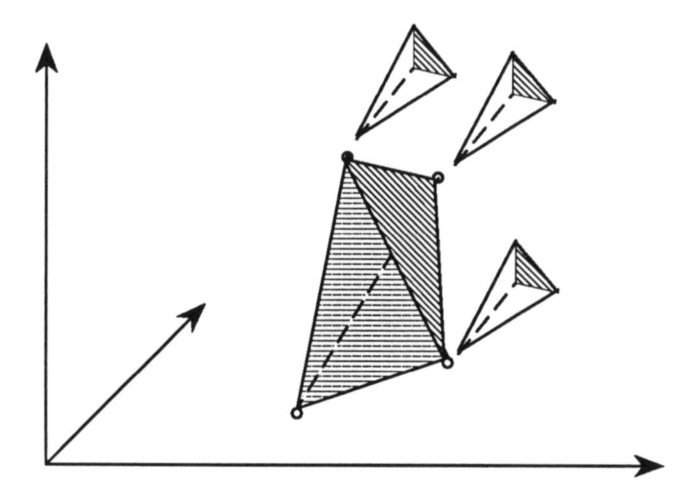

Figure 23 Adding the polyhedral cone $\mathcal{C}$ to the polyhedron $\mathcal{P}$.

polyhedral set. The set of boundary points again consists of different convex polyhedral sets, namely sides (two-dimensional), edges (one-dimensional) and vertices (zero-dimensional). The sides are called facets. In general, consider an arbitrary convex polyhedral set $\mathcal{B} \subset \mathbb{R}^n$. Without loss of generality, assume that $0 \in \mathcal{B}$ (if not, one could, for any fixed $z \in \mathcal{B}$, consider the transposition $\mathcal{B} - \{z\}$ obviously containing the origin). The dimension of $\mathcal{B}$, $\dim \mathcal{B}$, is the smallest dimension of all linear spaces (in $\mathbb{R}^n$) containing $\mathcal{B}$. Therefore $\dim \mathcal{B} \leq n$. For any linear subspace $\mathcal{U} \in \mathbb{R}^n$ and any $\hat{z} \in \mathcal{B}$ the intersection $\mathcal{B}_{\hat{z},\mathcal{U}} := [\{\hat{z}\} + \mathcal{U}] \cap \mathcal{B} \neq \emptyset$ is again a convex polyhedral set. This set is called a *facet* if

- $\hat{z}$ is a boundary point of $\mathcal{B}$ and $\mathcal{B}_{\hat{z},\mathcal{U}}$ does not contain interior points of $\mathcal{B}$;
- $\dim \mathcal{U} = \dim \mathcal{B}_{\hat{z},\mathcal{U}} = \dim \mathcal{B} - 1$.

In other words, a facet of $\mathcal{B}$ is a (maximal) piece of the boundary of $\mathcal{B}$ having the dimension $\dim \mathcal{B} - 1$.

The description of the feasible set of (6.1) given so far enables us to understand immediately under which conditions the linear program (6.1) is solvable and how the solution(s) may look.

Proposition 1.14 *The linear program (6.1) is solvable iff*

$$\mathcal{B} = \{x \mid Ax = b,\ x \geq 0\} \neq \emptyset \tag{6.8}$$

and

$$c^{\mathrm{T}} y \geq 0 \quad \forall y \in \mathcal{C} = \{y \mid Ay = 0,\ y \geq 0\}. \tag{6.9}$$

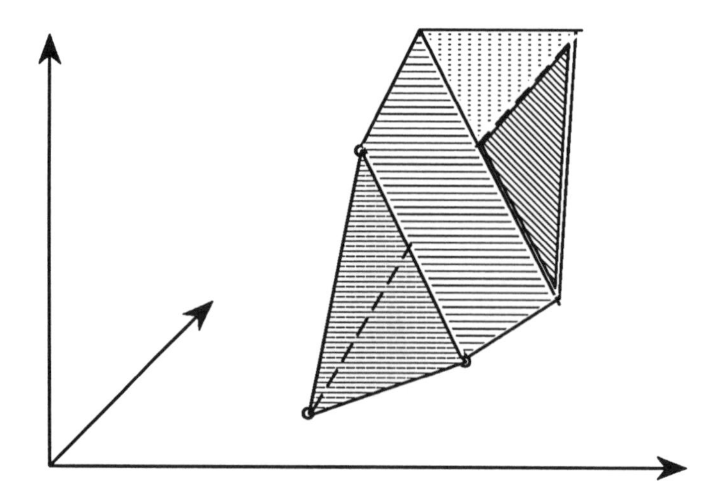

Figure 24 LP: unbounded feasible set.

Given that these two conditions are satisfied, there is at least one feasible basic solution that is an optimal solution.

Proof Obviously condition (6.9) is necessary for the existence of an optimal solution. If $\mathcal{B} \neq \emptyset$ then we know from Proposition 1.13 that $x \in \mathcal{B}$ iff

$$\begin{array}{l} x = \sum_{i=1}^{r} \lambda_i x^{\{i\}} + \sum_{j=1}^{s} \mu_j y^{\{j\}} \\ \text{with } \lambda_i \geq 0 \ \forall i,\ \mu_j \geq 0 \ \forall j \text{ and } \sum_{i=1}^{r} \lambda_i = 1 \end{array}$$

where $\{x^{\{1\}}, \cdots, x^{\{r\}}\}$ is the set of all feasible basic solutions in $\mathcal{B}$ and $\{y^{\{1\}}, \cdots, y^{\{s\}}\}$ is a set of elements generating $\mathcal{C}$, for instance as described in Proposition 1.12. Hence solving

$$\begin{array}{ll} \min c^{\mathrm{T}} x & \\ \text{s.t. } & Ax = b, \\ & x \geq 0 \end{array}$$

is equivalent to solving the problem

$$\begin{array}{ll} \min\{\sum_{i=1}^{r} \lambda_i c^{\mathrm{T}} x^{\{i\}} + \sum_{j=1}^{s} \mu_j c^{\mathrm{T}} y^{\{j\}}\} & \\ \text{s.t. } \sum_{i=1}^{r} \lambda_i = 1 & \\ \qquad \lambda_i \geq 0 \ \ \forall i, & \\ \qquad \mu_j \geq 0 \ \ \forall j. & \end{array}$$

The objective value of this latter program can be driven to $-\infty$ if and only if we have $c^{\mathrm{T}} y^{\{j\}} < 0$ for at least one $j \in \{1, \cdots, s\}$; otherwise, i.e. if

$c^{\mathrm{T}}y^{\{j\}} \geq 0 \ \forall j \in \{1, \cdots, s\}$ and hence $c^{\mathrm{T}}y \geq 0 \ \forall y \in \mathcal{C}$, the objective is minimized by setting $\mu_j = 0 \ \forall j$, and choosing $\lambda_{i_0} = 1$ and $\lambda_i = 0 \ \forall i \neq i_0$ for $x^{\{i_0\}}$ solving $\min_{1 \leq i \leq r}\{c^{\mathrm{T}}x^{\{i\}}\}$. □

Observe that in general the solution of a linear program need not be unique. Given the solvability conditions of Proposition 1.14 and the notation of its proof, if $c^{\mathrm{T}}y^{\{j_0\}} = 0$, we may choose $\mu_{j_0} > 0$, and $x^{\{i_0\}} + \mu_{j_0}y^{\{j_0\}}$ is a solution as well; and obviously it also may happen that $\min_{1 \leq i \leq r}\{c^{\mathrm{T}}x^{\{i\}}\}$ is assumed by more than just one feasible basic solution. In any case, if there is more than one (different) solution for our linear program then there are infinitely many owing to the fact that, given the optimal value γ, the set Γ of optimal solutions is characterized by the linear constraints

$$\begin{aligned} Ax &= b \\ c^{\mathrm{T}}x &\leq \gamma \\ x &\geq 0 \end{aligned}$$

and therefore Γ is itself a convex polyhedral set.

1.6.2 The Simplex Algorithm

If we have the task of solving a linear program of the form (6.1) then, by Proposition 1.14, we may restrict ourselves to feasible basic solutions. Let $\hat{x} \in \mathcal{B}$ be any basic solution and, as before, $I(\hat{x}) = \{i \mid \hat{x}_i > 0\}$. Under the assumption (6.3), the feasible basic solution is called

- *nondegenerate* if $|I(\hat{x})| = m$, and
- *degenerate* if $|I(\hat{x})| < m$.

To avoid lengthy discussions, we assume in this section that for all feasible basic solutions $x^{\{1\}}, \cdots, x^{\{r\}}$ of the linear program (6.1) we have

$$|I(x^{\{i\}})| = m, \ i = 1, \cdots, r, \tag{6.10}$$

i.e. that all feasible basic solutions are nondegenerate. For the case of degenerate basic solutions, and the adjustments that might be necessary in this case, the reader may consult the wide selection of books devoted to linear programming in particular. Referring to our former presentation (6.5), we have, owing to (6.10), that $I_B(\hat{x}) = I(\hat{x})$, and, with the basic part $B = (A_i \mid i \in I(\hat{x}))$ and the nonbasic part $N = (A_i \mid i \notin I(\hat{x}))$ of the matrix A, the constraints of (6.1) may be rewritten—using the basic and nonbasic variables as introduced in (6.4)—as

$$\left.\begin{aligned} x^{\{B\}} &= B^{-1}b - B^{-1}Nx^{\{NB\}}, \\ x^{\{B\}} &\geq 0, \\ x^{\{NB\}} &\geq 0. \end{aligned}\right\} \tag{6.11}$$

Obviously this system yields our feasible basic solution $\hat{x}$ iff $x^{\{NB\}} = 0$, and then we have, by our assumption (6.10), that $x^{\{B\}} = B^{-1}b > 0$. Rearranging the components of c analogously to (6.4) into the two vectors

$$\begin{aligned} c_k^{\{B\}} &= c_i, \ i \text{ the } k\text{th element of } I(\hat{x}), \ k = 1, \cdots, m, \\ c_l^{\{NB\}} &= c_i, \ i \text{ the } l\text{th element of } \{1, \cdots, n\} - I(\hat{x}), \ l = 1, \cdots, n-m, \end{aligned}$$

owing to (6.11), the objective may now be expressed as a function of the nonbasic variables:

$$\begin{aligned} c^{\mathrm{T}}x &= (c^{\{B\}})^{\mathrm{T}}x^{\{B\}} + (c^{\{NB\}})^{\mathrm{T}}x^{\{NB\}} \\ &= (c^{\{B\}})^{\mathrm{T}}B^{-1}b + [(c^{\{NB\}})^{\mathrm{T}} - (c^{\{B\}})^{\mathrm{T}}B^{-1}N]\, x^{\{NB\}}. \end{aligned} \tag{6.12}$$

This representation of the objective connected to the particular feasible basic solution $\hat{x}$ implies the optimality condition for linear programming—the so-called *simplex criterion.*

Proposition 1.15 *Under the assumption (6.10), the feasible basic solution resulting from (6.11) for $x^{\{NB\}} = 0$ is optimal iff*

$$[(c^{\{NB\}})^{\mathrm{T}} - (c^{\{B\}})^{\mathrm{T}}B^{-1}N]^{\mathrm{T}} \geq 0. \tag{6.13}$$

Proof By assumption (6.10), the feasible basic solution given by

$$\begin{aligned} x^{\{B\}} &= B^{-1}b - B^{-1}Nx^{\{NB\}}, \\ x^{\{NB\}} &= 0 \end{aligned}$$

satisfies $x^{\{B\}} = B^{-1}b > 0$. Therefore any nonbasic variable $x_l^{\{NB\}}$ may be increased to some positive amount without violating the constraints $x^{\{B\}} \geq 0$. Furthermore, increasing the nonbasic variables is the only feasible change applicable to them, owing to the constraints $x^{\{NB\}} \geq 0$. From the objective presentation in (6.12), we see immediately that

$$\begin{aligned} c^{\mathrm{T}}\hat{x} &= (c^{\{B\}})^{\mathrm{T}}B^{-1}b \\ &\leq (c^{\{B\}})^{\mathrm{T}}B^{-1}b + [(c^{\{NB\}})^{\mathrm{T}} - (c^{\{B\}})^{\mathrm{T}}B^{-1}N]\, x^{\{NB\}} \quad \forall x^{\{NB\}} \geq 0 \end{aligned}$$

iff $[(c^{\{NB\}})^{\mathrm{T}} - (c^{\{B\}})^{\mathrm{T}}B^{-1}N]^{\mathrm{T}} \geq 0$. □

Motivated by the above considerations, we call any nonsingular $m \times m$ submatrix $B = (A_i \mid i \in I_B)$ of A a *feasible basis* for the linear program (6.1) if $B^{-1}b \geq 0$. Obviously, on rearranging the variables as before into basic variables $x^{\{B\}}$—belonging to B—and nonbasic variables $x^{\{NB\}}$—belonging to $N = (A_i \mid i \notin I_B)$—the objective γ and the constraints of (6.1) read (see (6.11) and (6.12)) as

$$\begin{aligned} \gamma &= (c^{\{B\}})^{\mathrm{T}}B^{-1}b + [(c^{\{NB\}})^{\mathrm{T}} - (c^{\{B\}})^{\mathrm{T}}B^{-1}N]\, x^{\{NB\}}, \\ x^{\{B\}} &= B^{-1}b - B^{-1}Nx^{\{NB\}}, \\ x^{\{B\}} &\geq 0, \\ x^{\{NB\}} &\geq 0, \end{aligned}$$

and $x^{\{NB\}} = 0$ corresponds to a feasible basic solution—under our assumption (6.10), satisfying even $x^{\{B\}} = B^{-1}b > 0$ instead of only $x^{\{B\}} = B^{-1}b \geq 0$ in general.

Now we are ready—using the above notation—to formulate the classical solution procedure of linear programming: the simplex method

Simplex method.

Step 1 Determine a feasible basis $B = (A_i \mid i \in I_B)$ for (6.1) and $N = (A_i \mid i \notin I_B)$.

Step 2 If the simplex criterion (6.13) is satisfied then stop with

$$x^{\{B\}} = B^{-1}b, \; x^{\{NB\}} = 0$$

being an optimal solution; otherwise, there is some $\rho \in \{1, \cdots, n-m\}$ such that for the ρth component of $[(c^{\{NB\}})^{\mathrm{T}} - (c^{\{B\}})^{\mathrm{T}} B^{-1} N]^{\mathrm{T}}$ we have

$$[(c^{\{NB\}})^{\mathrm{T}} - (c^{\{B\}})^{\mathrm{T}} B^{-1} N]_{\rho}^{\mathrm{T}} < 0,$$

and we increase the ρ-th nonbasic variable $x_{\rho}^{\{NB\}}$.

If increasing $x_{\rho}^{\{NB\}}$ is not "blocked" by the constraints $x^{\{B\}} \geq 0$, i.e. if $x_{\rho}^{\{NB\}} \to \infty$ is feasible, then $\inf_{\mathcal{B}} \gamma = -\infty$ such that our problem has no (finite) optimal solution.

If, on the other hand, increasing $x_{\rho}^{\{NB\}}$ is "blocked" by one of the constraints $x_i^{\{B\}} \geq 0$, $i = 1, \cdots, m$, such that, for instance, for some $\mu \in \{1, \cdots, m\}$ the basic variable $x_{\mu}^{\{B\}}$ is the first one to become $x_{\mu}^{\{B\}} = 0$ while increasing $x_{\rho}^{\{NB\}}$, then go to step 3.

Step 3 Exchange the μth column of B with the ρth column of N, yielding new basic and nonbasic parts $\tilde{B}$ and $\tilde{N}$ of A such that $\tilde{B}$ contains N_{ρ} as its μth column and $\tilde{N}$ contains B_{μ} as its ρth column. Redefine $B := \tilde{B}$ and $N := \tilde{N}$, and rearrange $x^{\{B\}}, x^{\{NB\}}, c^{\{B\}}$ and $c^{\{NB\}}$ correspondingly, and then return to step 2.

Remark 1.5 The following comments on the single steps of the simplex method may be helpful for a better understanding of this procedure:

Step 1 Obviously we assume that $\mathcal{B} \neq \emptyset$. The existence of a feasible basis B follows from Propositions 1.9 and 1.8. Because of our assumption (6.10), we have $B^{-1}b > 0$.

Step 2 (a) If for a feasible basis B we have

$$[(c^{\{NB\}})^{\mathrm{T}} - (c^{\{B\}})^{\mathrm{T}} B^{-1} N]^{\mathrm{T}} \geq 0$$

then by Proposition 1.15 this basis (i.e. the corresponding basic solution) is optimal.

(b) If the simplex criterion is violated for the feasible basic solution belonging to B given by $x^{\{B\}} = B^{-1}b$, $x^{\{NB\}} = 0$, then there must be an index $\rho \in \{1, \cdots, n-m\}$ such that $\alpha_{0\rho} := [(c^{\{NB\}})^{\mathrm{T}} - (c^{\{B\}})^{\mathrm{T}} B^{-1} N]^{\mathrm{T}}_{\rho} < 0$, and, keeping all but the ρth nonbasic variables on their present values $x_j^{\{NB\}} = 0$, $j \neq \rho$, with $\alpha_{\cdot\rho} := -B^{-1}N_\rho$, the objective and the basic variables have the representations

$$\begin{aligned} \gamma &= (c^{\{B\}})^{\mathrm{T}} B^{-1} b + \alpha_{0\rho} x_\rho^{\{NB\}}, \\ x^{\{B\}} &= B^{-1}b \qquad\qquad + \alpha_{\cdot\rho} x_\rho^{\{NB\}}. \end{aligned}$$

According to these formulae, we conclude immediately that for $\alpha_{\cdot\rho} \geq 0$ the nonnegativity of the basic variables would never be violated by increasing $x_\rho^{\{NB\}}$ arbitrarily such that we had $\inf_B \gamma = -\infty$, whereas for $\alpha_{\cdot\rho} \not\geq 0$ it would follow that the set of rows $\{i \mid \alpha_{i\rho} < 0,\ 1 \leq i \leq m\} \neq \emptyset$, and consequently, with $\beta := B^{-1}b$, the constraints $x^{\{B\}} = \beta + \alpha_{\cdot\rho} x_\rho^{\{NB\}} \geq 0$ would "block" the increase of $x_\rho^{\{NB\}}$ at some positive value (remember that, by the assumption (6.10), we have $\beta > 0$). More precisely, we now have to observe the constraints

$$\beta_i + \alpha_{i\rho} x_\rho^{\{NB\}} \geq 0 \text{ for } i \in \{i \mid \alpha_{i\rho} < 0,\ 1 \leq i \leq m\}$$

or equivalently

$$x_\rho^{\{NB\}} \leq \frac{\beta_i}{-\alpha_{i\rho}} \text{ for } i \in \{i \mid \alpha_{i\rho} < 0,\ 1 \leq i \leq m\}.$$

Hence, with $\mu \in \{i \mid \alpha_{i\rho} < 0,\ 1 \leq i \leq m\}$ denoting a row for which

$$\frac{\beta_\mu}{-\alpha_{\mu\rho}} = \min\left\{\frac{\beta_i}{-\alpha_{i\rho}} \,\middle|\, \alpha_{i\rho} < 0,\ 1 \leq i \leq m\right\},$$

$x_\mu^{\{B\}}$ is the first basic variable to decrease to zero if $x_\rho^{\{NB\}}$ is increased to the value $\beta_\mu/(-\alpha_{\mu\rho})$, and we observe that at the same time the objective value is changed to

$$\begin{aligned} \gamma &= (c^{\{B\}})^{\mathrm{T}}\beta + \underbrace{\alpha_{0\rho}}_{<0} \underbrace{\overbrace{\frac{\beta_\mu}{-\alpha_{\mu\rho}}}^{>0}}_{>0} \\ &< (c^{\{B\}})^{\mathrm{T}}\beta \end{aligned}$$

such that we have a strict decrease of the objective.

Step 3 The only point to understand here is that $\tilde{B}$ as constructed in this step is again a basis. By assumption, B was a basis, i.e. the column set $(B_1, \cdots, B_\mu, \cdots, B_m)$ was linearly independent. Entering step 3 according to step 2 asserts that for $\alpha_{\cdot\rho} = -B^{-1}N_\rho$ we have $\alpha_{\mu\rho} < 0$, i.e. in the representation of the column N_ρ by the basic columns, $N_\rho = -\sum_{i=1}^{m} B_i \alpha_{i\rho}$ the column B_μ appears with a nonzero coefficient. In this case it is well known from linear algebra that the column set $(B_1, \cdots, N_\rho, \cdots, B_m)$ is linearly independent as well, and hence $\tilde{B}$ is a basis.

The operation of changing the basis by exchanging one column (step 3) is usually called a *pivot step.* □

Summarizing the above remarks immediately yields the following.

Proposition 1.16 *If the linear program (6.1) is feasible then the simplex method yields—under the assumption of nondegeneracy (6.10)—after finitely many steps either a solution or else the information that there is no finite solution, i.e. that* $\inf_{\mathcal{B}} \gamma = -\infty$.

Proof As mentioned in Remark 1.5, step 3, the objective strictly decreases in every pivot step. During the cycles (steps 2 and 3) of the method, we only consider feasible bases. Since there are no more than finitely many feasible bases for any linear program of the form (6.1), the simplex method must end after finitely many cycles. □

Remark 1.6 In step 2 of the simplex method it may happen that the simplex criterion is not satisfied and that we discover that $\inf_{\mathcal{B}} \gamma = -\infty$. It is worth mentioning that in this situation we may easily find a generating element of the cone $\mathcal{C}$ associated with $\mathcal{B}$, as discussed in Proposition 1.12. With the above notation, we then have a feasible basis B, and for some column $N_\rho \neq 0$ we have $B^{-1}N_\rho \leq 0$. Then, with $e = (1, \cdots, 1)^{\mathrm{T}}$ of appropriate dimensions, for $(\hat{y}^{\{B\}}, \hat{y}^{\{NB\}})$ satisfying

$$
\begin{aligned}
\hat{y}^{\{B\}} &= -B^{-1}N_\rho \hat{y}_\rho^{\{NB\}}, \\
\hat{y}_\rho^{\{NB\}} &= \frac{1}{-e^{\mathrm{T}}B^{-1}N_\rho + 1}, \\
\hat{y}_l^{\{NB\}} &= 0 \text{ for } l \neq \rho
\end{aligned}
$$

it follows that

$$\begin{aligned} B\hat{y}^{\{B\}} + N\hat{y}^{\{NB\}} &= 0 \\ e^{\mathrm{T}}\hat{y}^{\{B\}} + e^{\mathrm{T}}\hat{y}^{\{NB\}} &= -e^{\mathrm{T}}B^{-1}N_\rho\hat{y}_\rho^{\{NB\}} + \hat{y}_\rho^{\{NB\}} \\ &= (-e^{\mathrm{T}}B^{-1}N_\rho + 1)\hat{y}_\rho^{\{NB\}} \\ &= 1, \\ \hat{y}^{\{B\}} &\geq 0, \\ \hat{y}^{\{NB\}} &\geq 0. \end{aligned}$$

Observe that, with $B = (B_1, \cdots, B_m)$ a basis of $\mathbb{R}^m$, owing to

$$v = B^{-1}N_\rho \leq 0, \text{ and hence } 1 - e^{\mathrm{T}}v \geq 1,$$

we have

$$\begin{aligned} \mathrm{rk}\begin{pmatrix} B_1 & \cdots & B_m & 0 \\ 1 & \cdots & 1 & 1 - e^{\mathrm{T}}v \end{pmatrix} &= \mathrm{rk}\begin{pmatrix} B_1 & \cdots & B_m & 0 \\ 1 & \cdots & 1 & 1 \end{pmatrix} \\ &= \mathrm{rk}\begin{pmatrix} B_1 & \cdots & B_m & 0 \\ 0 & \cdots & 0 & 1 \end{pmatrix}. \end{aligned}$$

It follows that

$$\begin{pmatrix} B_1 & B_2 & \cdots & B_m & N_\rho \\ 1 & 1 & \cdots & 1 & 1 \end{pmatrix}$$

is a basis of $\mathbb{R}^{m+1}$. Hence $(\hat{y}^{\{B\}}, \hat{y}^{\{NB\}})$ is one of the generating elements of the convex polyhedral cone $\{(y^{\{B\}}, y^{\{NB\}}) \mid By^{\{B\}} + Ny^{\{NB\}} = 0,\ y^{\{B\}} \geq 0, y^{\{NB\}} \geq 0\}$, as derived in Proposition 1.12. □

1.6.3 Duality Statements

Given the linear program (6.1) as so-called *primal program*

$$\left.\begin{aligned} &\min c^{\mathrm{T}}x \\ &\text{s.t. } Ax = b, \\ &\qquad x \geq 0, \end{aligned}\right\} \tag{6.14}$$

the corresponding *dual program* is formulated as

$$\left.\begin{aligned} &\max b^{\mathrm{T}}u \\ &\text{s.t. } A^{\mathrm{T}}u \leq c. \end{aligned}\right\} \tag{6.15}$$

Remark 1.7 Instead of stating a whole bunch of rules on how to assign the correct dual program to any of the various possible formulations of the primal linear program, we might recommend transformation of the primal program to the standard form (6.14), followed by the assignment of the linear program (6.15) as its dual. Let us just give some examples.

Example 1.5 Assume that our primal program is of the form

$$\begin{aligned} &\min c^{\mathrm{T}} x \\ &\text{s.t. } Ax \geq b, \\ &\qquad x \geq 0, \end{aligned}$$

which, by transformation to the standard form, is equivalent to

$$\begin{aligned} &\min c^{\mathrm{T}} x \\ &\text{s.t. } Ax - Iy = b, \\ &\qquad x \geq 0, \\ &\qquad y \geq 0, \end{aligned}$$

I being the $m \times m$ identity matrix, and, according to the above definition, has the dual program

$$\begin{aligned} &\max b^{\mathrm{T}} u \\ &\text{s.t. } A^{\mathrm{T}} u \leq c, \\ &\qquad -Iu \leq 0, \end{aligned}$$

or equivalently

$$\begin{aligned} &\max b^{\mathrm{T}} u \\ &\text{s.t. } A^{\mathrm{T}} u \leq c, \\ &\qquad u \geq 0. \end{aligned}$$

Hence for this case the pair of the primal and its dual program looks like

$$\begin{aligned} &\min c^{\mathrm{T}} x & \qquad &\max b^{\mathrm{T}} u \\ &\text{s.t. } Ax \geq b, & &\text{s.t. } A^{\mathrm{T}} u \leq c, \\ &\qquad x \geq 0; & &\qquad u \geq 0. \end{aligned}$$

□

Example 1.6 Considering the primal program

$$\begin{aligned} &\min c^{\mathrm{T}} x \\ &\text{s.t. } Ax \leq b, \\ &\qquad x \geq 0 \end{aligned}$$

in its standard form

$$\begin{aligned} &\min c^{\mathrm{T}} x \\ &\text{s.t. } Ax + Iy = b, \\ &\qquad x \geq 0, \\ &\qquad y \geq 0 \end{aligned}$$

would yield the dual program

$$\begin{aligned}\max\ & b^{\mathrm{T}}u \\ \text{s.t.}\ & A^{\mathrm{T}}u \le c, \\ & u \le 0,\end{aligned}$$

or equivalently, with $v := -u$,

$$\begin{aligned}\max\ & (-b^{\mathrm{T}}v) \\ \text{s.t.}\ & A^{\mathrm{T}}v \ge -c, \\ & v \ge 0.\end{aligned}$$

Therefore we now have the following pair of a primal and the corresponding dual program:

$$\begin{aligned}\min\ & c^{\mathrm{T}}x \\ \text{s.t.}\ & Ax \le b, \\ & x \ge 0;\end{aligned} \qquad\qquad \begin{aligned}\max\ & (-b^{\mathrm{T}}v) \\ \text{s.t.}\ & A^{\mathrm{T}}v \ge -c, \\ & v \ge 0.\end{aligned}$$

□

Example 1.7 Finally consider the primal program

$$\begin{aligned}\max\ & g^{\mathrm{T}}x \\ \text{s.t.}\ & Dx \le f.\end{aligned}$$

This program is of the same form as the dual of our standard linear program (6.14) and—using the fact that for any function φ defined on some set $\mathcal{M}$ we have $\sup_{x\in\mathcal{M}} \varphi(x) = -\inf_{x\in\mathcal{M}}\{-\varphi(x)\}$—its standard form is written as

$$\begin{aligned}-\min\ & (-g^{\mathrm{T}}x^{+} + g^{\mathrm{T}}x^{-}) \\ \text{s.t.}\ & Dx^{+} - Dx^{-} + Iy = f, \\ & x^{+} \ge 0, \\ & x^{-} \ge 0, \\ & y \ge 0,\end{aligned}$$

with the dual program

$$\begin{aligned}-\max\ & f^{\mathrm{T}}z \\ \text{s.t.}\ & D^{\mathrm{T}}z \le -g, \\ & -D^{\mathrm{T}}z \le g, \\ & Iz \le 0\end{aligned}$$

which is (with $w := -z$) equivalent to

$$\begin{aligned}\min\ & f^{\mathrm{T}}w \\ \text{s.t.}\ & D^{\mathrm{T}}w = g, \\ & w \ge 0,\end{aligned}$$

such that we have the dual pair

$$\begin{array}{ll} \max g^{\mathrm{T}} x & \min f^{\mathrm{T}} w \\ \text{s.t. } Dx \le f; & \text{s.t. } D^{\mathrm{T}} w = g, \\ & \qquad w \ge 0. \end{array}$$

□

Hence, by comparison with our standard forms of the primal program (6.14) and the dual program (6.15), it follows that the dual of the dual is the primal program. □

There are close relations between a primal linear program and its dual program. Let us denote the feasible set of the primal program (6.14) by $\mathcal{B}$ and that of its dual program by $\mathcal{D}$. Furthermore, let us introduce the convention that

$$\begin{array}{ll} \inf_{x\in\mathcal{B}} c^{\mathrm{T}} x = +\infty & \text{if } \mathcal{B} = \emptyset, \\ \sup_{u\in\mathcal{D}} b^{\mathrm{T}} u = -\infty & \text{if } \mathcal{D} = \emptyset. \end{array} \tag{6.16}$$

Then we have as a first statement the following so-called *weak duality theorem*:

Proposition 1.17 *For the primal linear program (6.14) and its dual (6.15)*

$$\inf_{x\in\mathcal{B}} c^{\mathrm{T}} x \ge \sup_{u\in\mathcal{D}} b^{\mathrm{T}} u.$$

Proof If either $\mathcal{B} = \emptyset$ or $\mathcal{D} = \emptyset$ then the proposition is trivial owing to our convention (6.16). Assume therefore that both feasible sets are nonempty and choose arbitrarily an element $\hat{x} \in \mathcal{B}$ and an element $\hat{u} \in \mathcal{D}$. Then, from (6.15), we have

$$c - A^{\mathrm{T}}\hat{u} \ge 0,$$

and, by scalar multiplication with $\hat{x} \ge 0$,

$$\hat{x}^{\mathrm{T}}(c - A^{\mathrm{T}}\hat{u}) \ge 0,$$

which, observing that $A\hat{x} = b$ by (6.14), implies

$$\hat{x}^{\mathrm{T}} c - b^{\mathrm{T}}\hat{u} \ge 0.$$

Since $\hat{x} \in \mathcal{B}$ and $\hat{u} \in \mathcal{D}$ were arbitrarily chosen, we have

$$c^{\mathrm{T}} x \ge b^{\mathrm{T}} u \quad \forall x \in \mathcal{B}, u \in \mathcal{D},$$

and hence

$$\inf_{x\in\mathcal{B}} c^{\mathrm{T}} x \ge \sup_{u\in\mathcal{D}} b^{\mathrm{T}} u.$$

□

In view of this proposition, the question arises as to whether or when it might happen that

$$\inf_{x\in\mathcal{B}} c^{\mathrm{T}} x > \sup_{u\in\mathcal{D}} b^{\mathrm{T}} u.$$

Example 1.8 Consider the following primal linear program:

$$\begin{array}{ll}\min\{3x_1 + 3x_2 - 16x_3\} & \\ \text{s.t.} \quad 5x_1 + 3x_2 - 8x_3 = 2, & \\ \quad -5x_1 + 3x_2 - 8x_3 = 4, & \\ \qquad\qquad x_i \geq 0, \quad i = 1, 2, 3, & \end{array}$$

and its dual program

$$\begin{array}{ll}\max\{2u_1 + 4u_2\} & \\ \text{s.t.} \quad 5u_1 - 5u_2 \leq 3, & \\ \quad 3u_1 + 3u_2 \leq 3, & \\ \quad -8u_1 - 8u_2 \leq -16. & \end{array}$$

Adding the equations of the primal program, we get

$$6x_2 - 16x_3 = 6,$$

and hence

$$x_2 = 1 + \tfrac{8}{3}x_3,$$

which, on insertion into the first equation, yields

$$\begin{aligned} x_1 &= \tfrac{1}{5}(2 - 3 - 8x_3 + 8x_3) \\ &= -\tfrac{1}{5}, \end{aligned}$$

showing that the primal program is not feasible.

Looking at the dual constraints, we get from the second and third inequalities that

$$\begin{aligned} u_1 + u_2 &\leq 1, \\ u_1 + u_2 &\geq 2, \end{aligned}$$

such that also the dual constraints do not allow a feasible solution. Hence, by our convention (6.16), we have for this dual pair

$$\inf_{x\in\mathcal{B}} c^{\mathrm{T}}x = +\infty > \sup_{u\in\mathcal{D}} b^{\mathrm{T}}u = -\infty.$$

□

However, the so-called *duality gap* in the above example does not occur so long as at least one of the two problems is feasible, as is asserted by the following *strong duality theorem* of linear programming.

Proposition 1.18 *Consider the feasible sets $\mathcal{B}$ and $\mathcal{D}$ of the dual pair of linear programs (6.14) and (6.15) respectively. If either $\mathcal{B} \neq \emptyset$ or $\mathcal{D} \neq \emptyset$ then it follows that*

$$\inf_{x\in\mathcal{B}} c^{\mathrm{T}}x = \sup_{u\in\mathcal{D}} b^{\mathrm{T}}u.$$

If one of these two problems is solvable then so is the other, and we have

$$\min_{x \in \mathcal{B}} c^{\mathrm{T}} x = \max_{u \in \mathcal{D}} b^{\mathrm{T}} u.$$

Proof Assume that $\mathcal{B} \neq \emptyset$.

If $\inf_{x \in \mathcal{B}} c^{\mathrm{T}} x = -\infty$ then it follows from the weak duality theorem that $\sup_{u \in \mathcal{D}} b^{\mathrm{T}} u = -\infty$ as well, i.e. that the dual program (6.15) is infeasible.

If the primal program (6.14) is solvable then we know from Proposition 1.14 that there is an optimal feasible basis B such that the primal program may be rewritten as

$$\begin{aligned} \min\{(c^{\{B\}})^{\mathrm{T}} x^{\{B\}} + (c^{\{NB\}})^{\mathrm{T}} x^{\{NB\}}\} \\ \text{s.t.}\quad B x^{\{B\}} + N x^{\{NB\}} &= b, \\ x^{\{B\}} &\geq 0, \\ x^{\{NB\}} &\geq 0, \end{aligned}$$

and therefore the dual program reads as

$$\begin{aligned} \max\, & b^{\mathrm{T}} u \\ \text{s.t.}\quad & B^{\mathrm{T}} u \leq c^{\{B\}}, \\ & N^{\mathrm{T}} u \leq c^{\{NB\}}. \end{aligned}$$

For B an optimal feasible basis, owing to Proposition 1.15, the simplex criterion

$$[(c^{\{NB\}})^{\mathrm{T}} - (c^{\{B\}})^{\mathrm{T}} B^{-1} N]^{\mathrm{T}} \geq 0$$

has to hold. Hence it follows immediately that $\hat{u} := (B^{\mathrm{T}})^{-1} c^{\{B\}}$ satisfies the dual constraints. Additionally, the dual objective value $b^{\mathrm{T}} \hat{u} = b^{\mathrm{T}} (B^{\mathrm{T}})^{-1} c^{\{B\}}$ is equal to the primal optimal value $(c^{\{B\}})^{\mathrm{T}} B^{-1} b$. In view of Proposition 1.17, it follows that $\hat{u}$ is an optimal solution of the dual program. □

An immediate consequence of the strong duality theorem is *Farkas' lemma*, which yields a necessary and sufficient condition for the feasibility of a system of linear constraints, and may be stated as follows.

Proposition 1.19 *The set*

$$\{x \mid Ax = b,\ x \geq 0\} \neq \emptyset$$

if and only if

$$A^{\mathrm{T}} u \geq 0 \text{ implies that } b^{\mathrm{T}} u \geq 0.$$

Proof Assume that $\tilde{u}$ satisfies $A^{\mathrm{T}} \tilde{u} \geq 0$ and that $\{x \mid Ax = b,\ x \geq 0\} \neq \emptyset$. Then let $\hat{x}$ be a feasible solution, i.e. we have

$$A\hat{x} = b, \quad \hat{x} \geq 0,$$

and, by scalar multiplication with $\tilde{u}$, we get

$$\tilde{u}^{\mathrm{T}} b = \underbrace{\tilde{u}^{\mathrm{T}} A}_{\geq 0} \underbrace{\hat{x}}_{\geq 0} \geq 0,$$

so that the condition is necessary.

Assume now that the following condition holds:

$$A^{\mathrm{T}} u \geq 0 \text{ implies that } b^{\mathrm{T}} u \geq 0.$$

Choosing any $\hat{u} \neq 0$ and defining $c := A^{\mathrm{T}} \hat{u}$, it follows from Proposition 1.14 that the linear program

$$\begin{aligned} &\min b^{\mathrm{T}} u \\ &\text{s.t. } A^{\mathrm{T}} u \geq c \end{aligned}$$

is solvable. Then its dual program

$$\begin{aligned} &\max c^{\mathrm{T}} x \\ &\text{s.t. } Ax = b, \\ &\qquad\ x \geq 0 \end{aligned}$$

is solvable and hence feasible. □

1.6.4 A Dual Decomposition Method

In Section 1.4 we discussed stochastic linear programs with linear recourse and mentioned in particular the case of a finite support Ξ of the probability distribution. We saw that the deterministic equivalent—the linear program (4.2)—has a dual decomposition structure. We want to sketch a solution method that makes use of this structure. For simplicity, and just to present the essential ideas, we restrict ourselves to a support Ξ containing just one realization such that the problem to discuss is reduced to

$$\left.\begin{aligned} &\min\{c^{\mathrm{T}} x + q^{\mathrm{T}} y\} \\ &\text{s.t. } Ax \qquad\quad = b, \\ &\qquad Tx + Wy = h, \\ &\qquad\qquad\quad x \geq 0, \\ &\qquad\qquad\quad y \geq 0. \end{aligned}\right\} \tag{6.17}$$

In addition, we assume that the problem is solvable and that the set $\{x \mid Ax = b,\ x \geq 0\}$ is bounded. The above problem may be restated as

$$\begin{aligned} &\min\{c^{\mathrm{T}} x + f(x)\} \\ &\text{s.t. } Ax = b, \\ &\qquad\ x \geq 0, \end{aligned}$$

with

$$f(x) := \min\{q^{\mathrm{T}}y \mid Wy = h - Tx,\ y \geq 0\}.$$

Our recourse function $f(x)$ is easily seen to be piecewise linear and convex. It is also immediate that the above problem can be replaced by the equivalent problem

$$\begin{aligned} &\min\{c^{\mathrm{T}}x + \theta\} \\ &\text{s.t.} \quad Ax = b \\ &\quad \theta - f(x) \geq 0 \\ &\quad x \geq 0; \end{aligned}$$

however, this would require that we know the function $f(x)$ explicitly in advance. This will not be the case in general. Therefore we may try to construct a sequence of new (additional) linear constraints that can be used to define a monotonically decreasing feasible set $\mathcal{B}_1$ of $(n+1)$-vectors $(x_1, \cdots, x_n, \theta)^{\mathrm{T}}$ such that finally, with $\mathcal{B}_0 := \{(x^{\mathrm{T}}, \theta)^{\mathrm{T}} \mid Ax = b,\ x \geq 0,\ \theta \in \mathbb{R}\}$, the problem $\min_{(x,\theta)\in\mathcal{B}_0\cap\mathcal{B}_1}\{c^{\mathrm{T}}x + \theta\}$ yields a (first-stage) solution of our problem (6.17).

After these preparations, we may describe the following particular method.

Dual decomposition method

Step 1 With θ_0 a lower bound for

$$\min\{q^{\mathrm{T}}y \mid Ax = b,\ Tx + Wy = h,\ x \geq 0,\ y \geq 0\},$$

solve the program

$$\min\{c^{\mathrm{T}}x + \theta \mid Ax = b,\ \theta \geq \theta_0,\ x \geq 0\}$$

yielding a solution $(\hat{x}, \hat{\theta})$. Let $\mathcal{B}_1 := \{\mathbb{R}^n \times \{\theta\} \mid \theta \geq \theta_0\}$.

Step 2 Using the last first-stage solution $\hat{x}$, evaluate the recourse function

$$\begin{aligned} f(\hat{x}) &= \min\{q^{\mathrm{T}}y \mid Wy = h - T\hat{x},\ y \geq 0\} \\ &= \max\{(h - T\hat{x})^{\mathrm{T}}u \mid W^{\mathrm{T}}u \leq q\}. \end{aligned}$$

Now we have to distinguish two cases.

(a) If $f(\hat{x}) = +\infty$ then $\hat{x}$ is not feasible with respect to all constraints of (6.17) (i.e. $\hat{x}$ does not satisfy the induced constraints discussed in Proposition 1.3), and by Proposition 1.14 we have a $\tilde{u}$ such that $W^{\mathrm{T}}\tilde{u} \leq 0$ and $(h - T\hat{x})^{\mathrm{T}}\tilde{u} > 0$. On the other hand, for any feasible x there must exist a $y \geq 0$ such that $Wy = h - Tx$. Scalar multiplication of this equation by $\tilde{u}$ yields

$$\tilde{u}^{\mathrm{T}}(h - Tx) = \underbrace{\tilde{u}^{\mathrm{T}}W}_{\leq 0}\ \underbrace{y}_{\geq 0} \leq 0,$$

and hence

$$\tilde{u}^{\mathrm{T}} h \leq \tilde{u}^{\mathrm{T}} T x,$$

which has to hold for any feasible x, and obviously does not hold for $\hat{x}$, since $\tilde{u}^{\mathrm{T}}(h - T\hat{x}) > 0$. Therefore we introduce the *feasibility cut*, cutting off the infeasible solution $\hat{x}$:

$$\tilde{u}^{\mathrm{T}}(h - Tx) \leq 0.$$

Then we redefine $\mathcal{B}_1 := \mathcal{B}_1 \bigcap \{(x^{\mathrm{T}}, \theta) \mid \tilde{u}^{\mathrm{T}}(h - Tx) \leq 0\}$ and go on to step 3.

(b) Otherwise, if $f(\hat{x})$ is finite, we have for the recourse problem (see the proof of Proposition 1.18) simultaneously—for $\hat{x}$—a primal optimal basic solution $\hat{y}$ and a dual optimal basic solution $\hat{u}$. From the dual formulation of the recourse problem, it is evident that

$$f(\hat{x}) = (h - T\hat{x})^{\mathrm{T}} \hat{u},$$

whereas for any x we have

$$\begin{aligned} f(x) &= \sup\{(h - Tx)^{\mathrm{T}} u \mid W^{\mathrm{T}} u \leq q\} \\ &\geq (h - Tx)^{\mathrm{T}} \hat{u} \\ &= \hat{u}^{\mathrm{T}}(h - Tx). \end{aligned}$$

The intended constraint $\theta \geq f(x)$ implies the linear constraint

$$\theta \geq \hat{u}^{\mathrm{T}}(h - Tx),$$

which is violated by $(\hat{x}^{\mathrm{T}}, \hat{\theta})^{\mathrm{T}}$ iff $(h - T\hat{x})^{\mathrm{T}} \hat{u} > \hat{\theta}$; in this case we introduce the ***optimality cut*** (see Figure 25), cutting off the nonoptimal solution $(\hat{x}^{\mathrm{T}}, \hat{\theta})^{\mathrm{T}}$:

$$\theta \geq \hat{u}^{\mathrm{T}}(h - Tx).$$

Correspondingly, we redefine $\mathcal{B}_1 := \mathcal{B}_1 \bigcap \{(x^{\mathrm{T}}, \theta) \mid \theta \geq \hat{u}^{\mathrm{T}}(h - Tx)\}$ and continue with step 3; otherwise, i.e. if $f(\hat{x}) \leq \hat{\theta}$, we stop, with $\hat{x}$ being an optimal first-stage solution.

Step 3 Solve the updated problem

$$\min\{c^{\mathrm{T}} x + \theta \mid (x^{\mathrm{T}}, \theta) \in \mathcal{B}_0 \cap \mathcal{B}_1\},$$

yielding the optimal solution $(\tilde{x}^{\mathrm{T}}, \tilde{\theta})^{\mathrm{T}}$.

With $(\hat{x}^{\mathrm{T}}, \hat{\theta})^{\mathrm{T}} := (\tilde{x}^{\mathrm{T}}, \tilde{\theta})^{\mathrm{T}}$, we return to step 2.

Remark 1.8 We briefly sketch the arguments regarding the proper functioning of this method.

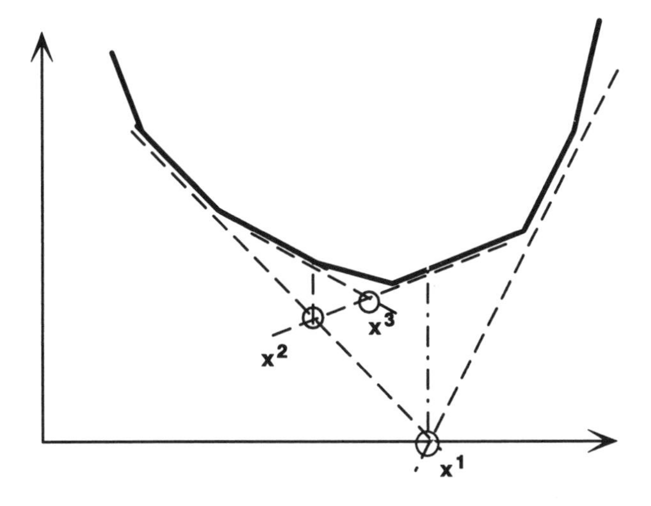

Figure 25 Dual decomposition: optimality cuts.

Step 1 We have assumed problem (6.17) to be solvable, which implies, by Proposition 1.14, that

$$\{(x,y) \mid Ax = b,\ Tx + Wy = h,\ x \geq 0,\ y \geq 0\} \neq \emptyset,$$

$$\{v \mid Wv = 0, q^{\mathrm{T}}v < 0,\quad v \geq 0\} = \emptyset.$$

In addition, we have assumed $\{x \mid Ax = b,\ x \geq 0\}$ to be bounded. Hence $\inf\{f(x) \mid Ax = b,\ x \geq 0\}$ is finite such that the lower bound θ_0 exists. This (and the boundedness of $\{x \mid Ax = b,\ x \geq 0\}$) implies that

$$\min\{c^{\mathrm{T}}x + \theta \mid Ax = b,\ \theta \geq \theta_0,\ x \geq 0\}$$

is solvable.

Step 2 If $f(\hat{x}) = +\infty$, we know from Proposition 1.14 that $\{u \mid W^{\mathrm{T}}u \leq 0,\ (h - T\hat{x})^{\mathrm{T}}u > 0\} \neq \emptyset$, and, according to Remark 1.6, for the convex polyhedral cone $\{u \mid W^{\mathrm{T}}u \leq 0\}$ we may find with the simplex method one of the generating elements $\tilde{u}$ mentioned in Proposition 1.12 that satisfies $(h - T\hat{x})^{\mathrm{T}}\tilde{u} > 0$. By Proposition 1.12, we have finitely many generating elements for the cone $\{u \mid W^{\mathrm{T}}u \leq 0\}$ such that, after having used all of them to construct feasibility cuts, for all feasible x we should have $(h - Tx)^{\mathrm{T}}u \leq 0\ \forall u \in \{u \mid W^{\mathrm{T}}u \leq 0\}$ and hence solvability of the recourse problem. This shows that $f(\hat{x}) = +\infty$ may appear only finitely many times within this method.

If $f(\hat{x})$ is finite, the simplex method yields primal and dual optimal feasible basic solutions $\hat{y}$ and $\hat{u}$ respectively. Assume that we already had the same dual basic solution $\tilde{u} := \hat{u}$ in a previous step to construct an optimality cut

$$\theta \geq \tilde{u}^{\mathrm{T}}(h - Tx);$$

then our present $\hat{\theta}$ has to satisfy this constraint for $x = \hat{x}$ such that

$$\begin{aligned}\hat{\theta} &\geq \tilde{u}^{\mathrm{T}}(h - T\hat{x}) \\ &= \hat{u}^{\mathrm{T}}(h - T\hat{x})\end{aligned}$$

holds, or equivalently we have $f(\hat{x}) \leq \hat{\theta}$ and stop the procedure. From the above inequalities, it follows that

$$\hat{\theta} \geq (h - T\hat{x})^{\mathrm{T}} u^{\{i\}}, \quad i = 1, \cdots, k,$$

if $u^{\{1\}}, \cdots, u^{\{k\}}$ denote the feasible basic solutions in $\{u \mid W^{\mathrm{T}} u \leq q\}$ used so far for optimality cuts. Observing that in step 3 for any x we minimize θ with respect to $\mathcal{B}_1$ this implies that

$$\hat{\theta} = \max_{1 \leq i \leq k} (h - T\hat{x})^{\mathrm{T}} u^{\{i\}}.$$

Given our stopping rule $f(\hat{x}) \leq \hat{\theta}$, with the set of *all* feasible basic solutions, $\{u^{\{1\}}, \cdots, u^{\{k\}}, \cdots, u^{\{r\}}\}$, of $\{u \mid W^{\mathrm{T}} u \leq q\}$, it follows that

$$\begin{aligned}\hat{\theta} &= \max\nolimits_{1 \leq i \leq k} (h - T\hat{x})^{\mathrm{T}} u^{\{i\}} \\ &\leq \max\nolimits_{1 \leq i \leq r} (h - T\hat{x})^{\mathrm{T}} u^{\{i\}} \\ &= f(\hat{x}) \\ &\leq \hat{\theta}\end{aligned}$$

and hence $\hat{\theta} = f(\hat{x})$, which implies the optimality of $\hat{x}$.

□

Summarizing the above remarks we have the following.

Proposition 1.20 *Provided that the program (6.17) is solvable and* $\{x \mid Ax = b,\ x \geq 0\}$ *is bounded, the dual decomposition method yields an optimal solution after finitely many steps.*

We have described this method for the data structure of the linear program (6.17) that would result if a stochastic linear program with recourse had just one realization of the random data. To this end, we introduced the feasibility and optimality cuts for the recourse function $f(x) := \min\{q^{\mathrm{T}} y \mid Wy = h - Tx,\ y \geq 0\}$. The modification for a finite discrete distribution with K

realizations is immediate. From the discussion in Section 1.4, our problem is of the form

$$\begin{aligned} \min & \left\{c^{\mathrm{T}}x + \sum_{i=1}^{K} q^{i\mathrm{T}}y^i\right\} \\ \text{s.t.} \quad & Ax = b \\ & T^i x + W y^i = h^i, \quad i = 1, \cdots, K \\ & x \geq 0, \\ & y^i \geq 0, \quad i = 1, \cdots, K. \end{aligned}$$

Thus we may simply introduce feasibility and optimality cuts for all the recourse functions $f_i(x) := \min\{q^{i\mathrm{T}}y^i \mid Wy^i = h^i - T^i x,\ y^i \geq 0\}$, $i = 1, \cdots, K$, yielding the so-called multicut version of the dual decomposition method. Alternatively, combining the single cuts corresponding to the particular blocks $i = 1, \cdots, K$ with their respective probabilities leads to the so-called L-shaped method.

1.7 Nonlinear Programming

In this section we summarize some basic facts about nonlinear programming problems written in the standard form

$$\left.\begin{aligned} & \min f(x) \\ & \text{s.t. } g_i(x) \leq 0, \quad i = 1, \cdots, m. \end{aligned}\right\} \tag{7.1}$$

The feasible set is again denoted by $\mathcal{B}$:

$$\mathcal{B} := \{x \mid g_i(x) \leq 0, \quad i = 1, \cdots, m\}.$$

As in the previous section, any other nonlinear program, for instance

$$\begin{aligned} & \min f(x) \\ & \text{s.t. } g_i(x) \leq 0, \quad i = 1, \cdots, m, \\ & \qquad x \geq 0 \end{aligned}$$

or

$$\begin{aligned} & \min f(x) \\ & \text{s.t. } g_i(x) \leq 0, \quad i = 1, \cdots, m_1, \\ & \qquad g_i(x) = 0, \quad i = m_{1+1}, \cdots, m, \\ & \qquad x \geq 0 \end{aligned}$$

or

$$\begin{aligned} & \min f(x) \\ & \text{s.t. } g_i(x) \geq 0, \quad i = 1, \cdots, m, \\ & \qquad x \geq 0, \end{aligned}$$

may be transformed into the standard form (7.1).

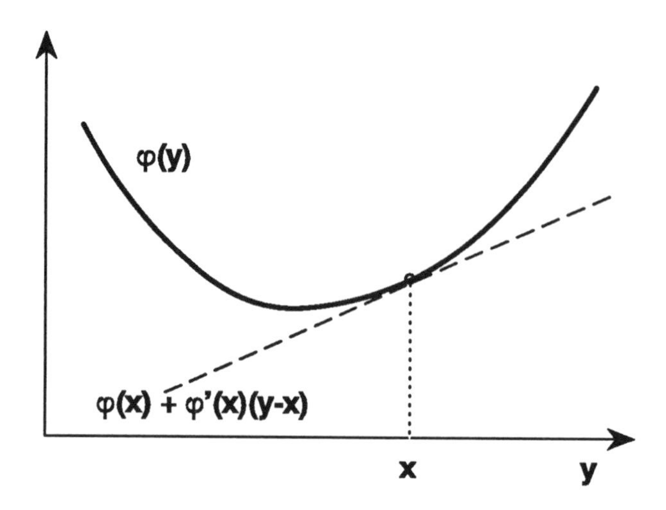

Figure 26 Convex function with tangent as supporting hyperplane.

We assume throughout this section that the functions $f, g_i : \mathbb{R}^n \longrightarrow \mathbb{R}$ are given, that at least one of them is not a linear function, and that all of them are continuously (partially) differentiable (i.e. $\frac{\partial f}{\partial x_j}$ and $\frac{\partial g_i}{\partial x_j}$ are continuous). Occasionally we restrict ourselves to the case that the functions are convex, since we shall not widely deal with nonconvex problems in this book. This implies, according to Lemma 1.1 that any local minimum of program (7.1) is a global minimum.

First of all, we have to refer to a well known fact from analysis.

Proposition 1.21 *The function $\varphi : \mathbb{R}^n \longrightarrow \mathbb{R}$ is convex iff for all arbitrarily chosen $x, y \in \mathbb{R}^n$ we have*

$$(y - x)^{\mathrm{T}} \nabla \varphi(x) \le \varphi(y) - \varphi(x).$$

In other words, for a convex function, a tangent (hyperplane) at any arbitrary point (of its graph) supports the function everywhere from below; a hyperplane with this property is called a *supporting hyperplane* for this function (see Figure 26).

We know from calculus that for some $\hat{x} \in \mathbb{R}^n$ to yield a local minimum for a differentiable function $\varphi : \mathbb{R}^n \longrightarrow \mathbb{R}$ we have the necessary condition

$$\nabla \varphi(\hat{x}) = 0.$$

If, moreover, the function φ is convex then, owing to Proposition 1.21, this condition is also sufficient for $\hat{x}$ to be a global minimum, since then for any arbitrary $x \in \mathbb{R}^n$ we have

$$0 = (x - \hat{x})^{\mathrm{T}} \nabla \varphi(\hat{x}), \leq \varphi(x) - \varphi(\hat{x})$$

and hence

$$\varphi(\hat{x}) \leq \varphi(x) \quad \forall x \in \mathbb{R}^n.$$

Whereas the above optimality condition is necessary for *unconstrained minimization*, the situation may become somewhat different for *constrained minimization*.

Example 1.9 For $x \in \mathbb{R}$ consider the simple problem

$$\begin{aligned} &\min \psi(x) = x^2 \\ &\text{s.t. } \; x \geq 1, \end{aligned}$$

with the obvious unique solution

$$\hat{x} = 1, \quad \text{with } \nabla \psi(\hat{x}) = \frac{d\psi}{dx}(\hat{x}) = 2.$$

Hence we cannot just transfer the optimality conditions for unconstrained optimization to the constrained case. □

Therefore we shall first deal with the necessary and/or sufficient conditions for some $\hat{x} \in \mathbb{R}^n$ to be a local or global solution of the program (7.1).

1.7.1 The Kuhn–Tucker Conditions

Remark 1.9 To get an idea of what kind of optimality conditions we may expect for problems of the type (7.1), let us first—contrary to our general assumption—consider the case where $f, g_i,\ i = 1, \cdots, m$, are linear functions

$$\left.\begin{aligned} f(x) &:= c^{\mathrm{T}} x, \\ g_i(x) &:= a_i^{\mathrm{T}} x - b_i, \quad i = 1, \cdots, m, \end{aligned}\right\} \tag{7.2}$$

such that we have the gradients

$$\left.\begin{aligned} \nabla f(x) &= c, \\ \nabla g_i(x) &= a_i, \end{aligned}\right\} \tag{7.3}$$

and problem (7.1) becomes the linear program

$$\left.\begin{aligned} &\min c^{\mathrm{T}} x \\ &\text{s.t. } \; a_i^{\mathrm{T}} x \leq b_i, \quad i = 1, \cdots, m. \end{aligned}\right\} \tag{7.4}$$

Although we did not explicitly discuss optimality conditions for linear programs in the previous section, they are implicitly available in the duality statements discussed there. The dual problem of (7.4) is

$$\left.\begin{array}{rl} \max\{-b^{\mathrm{T}}u\} & \\ \text{s.t.} \ -\displaystyle\sum_{i=1}^{m} a_i u_i = c, & \\ u \geq 0. & \end{array}\right\} \tag{7.5}$$

Let A be the $m \times n$ matrix having a_i^{T}, $i = 1, \cdots, m$, as rows. The difference of the primal and the dual objective functions can then be written as

$$\begin{aligned} c^{\mathrm{T}}x + b^{\mathrm{T}}u &= c^{\mathrm{T}}x + u^{\mathrm{T}}Ax - u^{\mathrm{T}}Ax + b^{\mathrm{T}}u \\ &= (c + A^{\mathrm{T}}u)^{\mathrm{T}}x + (b - Ax)^{\mathrm{T}}u \\ &= [\nabla f(x) + \sum_{i=1}^{m} u_i \nabla g_i(x)]^{\mathrm{T}}x - \sum_{i=1}^{m} u_i g_i(x). \end{aligned} \tag{7.6}$$

From the duality statements for linear programming (Propositions 1.17 and 1.18), we know the following.

(a) If $\hat{x}$ is an optimal solution of the primal program (7.4) then, by the strong duality theorem (Proposition 1.18), there exists a solution $\hat{u}$ of the dual program (7.5) such that the difference of the primal and dual objective vanishes. For the pair of dual problems (7.4) and (7.5) this means that $c^{\mathrm{T}}\hat{x} - (-b^{\mathrm{T}}\hat{u}) = c^{\mathrm{T}}\hat{x} + b^{\mathrm{T}}\hat{u} = 0$. In view of (7.6) this may also be stated as the necessary condition

$$\begin{aligned} \exists \hat{u} \geq 0 \text{ such that } \nabla f(\hat{x}) + \sum_{i=1}^{m} \hat{u}_i \nabla g_i(\hat{x}) &= 0, \\ \sum_{i=1}^{m} \hat{u}_i g_i(\hat{x}) &= 0. \end{aligned}$$

(b) if we have a primal feasible and a dual feasible solution $\tilde{x}$ and $\tilde{u}$ respectively, such that the difference of the respective objectives is zero then, by the weak duality theorem (Proposition 1.17), $\tilde{x}$ solves the primal problem; in other words, given a feasible $\tilde{x}$, the condition

$$\begin{aligned} \exists \tilde{u} \geq 0 \text{ such that } \nabla f(\tilde{x}) + \sum_{i=1}^{m} \tilde{u}_i \nabla g_i(\tilde{x}) &= 0, \\ \sum_{i=1}^{m} \tilde{u}_i g_i(\tilde{x}) &= 0 \end{aligned}$$

is sufficient for $\tilde{x}$ to be a solution of the program (7.4).

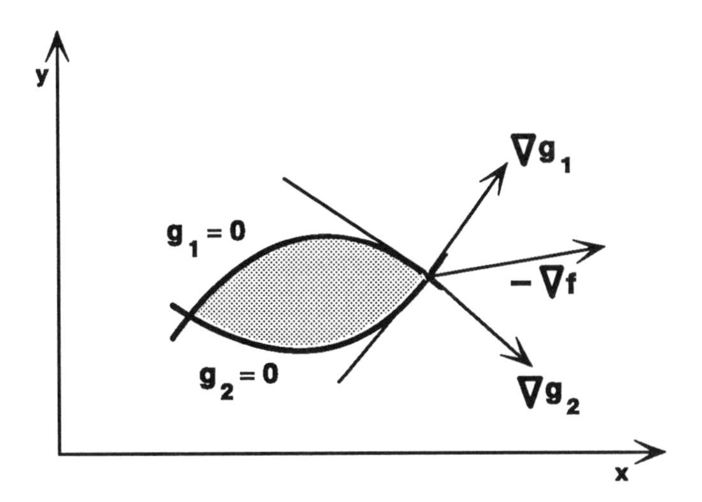

Figure 27 Kuhn–Tucker conditions.

□

Remark 1.10 The optimality condition derived in Remark 1.9 for the linear case could be formulated as follows:

(1) For the feasible $\hat{x}$ the negative gradient of the objective f—i.e. the direction of the greatest (local) descent of f—is equal (with the *multipliers* $\hat{u}_i \geq 0$) to a nonnegative linear combination of the gradients of those constraint functions g_i that are *active* at $\hat{x}$, i.e. that satisfy $g_i(\hat{x}) = 0$.

(2) This corresponds to the fact that the multipliers satisfy the *complementarity conditions* $\hat{u}_i g_i(\hat{x}) = 0,\ i = 1, \cdots, m$, stating that the multipliers $\hat{u}_i$ are zero for those constraints that are not active at $\hat{x}$, i.e. that satisfy $g_i(\hat{x}) < 0$.

In conclusion, this optimality condition says that $-\nabla f(\hat{x})$ must be contained in the convex polyhedral cone generated by the gradients $\nabla g_i(\hat{x})$ of the constraints being active in $\hat{x}$. This is one possible formulation of the *Kuhn–Tucker conditions* illustrated in Figure 27. □

Let us now return to the more general nonlinear case and consider the following question. Given that $\hat{x}$ is a (local) solution, under what assumption does this imply that the above optimality conditions,

$$\left.\begin{aligned} \exists \hat{u} \geq 0 \text{ such that } \nabla f(\hat{x}) + \sum_{i=1}^{m} \hat{u}_i \nabla g_i(\hat{x}) &= 0, \\ \sum_{i=1}^{m} \hat{u}_i g_i(\hat{x}) &= 0, \end{aligned}\right\} \tag{7.7}$$

hold? Hence we ask under what assumption are the conditions (7.7) necessary for $\hat{x}$ to be a (locally) optimal solution of the program (7.1). To answer this question, let $I(\hat{x}) := \{i \mid g_i(\hat{x}) = 0\}$, such that the optimality conditions (7.7) are equivalent to

$$\left\{u \,\middle|\, \sum_{i \in I(\hat{x})} u_i \nabla g_i(\hat{x}) = -\nabla f(\hat{x}),\ u_i \geq 0 \text{ for } i \in I(\hat{x})\right\} \neq \emptyset.$$

Observing that $\nabla g_i(\hat{x})$ and $\nabla f(\hat{x})$ are constant vectors when x is fixed at $\hat{x}$, the condition of Farkas' lemma (Proposition 1.19) is satisfied if and only if the following *regularity condition* holds in $\hat{x}$:

<u>$\mathcal{RC}\_0$</u>

$$z^{\mathrm{T}} \nabla g_i(\hat{x}) \leq 0,\ i \in I(\hat{x}) \text{ implies that } z^{\mathrm{T}} \nabla f(\hat{x}) \geq 0. \tag{7.8}$$

Hence we have the rigorous formulation of the Kuhn–Tucker conditions:

Proposition 1.22 *Given that $\hat{x}$ is a (local) solution of the nonlinear program (7.1), under the assumption that the regularity condition <u>$\mathcal{RC}\_0$</u> is satisfied in $\hat{x}$ it necessarily follows that*

$$\begin{aligned} \exists \hat{u} \geq 0 \text{ such that } \nabla f(\hat{x}) + \sum_{i=1}^{m} \hat{u}_i \nabla g_i(\hat{x}) &= 0, \\ \sum_{i=1}^{m} \hat{u}_i g_i(\hat{x}) &= 0. \end{aligned}$$

Example 1.10 The Kuhn–Tucker conditions need not hold if the regularity condition cannot be asserted. Consider the following simple problem ($x \in \mathbb{R}^1$):

$$\min\{x \mid x^2 \leq 0\}.$$

Its unique solution is $\hat{x} = 0$. Obviously we have

$$\nabla f(\hat{x}) = (1), \quad \nabla g(\hat{x}) = (0),$$

and there is no way to represent $\nabla f(\hat{x})$ as (positive) multiple of $\nabla g(\hat{x})$. (Needless to say, the regularity condition <u>$\mathcal{RC}\_0$</u> is not satisfied in $\hat{x}$.) □

We just mention that for the case of linear constraints the Kuhn–Tucker conditions are necessary for optimality, without the addition of any regularity condition.

Instead of condition $\underline{\mathcal{RC}\_0}$, there are various other regularity conditions popular in optimization theory, only two of which we shall mention here. The first is stated as

$\underline{\mathcal{RC}\_1}$

$$\forall z \neq 0 \text{ s.t. } z^{\mathrm{T}} \nabla g_i(\hat{x}) \leq 0,\ i \in I(\hat{x}),\ \exists \{x^k \mid x^k \neq \hat{x},\ k = 1, 2, \cdots\} \subset \mathcal{B}$$

such that

$$\lim_{k \to \infty} x^k = \hat{x}, \quad \lim_{k \to \infty} \frac{x^k - \hat{x}}{||x^k - \hat{x}||} = \frac{z}{||z||}.$$

The second—used frequently for the convex case, i.e. if the functions g_i are convex—is the *Slater condition*

$\underline{\mathcal{RC}\_2}$

$$\exists \tilde{x} \in \mathcal{B} \text{ such that } g_i(\tilde{x}) < 0\ \forall i. \tag{7.9}$$

Observe that there is an essential difference among these regularity conditions: to verify $\underline{\mathcal{RC}\_0}$ or $\underline{\mathcal{RC}\_1}$, we need to know the (locally) optimal point for which we want the Kuhn–Tucker conditions (7.7) to be necessary, whereas the Slater condition $\mathcal{RC}\_2$—for the convex case—requires the existence of an $\tilde{x}$ such that $g_i(\tilde{x}) < 0\ \forall i$, but does not refer to any optimal solution. Without proof we might mention the following.

Proposition 1.23

(a) The regularity condition $\underline{\mathcal{RC}\_1}$ (in any locally optimal solution) implies the regularity condition $\underline{\mathcal{RC}\_0}$.

(b) For the convex case the Slater condition $\underline{\mathcal{RC}\_2}$ implies the regularity condition $\underline{\mathcal{RC}\_1}$ (for every feasible solution).

In Figure 28 we indicate how the proof of the implication $\underline{\mathcal{RC}\_2} \Longrightarrow \underline{\mathcal{RC}\_1}$ can be constructed.

Based on these facts we immediately get the following.

Proposition 1.24

(a) If $\hat{x}$ (locally) solves problem (7.1) and satisfies $\underline{\mathcal{RC}\_0}$ then the Kuhn–Tucker conditions (7.7) necessarily hold in $\hat{x}$.

(b) If the functions $f, g_i,\ i = 1, \cdots, m$, are convex and the Slater condition $\underline{\mathcal{RC}\_2}$ holds, then $\hat{x} \in \mathcal{B}$ (globally) solves problem (7.1) if and only if the Kuhn–Tucker conditions (7.7) are satisfied for $\hat{x}$.

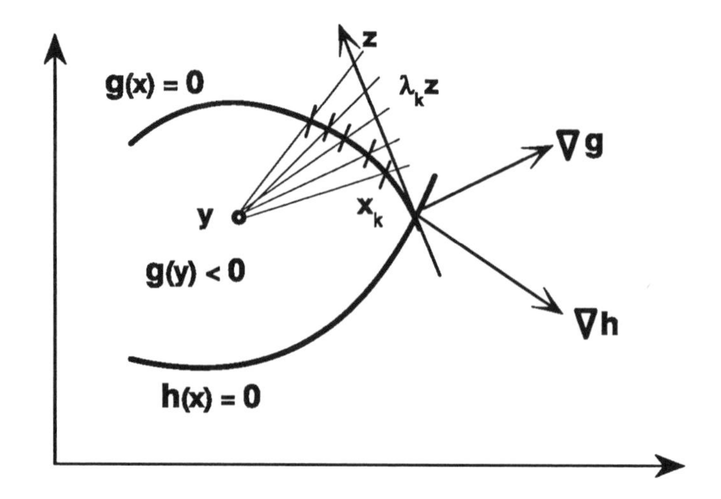

Figure 28 The Slater condition implies $\underline{\mathcal{RC}\_1}$.

Proof: Referring to Proposition 1.23, the necessity of the Kuhn–Tucker conditions has already been demonstrated. Hence we need only show that in the convex case the Kuhn–Tucker conditions are also sufficient for optimality. Assume therefore that we have an $\hat{x} \in \mathcal{B}$ and a $\hat{u} \geq 0$ such that

$$\nabla f(\hat{x}) + \sum_{i=1}^{m} \hat{u}_i \nabla g_i(\hat{x}) = 0,$$
$$\sum_{i=1}^{m} \hat{u}_i g_i(\hat{x}) = 0.$$

Then, with $I(\hat{x}) = \{i \mid g_i(\hat{x}) = 0\}$, we have

$$\nabla f(\hat{x}) = - \sum_{i \in I(\hat{x})} \hat{u}_i \nabla g_i(\hat{x})$$

and owing to $\hat{u} \geq 0$ and the convexity of f and g_i, it follows from Proposition 1.21 that for any arbitrary $x \in \mathcal{B}$

$$\begin{aligned} f(x) - f(\hat{x}) \geq\ & (x - \hat{x})^{\mathrm{T}} \nabla f(\hat{x}) \\ =& - \sum_{i \in I(\hat{x})} \hat{u}_i (x - \hat{x})^{\mathrm{T}} \nabla g_i(\hat{x}) \\ \geq& - \sum_{i \in I(\hat{x})} \underbrace{\hat{u}_i}_{\geq 0} \underbrace{[g_i(x) - g_i(\hat{x})]}_{\leq 0\ \forall x \in \mathcal{B},\ i \in I(\hat{x})} \\ \geq& \ 0 \end{aligned}$$

such that $f(x) \geq f(\hat{x})\ \forall x \in \mathcal{B}$. □

Observe that

- to show the necessity of the Kuhn–Tucker conditions we had to use the regularity condition $\underline{\mathcal{RC}\_0}$ (or one of the other two, being stronger), but we did not need any convexity assumption;
- to demonstrate that in the convex case the Kuhn–Tucker conditions are sufficient for optimality we have indeed used the assumed convexity, but we did not need any regularity condition at all.

Defining the Lagrange function for problem (7.1),

$$L(x, u) := f(x) + \sum_{i=1}^{m} u_i g_i(x)$$

we may restate our optimality conditions. With the notation

$$\begin{aligned} \nabla_x L(x, u) &:= \left(\frac{\partial L(x,u)}{\partial x_1}, \cdots, \frac{\partial L(x,u)}{\partial x_n}\right)^{\mathrm{T}}, \\ \nabla_u L(x, u) &:= \left(\frac{\partial L(x,u)}{\partial u_1}, \cdots, \frac{\partial L(x,u)}{\partial u_m}\right)^{\mathrm{T}}, \end{aligned}$$

and observing that $\nabla_u L(x, u) \leq 0$ simply repeats the constraints $g_i(x) \leq 0\ \forall i$ of our original program (7.1), the Kuhn–Tucker conditions now read as

$$\left.\begin{aligned} \nabla_x L(\hat{x}, \hat{u}) &= 0, \\ \nabla_u L(\hat{x}, \hat{u}) &\leq 0, \\ \hat{u}^{\mathrm{T}} \nabla_u L(\hat{x}, \hat{u}) &= 0, \\ \hat{u} &\geq 0. \end{aligned}\right\} \tag{7.10}$$

Assume now that the functions $f, g_i,\ i = 1, \cdots, m$, are convex. Then for any fixed $u \geq 0$ the Lagrange function is obviously convex in x. For $(\hat{x}, \hat{u})$ satisfying the Kuhn–Tucker conditions, it follows by Proposition 1.21 that for any arbitrary x

$$L(x, \hat{u}) - L(\hat{x}, \hat{u}) \geq (x - \hat{x})^{\mathrm{T}} \nabla_x L(\hat{x}, \hat{u}) = 0$$

and hence

$$L(\hat{x}, \hat{u}) \leq L(x, \hat{u})\ \ \forall x \in \mathbb{R}^n.$$

On the other hand, since $\nabla_u L(\hat{x}, \hat{u}) \leq 0$ is equivalent to $g_i(\hat{x}) \leq 0\ \forall i$, and the Kuhn–Tucker conditions assert that $\hat{u}^{\mathrm{T}} \nabla_u L(\hat{x}, \hat{u}) = \sum_{i=1}^{m} \hat{u}_i g_i(\hat{x}) = 0$, it follows that

$$L(\hat{x}, u) \leq L(\hat{x}, \hat{u})\ \ \forall u \geq 0.$$

Hence we have the following.

Proposition 1.25 *Given that the functions f, g_i, $i = 1, \cdots, m$, in problem (7.1) are convex, any Kuhn–Tucker point, i.e. any pair $(\hat{x}, \hat{u})$ satisfying the Kuhn–Tucker conditions, is a* saddle point *of the Lagrange function, i.e. it satisfies*

$$\forall u \geq 0 \quad L(\hat{x}, u) \leq L(\hat{x}, \hat{u}) \leq L(x, \hat{u}) \quad \forall x \in \mathbb{R}^n.$$

Furthermore, it follows by the complementarity conditions that

$$L(\hat{x}, \hat{u}) = f(\hat{x}).$$

It is an easy exercise to show that for any saddle point $(\hat{x}, \hat{u})$, with $\hat{u} \geq 0$, of the Lagrange function, the Kuhn–Tucker conditions (7.10) are satisfied. Therefore, if we knew the right multiplier vector $\hat{u}$ in advance, the task to solve the constrained optimization problem (7.1) would be equivalent to that of solving the unconstrained optimization problem $\min_{x \in \mathbb{R}^n} L(x, \hat{u})$. This observation can be seen as the basic motivation for the development of a class of solution techniques known in the literature as *Lagrangian methods.*

1.7.2 Solution Techniques

When solving stochastic programs, we need to use known procedures from both linear and nonlinear programming, or at least adopt their underlying ideas. Unlike linear programs, nonlinear programs generally cannot be solved in finitely many steps. Instead, we shall have to deal with iterative procedures that we might expect to converge—in some reasonable sense—to a solution of the nonlinear program under consideration. For better readability of the subsequent chapters of this book, we sketch the basic ideas of some types of methods; for detailed technical presentations and convergence proofs the reader is referred to the extensive specialized literature on nonlinear programming. We shall discuss

- cutting-plane methods;
- methods of descent;
- penalty methods;
- Lagrangian methods

by presenting one particular variant of each of these methods.

1.7.2.1 Cutting-plane methods

Assume that for problem (7.1) the functions f and g_i, $i = 1, \cdots, m$, are convex and that the—convex—feasible set

$$\mathcal{B} = \{x \mid g_i(x) \leq 0, \ i = 1, \cdots, m\}$$

is bounded. Furthermore, assume that $\exists \hat{y} \in \operatorname{int} \mathcal{B}$—which for instance would be true if the Slater condition (7.9) held. Then, instead of the original problem

$$\min_{x \in \mathcal{B}} f(x),$$

we could consider the equivalent problem

$$\begin{array}{ll} \min \theta & \\ \text{s.t. } g_i(x) & \leq 0, \quad i = 1, \cdots, m, \\ \quad f(x) - \theta & \leq 0, \end{array}$$

with the feasible set $\overline{\mathcal{B}} = \{(x, \theta) \mid f(x) - \theta \leq 0,\ g_i(x) \leq 0, i = 1, \cdots, m\} \subset \mathbb{R}^{n+1}$ being obviously convex. With the assumption $\hat{y} \in \operatorname{int} \mathcal{B}$, we may further restrict the feasible solutions in $\overline{\mathcal{B}}$ to satisfy the inequality $\theta \leq f(\hat{y})$ without any effect on the solution set. The resulting problem can be interpreted as the minimization of the linear objective θ on the *bounded* convex set $\{(x, \theta) \in \overline{\mathcal{B}} \mid \theta \leq f(\hat{y})\}$, which is easily seen to contain an interior point $(\tilde{y}, \tilde{\theta})$ as well.

Hence, instead of the nonlinear program (7.1), we may consider—without loss of generality if the original feasible set $\mathcal{B}$ was bounded—the minimization of a linear objective on a bounded convex set

$$\min\{c^{\mathrm{T}} x \mid x \in \mathcal{B}\}, \tag{7.11}$$

where the bounded convex set $\mathcal{B}$ is assumed to contain an interior point $\hat{y}$.

Under the assumptions mentioned, it is possible to include the feasible set $\mathcal{B}$ of problem (7.11) in a convex polyhedron $\mathcal{P}$, which—after our discussions in Section 1.6—we may expect to be able to represent by linear constraints. Observe that the inclusion $\mathcal{P} \supset \mathcal{B}$ implies the inequality

$$\min_{x \in \mathcal{P}} c^{\mathrm{T}} x \leq \min_{x \in \mathcal{B}} c^{\mathrm{T}} x.$$

The cutting-plane method for problem (7.11) proceeds as follows.

Step 1 Determine a $\hat{y} \in \operatorname{int} \mathcal{B}$ and a convex polyhedron $\mathcal{P}_0 \supset \mathcal{B}$; let $k := 0$.

Step 2 Solve the linear program

$$\min\{c^{\mathrm{T}} x \mid x \in \mathcal{P}_k\},$$

yielding the solution $\hat{x}^k$.

If $\hat{x}^k \in \mathcal{B}$ then stop ($\hat{x}^k$ solves problem (7.11)); otherwise, i.e. if $\hat{x}^k \notin \mathcal{B}$, determine

$$\lambda_k := \min\{\lambda \mid \lambda \hat{y} + (1 - \lambda)\hat{x}^k \in \mathcal{B}\}$$

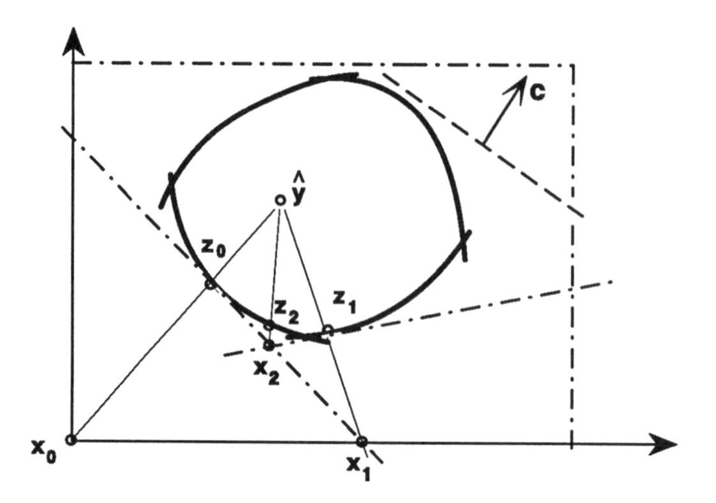

Figure 29 Cutting-plane method: iteration k.

and let

$$z^k := \lambda_k \hat{y} + (1 - \lambda_k)\hat{x}^k.$$

(Obviously we have $z^k \in \mathcal{B}$ and moreover z^k is a boundary point of $\mathcal{B}$ on the line segment between the interior point $\hat{y}$ of $\mathcal{B}$ and the point $\hat{x}^k$, which is "external" to $\mathcal{B}$.)

Step 3 Determine a "supporting hyperplane" of $\mathcal{B}$ in z^k (i.e. a hyperplane being tangent to $\mathcal{B}$ at the boundary point z^k). Let this hyperplane be given as

$$H_k := \{x \mid (a^k)^{\mathrm{T}} x = \alpha_k\}$$

such that the inequalities

$$(a^k)^{\mathrm{T}} \hat{x}^k > \alpha_k \geq (a^k)^{\mathrm{T}} x \quad \forall x \in \mathcal{B}$$

hold. Then define

$$\mathcal{P}_{k+1} := \mathcal{P}_k \bigcap \{x \mid (a^k)^{\mathrm{T}} x \leq \alpha_k\},$$

let $k := k + 1$, and return to step 2.

In Figure 29 we illustrate one step of the cutting-plane method.

Remark 1.11 By construction—see steps 1 and 3 of the above method—we have

$$\mathcal{P}_k \supset \mathcal{B}, \quad k = 0, 1, 2, \cdots,$$

and hence

$$c^{\mathrm{T}}\hat{x}^k \le \min_{x\in\mathcal{B}} c^{\mathrm{T}}x, \quad k = 0, 1, 2, \cdots,$$

such that as soon as $\hat{x}^k \in \mathcal{B}$ for some $k \ge 0$ we would have that $\hat{x}^k$ is an optimal solution of problem (7.11), as claimed in step 2. Furthermore, since $z^k \in \mathcal{B}\ \forall k$, we have

$$c^{\mathrm{T}}z^k \ge \min_{x\in\mathcal{B}} c^{\mathrm{T}}x \ \forall k$$

such that $c^{\mathrm{T}}z^k - c^{\mathrm{T}}\hat{x}^k$ could be taken after the kth iteration as an upper bound on the distance of either the feasible (but in general nonoptimal) objective value $c^{\mathrm{T}}z^k$ or the optimal (but in general nonfeasible) objective value $c^{\mathrm{T}}\hat{x}^k$ to the feasible optimal value $\min_{x\in\mathcal{B}} c^{\mathrm{T}}x$. Observe that in general the sequence $\{c^{\mathrm{T}}z^k\}$ need not be monotonically decreasing, whereas $\mathcal{P}_{k+1} \subset \mathcal{P}_k\ \forall k$ ensures that the sequence $\{c^{\mathrm{T}}\hat{x}^k\}$ is monotonically increasing. Thus we may enforce a monotonically decreasing error bound

$$\Delta_k := c^{\mathrm{T}}z^{l_k} - c^{\mathrm{T}}\hat{x}^k, \quad k = 0, 1, 2, \cdots,$$

by choosing z^{l_k} from the boundary points of $\mathcal{B}$ constructed in step 2 up to iteration k such that

$$c^{\mathrm{T}}z^{l_k} = \min_{l\in\{0,\cdots,k\}} c^{\mathrm{T}}z^l.$$

Finally we describe briefly how the "supporting hyperplane" of $\mathcal{B}$ in z^k of step 3 can be determined. By our assumptions, $\hat{y} \in \operatorname{int}\mathcal{B}$ and $\hat{x}^k \notin \mathcal{B}$, we get in step 2 that $0 < \lambda_k < 1$. Since $\lambda_k > 0$ is minimal under the condition $\lambda\hat{y} + (1-\lambda)\hat{x}^k \in \mathcal{B}$, there is at least one constraint i_0 active in z^k meaning that $g_{i_0}(z^k) = 0$. The convexity of g_{i_0} implies, owing to Proposition 1.21, that

$$0 > g_{i_0}(\hat{y}) = g_{i_0}(\hat{y}) - g_{i_0}(z^k) \ge (\hat{y} - z^k)^{\mathrm{T}}\nabla g_{i_0}(z^k),$$

and therefore that $a^k := \nabla g_{i_0}(z^k) \ne 0$.

Observing that $z_k = \lambda_k\hat{y} + (1-\lambda_k)\hat{x}^k$ with $0 < \lambda_k < 1$ is equivalent to

$$\hat{x}^k - z^k = -\frac{\lambda_k}{1-\lambda_k}(\hat{y} - z^k),$$

we conclude from the last inequality that

$$(\hat{x}^k - z^k)^{\mathrm{T}}\nabla g_{i_0}(z^k) > 0.$$

On the other hand, for any $x \in \mathcal{B}$, $g_{i_0}(x) \le 0$. Again by Proposition 1.21, it follows that

$$(x - z^k)^{\mathrm{T}}\nabla g_{i_0}(z^k) \le g_{i_0}(x) - g_{i_0}(z^k) = g_{i_0}(x) \le 0.$$

Therefore, with $a^k := \nabla g_{i_0}(z^k)$ and $\alpha_k := z^{k\mathrm{T}}\nabla g_{i_0}(z^k)$, we may define a supporting hyperplane as required in step 3; this hyperplane is then used in the definition of $\mathcal{P}_{k+1}$ to cut off the set $\{x \mid (a^k)^{\mathrm{T}}x > \alpha_k\}$—and hence in particular the infeasible solution $\hat{x}_k$—from further consideration. □

1.7.2.2 Descent methods

For the sake of simplicity, we consider the special case of minimizing a convex function under linear constraints

$$\left.\begin{array}{r} \min f(x) \\ \text{s.t. } Ax = b, \\ x \geq 0. \end{array}\right\} \tag{7.12}$$

Assume that we have a feasible point $z \in \mathcal{B} = \{x \mid Ax = b,\ x \geq 0\}$. Then there are two possibilities.

(a) If z is optimal then the Kuhn–Tucker conditions have to hold. For (7.12) these are

$$\begin{aligned} \nabla f(z) + A^{\mathrm{T}}u - w &= 0, \\ z^{\mathrm{T}}w &= 0, \\ w &\geq 0, \end{aligned}$$

or—with $J(z) := \{j \mid z_j > 0\}$—equivalently

$$\begin{aligned} A^{\mathrm{T}}u - w &= -\nabla f(z), \\ w_j &= 0 \quad \text{for } j \in J(z), \\ w &\geq 0. \end{aligned}$$

Applying Farkas' Lemma 1.19 tells us that this system (and hence the above Kuhn–Tucker system) is feasible if and only if

$$[\nabla f(z)]^{\mathrm{T}}d \geq 0 \quad \forall d \in \{d \mid Ad = 0, d_j \geq 0 \text{ for } j \notin J(z)\};$$

(b) If the feasible point z is not optimal then the Kuhn–Tucker conditions cannot hold, and, according to (a), there exists a direction d such that $Ad = 0, d_j \geq 0\ \forall j : z_j = 0$ and $[\nabla f(z)]^{\mathrm{T}}d < 0$. A direction like this is called a *feasible descent direction at* z, which has to satisfy the following two conditions: $\exists\lambda_0 > 0$ such that $z + \lambda d \in \mathcal{B}\ \forall\lambda \in [0, \lambda_0]$ and $[\nabla f(z)]^{\mathrm{T}}d < 0$. Hence, having at a feasible point z a feasible descent direction d (for which, by its definition, $d \neq 0$ is obvious), it is possible to move from z in direction d with some positive step length without leaving $\mathcal{B}$ and at the same time at least locally to decrease the objective's value.

From these brief considerations, we may state the following.

Conceptual method of descent directions

Step 1 Determine a feasible solution $z^{(0)}$, let $k := 0$.

Step 2 If there is no feasible descent direction at $z^{(k)}$ then stop ($z^{(k)}$ is optimal).
Otherwise, choose a feasible descent direction $d^{(k)}$ at $z^{(k)}$ and go to step 3.

Step 3 Solve the so-called *line search* problem

$$\min_{\lambda}\{f(z^{(k)} + \lambda d^{(k)}) \mid (z^{(k)} + \lambda d^{(k)}) \in \mathcal{B}\},$$

and with its solution λ_k define $z^{(k+1)} := z^{(k)} + \lambda_k d^{(k)}$. Let $k := k + 1$ and return to step 2.

Remark 1.12 It is worth mentioning that not every choice of feasible descent directions would lead to a well-behaved algorithm. By construction we should get—in any case—a sequence of feasible points $\{z^{(k)}\}$ with a monotonically (strictly) decreasing sequence $\{f(z^{(k)})\}$ such that for the case that f is bounded below on $\mathcal{B}$ the sequence $\{f(z^{(k)})\}$ has to converge to some value γ. However, there are examples in the literature showing that if we do not restrict the choice of the feasible descent directions in an appropriate way, it may happen that $\gamma > \inf_{\mathcal{B}} f(x)$, which is certainly not the kind of a result we want to achieve.

Let us assume that $\mathcal{B} \neq \emptyset$ is bounded, implying that our problem (7.12) is solvable. Then there are various possibilities of determining the feasible descent direction, each of which defines its own algorithm for which a "reasonable" convergence behaviour can be asserted in the sense that the sequence $\{f(z^{(k)})\}$ converges to the true optimal value and any accumulation point of the sequence $\{z^{(k)}\}$ is an optimal solution of our problem (7.12). Let us just mention two of those algorithms:

(a) *The feasible direction method* For this algorithm we determine in step 2 the direction $d^{(k)}$ as the solution of the following linear program:

$$\begin{aligned} &\min[\nabla f(z^{(k)})]^{\mathrm{T}} d \\ &\text{s.t. } Ad = 0 \\ &\qquad d_j \geq 0 \ \ \forall j : z_j^{(k)} = 0, \\ &\qquad d \leq e, \\ &\qquad d \geq -e, \end{aligned}$$

with $e = (1, \cdots, 1)^{\mathrm{T}}$. Then for $[\nabla f(z^{(k)})]^{\mathrm{T}} d^{(k)} < 0$ we have a feasible descent direction, whereas for $[\nabla f(z^{(k)})]^{\mathrm{T}} d^{(k)} = 0$ the point $z^{(k)}$ is an optimal solution of (7.12).

(b) *The reduced gradient method* Assume that $\mathcal{B}$ is bounded and every feasible basic solution of (7.12) is nondegenerate. Then for $z^{(k)}$ we find a basis B in A such that the components of $z^{(k)}$ belonging to B are strictly positive. Rewriting A—after the necessary rearrangements of columns—as (B, N) and correspondingly presenting $z^{(k)}$ as (x^B, x^{NB}), we have

$$Bx^B + Nx^{NB} = b,$$

or equivalently

$$x^B = B^{-1}b - B^{-1}Nx^{NB}.$$

We also may rewrite the gradient $\nabla f(z^{(k)})$ as $(\nabla_B f(z^{(k)}), \nabla_{NB} f(z^{(k)}))$. Then, rearranging d accordingly into (u, v), for a feasible direction we need to have

$$Bu + Nv = 0,$$

and hence

$$u = -B^{-1}Nv.$$

For the directional derivative $[\nabla f(z^{(k)})]^{\mathrm{T}} d$ it follows

$$\begin{aligned}
[\nabla f(z^{(k)})]^{\mathrm{T}} d &= [\nabla_B f(z^{(k)})]^{\mathrm{T}} u + [\nabla_{NB} f(z^{(k)})]^{\mathrm{T}} v \\
&= [\nabla_B f(z^{(k)})]^{\mathrm{T}} (-B^{-1}Nv) + [\nabla_{NB} f(z^{(k)})]^{\mathrm{T}} v \\
&= ([\nabla_{NB} f(z^{(k)})]^{\mathrm{T}} - [\nabla_B f(z^{(k)})]^{\mathrm{T}} B^{-1}N)v.
\end{aligned}$$

Defining the *reduced gradient* r by

$$\begin{aligned}
r^{\mathrm{T}} &= \begin{pmatrix} r^B \\ r^{NB} \end{pmatrix}^{\mathrm{T}} \\
&= ([\nabla_B f(z^{(k)})]^{\mathrm{T}}, [\nabla_{NB} f(z^{(k)})]^{\mathrm{T}}) - [\nabla_B f(z^{(k)})]^{\mathrm{T}} B^{-1}(B, N),
\end{aligned}$$

we have

$$r^B = 0, \quad r^{NB} = ([\nabla_{NB} f(z^{(k)})]^{\mathrm{T}} - [\nabla_B f(z^{(k)})]^{\mathrm{T}} B^{-1}N)^{\mathrm{T}},$$

and hence

$$\begin{aligned}
[\nabla f(z^{(k)})]^{\mathrm{T}} d &= (u^{\mathrm{T}}, v^{\mathrm{T}}) \begin{pmatrix} r^B \\ r^{NB} \end{pmatrix} \\
&= ([\nabla_{NB} f(z^{(k)})]^{\mathrm{T}} - [\nabla_B f(z^{(k)})]^{\mathrm{T}} B^{-1}N)v.
\end{aligned}$$

Defining v as

$$v_j := \begin{cases} -r_j^{NB} & \text{if } r_j^{NB} \leq 0, \\ -x_j^{NB} r_j^{NB} & \text{if } r_j^{NB} > 0, \end{cases}$$

and, as above, $u := B^{-1}Nv$, we have that $(u^{\mathrm{T}}, v^{\mathrm{T}})^{\mathrm{T}}$ is a feasible direction (observe that $v_j \geq 0$ if $x_j^{NB} = 0$ and $x^B > 0$ owing to our assumption) and it is a descent direction if $v \neq 0$. Furthermore, $z^{(k)}$ is a solution of problem (7.12) iff $v = 0$ (and hence $u = 0$), since then $r \geq 0$, and, with $w^{\mathrm{T}} := [\nabla_B f(z^{(k)})]^{\mathrm{T}} B^{-1}$, we have

$$\begin{aligned} r^B &= \nabla_B f(z^{(k)}) - B^{\mathrm{T}} w = 0, \\ r^{NB} &= \nabla_{NB} f(z^{(k)}) - N^{\mathrm{T}} w \geq 0, \end{aligned}$$

and

$$(r^B)^{\mathrm{T}} x^B = 0, \quad (r^{NB})^{\mathrm{T}} x^{NB} = 0,$$

i.e. $v = 0$ is equivalent to satisfying the Kuhn–Tucker conditions.

It is known that the reduced gradient method with the above definition of v may fail to converge to a solution (so-called "zigzagging"). However, we can perturb v as follows:

$$v_j := \begin{cases} -r_j^{NB} & \text{if } r_j^{NB} \leq 0, \\ -x_j^{NB} r_j^{NB} & \text{if } r_j^{NB} > 0 \text{ and } x_j^{NB} \geq \varepsilon, \\ 0 & \text{if } r_j^{NB} > 0 \text{ and } x_j^{NB} < \varepsilon. \end{cases}$$

Then a proper control of the perturbation $\varepsilon > 0$ during the procedure can be shown to enforce convergence.

□

The feasible direction and the reduced gradient methods have been extended to the case of nonlinear constraints. We omit the presentation of the general case here for the sake of better readability.

1.7.2.3 Penalty methods

The term "penalty" reflects the following attempt. Replace the original problem (7.1)

$$\begin{aligned} &\min f(x) \\ &\text{s.t. } g_i(x) \leq 0, \quad i = 1, \cdots, m, \end{aligned}$$

by appropriate free (i.e. unconstrained) optimization problems

$$\left.\begin{aligned} &\min_{x \in \mathbb{R}^n} F_{rs}(x), \\ &\text{the function } F_{rs} \text{ being defined as} \\ &F_{rs}(x) := f(x) + r \sum_{i \in I} \varphi(g_i(x)) + \frac{1}{s} \sum_{i \in J} \psi(g_i(x)), \end{aligned}\right\} \quad (7.13)$$

where $I, J \subset \{1, \cdots, m\}$ such that $I \cap J = \emptyset$, $I \cup J = \{1, \cdots, m\}$, and the parameters $r, s > 0$ are to be chosen or adapted in the course of the procedure. The role of the functions φ and ψ is to inhibit and to penalize respectively the violation of any one of the constraints. More precisely, for these functions we assume that

- φ, ψ are monotonically increasing and convex;
- the so-called *barrier function* satisfies

$$\begin{aligned} \varphi(\eta) < +\infty \quad & \forall \eta < 0, \\ \lim_{\eta \uparrow 0} \varphi(\eta) = +\infty; & \end{aligned}$$

- for the so-called *loss function* we have

$$\psi(\eta) \begin{cases} = 0 & \forall \eta \le 0, \\ > 0 & \forall \eta > 0. \end{cases}$$

Observe that the convexity of $f, g_i, i = 1, \cdots, m$, and the convexity and monotonicity of φ, ψ imply the convexity of F_{rs} for any choice of the parameters $r, s > 0$. Solving the free optimization problem (7.13) with parameters $r, s > 0$ would inhibit the violation of the constraints $i \in I$, whereas the violation of anyone of the constraints $i \in J$ would be penalized with a positive additive term. Intuitively, we might expect that the solutions of (7.13) will satisfy the constraints $i \in I$ and that for

$$s \downarrow 0, \quad \text{or equivalently} \quad \frac{1}{s} \uparrow +\infty,$$

they will eventually satisfy the constraints $i \in J$. Therefore it seems plausible to control the parameter s in such a way that it tends to zero. Now what about the parameter r of the barrier term in (7.13)? Imagine that for the (presumably unique) solution $\hat{x}$ of problem (7.1) some constraint $i_0 \in I$ is active, i.e. $g_{i_0}(\hat{x}) = 0$. For any fixed $r > 0$, minimization of (7.13) will not allow us to approach the solution $\hat{x}$, since obviously, by the definition of a barrier function, this would drive the new objective F_{rs} to $+\infty$. Hence it seems reasonable to drive the parameter r downwards to zero, as well.

With

$$\mathcal{B}_1 := \{x \mid g_i(x) \le 0, \ i \in I\}, \quad \mathcal{B}_2 := \{x \mid g_i(x) \le 0, \ i \in J\}$$

we have $\mathcal{B} = \mathcal{B}_1 \cap \mathcal{B}_2$, and for $r > 0$ we may expect finite values of F_{rs} only for $x \in \mathcal{B}_1^0 := \{x \mid g_i(x) < 0, \ i \in I\}$. We may close this short presentation of general penalty methods by a statement showing that, under mild assumptions, a method of this type may be controlled in such a way that it results in what we should like to experience.

Proposition 1.26 *Let f, g_i, $i = 1, \cdots, m$, be convex and assume that*

$$\mathcal{B}_1^0 \cap \mathcal{B}_2 \neq \emptyset$$

and that $\mathcal{B} = \mathcal{B}_1 \cap \mathcal{B}_2$ is bounded. Then for $\{r_k\}$ and $\{s_k\}$ strictly monotone sequences decreasing to zero there exists an index k_0 such that for all $k \geq k_0$ the modified objective function $F_{r_k s_k}$ attains its (free) minimum at some point $x^{(k)}$ where $x^{(k)} \in \mathcal{B}_1^0$.

The sequence $\{x^{(k)} \mid k \geq k_0\}$ is bounded, and any of its accumulation points is a solution of the original problem (7.1). With γ the optimal value of (7.1), the following relations hold:

$$\begin{aligned}
&\lim_{k\to\infty} f(x^{(k)}) &&= \gamma, \\
&\lim_{k\to\infty} r_k \sum_{i\in I} \varphi(g_i(x^{(k)})) &&= 0, \\
&\lim_{k\to\infty} \frac{1}{s_k} \sum_{i\in J} \psi(g_i(x^{(k)})) &&= 0.
\end{aligned}$$

1.7.2.4 Lagrangian methods

As mentioned at the end of Section 1.7.1, knowledge of the proper multiplier vector $\hat{u}$ in the Lagrange function $L(x, u) = f(x) + \sum_{i=1}^m u_i g_i(x)$ for problem (7.1) would allow us to solve the free optimization problem

$$\min_{x\in\mathbb{R}^n} L(x, \hat{u})$$

instead of the constrained problem

$$\begin{aligned}
&\min f(x) \\
&\text{s.t. } g_i(x) \leq 0, \quad i = 1, \cdots, m.
\end{aligned}$$

To simplify the description, let us first consider the optimization problem with equality constraints

$$\left.\begin{aligned}
&\min f(x) \\
&\text{s.t. } g_i(x) = 0, \quad i = 1, \cdots, m.
\end{aligned}\right\} \tag{7.14}$$

Knowing for this problem the proper multiplier vector $\hat{u}$ or at least a good approximate u of it, we should find

$$\min_{x\in\mathbb{R}^n} [f(x) + u^{\mathrm{T}} g(x)], \tag{7.15}$$

where $u^{\mathrm{T}} g(x) = \sum_{i=1}^m u_i g_i(x)$. However, at the beginning of any solution procedure we hardly have any knowledge about the numerical size of the multipliers in a Kuhn–Tucker point of problem (7.14), and using some guess for u might easily result in an unsolvable problem ($\inf_x L(x,u) = -\infty$).

On the other hand, we have just introduced penalty methods. Using for problem (7.14) a quadratic loss function for violating the equality constraints seems to be reasonable. Hence we could think of a penalty method using as modified objective

$$\min_{x \in \mathbb{R}^n} [f(x) + \tfrac{1}{2}\lambda \|g(x)\|^2] \tag{7.16}$$

and driving the parameter λ towards $+\infty$, with $\|g(x)\|$ being the Euclidean norm of $g(x) = (g_1(x), \cdots, g_m(x))^{\mathrm{T}}$.

One idea is to combine the two approaches (7.15) and (7.16) such that we are dealing with the so-called *augmented Lagrangian* as our modified objective:

$$\min_{x \in \mathbb{R}^n} [f(x) + u^{\mathrm{T}} g(x) + \tfrac{1}{2}\lambda \|g(x)\|^2].$$

The now obvious intention is to control the parameters u and λ in such a way that $\lambda \to \infty$—to eliminate infeasibilities—and that at the same time $u \to \hat{u}$, the proper Kuhn–Tucker multiplier vector. Although we are not yet in a position to appropriately adjust the parameters, we know at least the skeleton of the algorithm, which usually is referred to as *augmented Lagrange method*: With the augmented Lagrangian

$$L_\lambda(x,u) := f(x) + u^{\mathrm{T}} g(x) + \tfrac{1}{2}\lambda \|g(x)\|^2,$$

it may be loosely stated as follows:

$$\left.\begin{array}{l} \text{For} \\ \bullet\ \{u^{(k)}\} \subset \mathbb{R}^m \text{ bounded,} \\ \bullet\ \{\lambda_k\} \subset \mathbb{R} \text{ such that } 0 < \lambda_k < \lambda_{k+1}\ \forall k,\ \lambda_k \to \infty, \\ \text{solve successively } \min_{x \in \mathbb{R}^n} L_{\lambda_k}(x, u^{(k)}). \end{array}\right\} \tag{7.17}$$

Observe that for $u^{(k)} = 0\ \forall k$ we should get back the penalty method with a quadratic loss function, which, according to Proposition 1.26, is known to "converge" in the sense asserted there.

For the method (7.17) in general the following two statements can be proved, showing

(a) that we may expect a convergence behaviour as we know it already for penalty methods; and

(b) how we should successively adjust the multiplier vector $u^{(k)}$ to get the intended convergence to the proper Kuhn–Tucker multipliers.

Proposition 1.27 *If f and g_i, $i = 1, \cdots, m$, are continuous and $x^{(k)}$, $k = 1, 2, \cdots$, are global solutions of*

$$\min_x L_{\lambda_k}(x, u^{(k)})$$

then any accumulation point $\bar{x}$ of $\{x^{(k)}\}$ is a global solution of problem (7.14).

The following statement also shows that it would be sufficient to solve the free optimization problems $\min_x L_{\lambda_k}(x, u^{(k)})$ only approximately.

Proposition 1.28 *Let f and g_i, $i = 1, \cdots, m$, be continuously differentiable, and let the approximate solutions $x^{(k)}$ to the free minimization problems in (7.17) satisfy*

$$\|\nabla_x L_{\lambda_k}(x^{(k)}, u^{(k)})\| \leq \epsilon_k \; \forall k,$$

where $\epsilon_k \geq 0 \; \forall k$ and $\epsilon_k \longrightarrow 0$. For some $K \subset \mathbb{N}$ let $\{x^{(k)}, \; k \in K\}$ converge to some $x^\star$ (i.e. $x^\star$ is an accumulation point of $\{x^{(k)}, \; k \in \mathbb{N}\}$), and let $\{\nabla g_1(x^\star), \cdots, \nabla g_m(x^\star)\}$ be linearly independent. Then $\exists u^\star$ such that

$$\{u^{(k)} + \lambda_k g(x^{(k)}), \; k \in K\} \longrightarrow u^\star, \tag{7.18}$$

$$\nabla f(x^\star) + \sum_{i=1}^{m} u_i^\star \nabla g_i(x^\star) = 0, \; g(x^\star) = 0.$$

Choosing the parameters λ_k according to (7.17), for instance as

$$\lambda_1 := 1, \; \lambda_{k+1} := 1.1\lambda_k \; \forall k \geq 1,$$

the above statement suggests, by (7.18), that

$$u^{(k+1)} := u^{(k)} + \lambda_k g(x^{(k)}) \tag{7.19}$$

is an appropriate update formula for the multipliers in order to eventually get—together with $x^\star$—a Kuhn–Tucker point.

Now let us come back to our original nonlinear program (7.1) with inequality constraints and show how we can make use of the above results for the case of equality constraints. The key to this is the observation that our problem with inequality constraints

$$\begin{aligned} &\min f(x) \\ &\text{s.t. } g_i(x) \leq 0, \quad i = 1, \cdots, m \end{aligned}$$

is equivalent to the following one with equality constraints:

$$\begin{aligned} &\min f(x) \\ &\text{s.t. } g_i(x) + z_i^2 = 0, \quad i = 1, \cdots, m. \end{aligned}$$

Now applying the augmented Lagrangian method (7.17) to this equality-constrained problem requires that for

$$L_\lambda(x, z, u) := f(x) + \sum_{i=1}^{m} \left\{ u_i [g_i(x) + z_i^2] + \tfrac{1}{2}\lambda [g_i(x) + z_i^2]^2 \right\} \tag{7.20}$$

we solve successively the problem

$$\min_{x,\, z} L_{\lambda_k}(x, z, u^{(k)}).$$

The minimization with respect to z included in this problem may be carried through explicitly, observing that

$$\begin{aligned} &\min_{z \in \mathbb{R}^m} L_{\lambda_k}(x, z, u^{(k)}) \\ &\quad = \min_{z \in \mathbb{R}^m} \Big\{ f(x) + \sum_{i=1}^{m} \{ u_i^{(k)} [g_i(x) + z_i^2] + \tfrac{1}{2}\lambda_k [g_i(x) + z_i^2]^2 \} \Big\} \\ &\quad = f(x) + \sum_{i=1}^{m} \min_{z_i} \{ u_i^{(k)} [g_i(x) + z_i^2] + \tfrac{1}{2}\lambda_k [g_i(x) + z_i^2]^2 \}. \end{aligned}$$

Therefore the minimization of L with respect to z requires—with $y_i := z_i^2$—the solution of m problems of the form

$$\min_{y_i \geq 0} \{ u_i [g_i(x) + y_i] + \tfrac{1}{2}\lambda [g_i(x) + y_i]^2 \}, \tag{7.21}$$

i.e. the minimization of strictly convex ($\lambda > 0$) quadratic functions in y_i on $y_i \geq 0$. The free minima (i.e. $y_i \in \mathbb{R}$) of (7.21) have to satisfy

$$u_i + \lambda [g_i(x) + y_i] = 0,$$

yielding

$$\tilde{y}_i = -\Big[\frac{u_i}{\lambda} + g_i(x)\Big].$$

Hence we have for the solution of (7.21)

$$\left.\begin{aligned} y_i^\star &= \begin{cases} \tilde{y}_i & \text{if } \tilde{y}_i \geq 0, \\ 0 & \text{otherwise,} \end{cases} \\ &= \max\Big\{0, -\Big[\frac{u_i}{\lambda} + g_i(x)\Big]\Big\} \end{aligned}\right\} \tag{7.22}$$

implying

$$g_i(x) + y_i^\star = \max\Big[g_i(x), -\frac{u_i}{\lambda}\Big] \tag{7.23}$$

which, with $\hat{z}_i^2 = y_i^\star$, after an elementary algebraic manipulation reduces our extended Lagrangian (7.20) to

$$\begin{aligned} \tilde{L}_\lambda(x, u) &= L_\lambda(x, \hat{z}, u) \\ &= f(x) + \frac{1}{2\lambda} \sum_{i=1}^{m} \Big\{ \{ \max[0, u_i + \lambda g_i(x)] \}^2 - u_i^2 \Big\}. \end{aligned}$$

Minimization for some given u^k and λ_k of the Lagrangian (7.20) with respect to x and z will now be achieved by solving the problem

$$\min_x \tilde{L}_{\lambda_k}(x, u^{(k)}),$$

and, with a solution $x^{(k)}$ of this problem, our update formula (7.19) for the multipliers—recalling that we now have the equality constraints $g_i(x)+z_i^2 = 0$ instead of $g_i(x) = 0$ as before—becomes by (7.23)

$$\begin{aligned} u^{(k+1)} &:= u^{(k)} + \lambda_k \text{"max"} \left[g(x^{(k)}), -\frac{u^{(k)}}{\lambda_k}\right] \\ &= \text{"max"} \left[0, u^{(k)} + \lambda_k g(x^{(k)})\right], \end{aligned} \tag{7.24}$$

where "max" is to be understood componentwise.

1.8 Bibliographical Notes

The observation that some data in real life optimization problems could be random, i.e. the origin of stochastic programming, dates back to the 1950s. Without any attempt at completeness, we might mention from the early contributions to this field Avriel and Williams [3], Beale [5, 6], Bereanu [8], Dantzig [11], Dantzig and Madansky [13], Tintner [43] and Williams [49].

For more detailed discussions of the situation of the decision maker facing random parameters in an optimization problem we refer for instance to Dempster [14], Ermoliev and Wets [16], Frauendorfer [18], Kall [22], Kall and Prékopa [24], Kolbin [28], Sengupta [42] and Vajda [45].

Wait-and-see problems have led to investigations of the distribution of the optimal value (and the optimal solution); as examples of these efforts, we mention Bereanu [8] and King [26].

The linear programs resulting as deterministic equivalents in the recourse case may become (very) large in scale, but their particular block structure is amenable to specially designed algorithms, which are until now under investigation and for which further progress is to be expected in view of the possibilities given with parallel computers (see e.g. Zenios [51]). For those problems the particular decomposition method QDECOM—which will be described later—was proposed by Ruszczyński [41].

The idea of approximating stochastic programs with recourse (with a continuous type distribution) by discretizing the distribution, as mentioned in Section 1.2, is related to special convergence requirements for the (discretized) expected recourse functions, as discussed for example by Attouch and Wets [2] and Kall [23].

More on probabilistically constrained models and corresponding applications may be found for example in Dupačová et al. [15], Ermoliev and

Wets [16] and Prékopa et al. [36]. The convexity statement of Proposition 1.5 can be found in Wets [48]. The probalistically constrained program at the end of Section 1.2 (page 14) was solved by PROCON. This solution method for problems with a joint chance constraint (with normally distributed right-hand side) was described first by Mayer [30], and has its theoretical base in Prékopa [35].

Statements on the induced feasible set K and induced constraints are found in Rockafellar and Wets [39], Walkup and Wets [46] and Kall [22]. The requirement that the decision on x does not depend on the outcome of $\tilde{\xi}$ is denoted as *nonanticipativity*, and was discussed rigorously in Rockafellar and Wets [40]. The conditions for complete recourse matrices were proved in Kall [21], and may be found in [22].

Necessary and sufficient conditions for log-concave distributions were derived first in Prékopa [35]; later corresponding conditions for quasi-concave measures were derived in Borell [10] and Rinott [37].

More details on stochastic linear programs may be found in Kall [22]; multistage stochastic programs are still under investigation, and were discussed early by Olsen [31, 32, 33]; useful results on the deterministic equivalent of recourse problems and for the expectation functionals arising in Section 1.3 are due to Wets [47, 48].

There is a wide literature on linear programming, which cannot be listed here to any reasonable degree of completeness. Hence we restrict ourselves to mentioning the book of Dantzig [12] as a classic reference.

For a rigorous development of measure theory and the foundations of probability theory we mention the standard reference Halmos [19].

The idea of feasibility and optimality cuts in the dual decomposition method may be traced back to Benders [7].

There is a great variety of good textbooks on nonlinear programming (theory and methods) as well. Again we have to restrict ourselves, and just mention Bazaraa and Shetty [4] and Luenberger [29] as general texts.

Cutting-plane methods have been proposed in various publications, differing in the way the cuts (separating hyperplanes) are defined. An early version was published by Kelley [25]; the method we have presented is due to Kleibohm [27].

The method of feasible directions is due to Zoutendijk [52, 53]; an extension to nonlinear constraints was proposed by Topkis and Veinott [44].

The reduced gradient method can be found in Wolfe [50], and its extension to nonlinear constraints was developed by Abadie and Carpentier [1].

A standard reference for penalty methods is the monograph of Fiacco and McCormick [17].

The update formula (7.19) for the multipliers in the augmented Lagrangian method for equality constraints motivated by Proposition 1.28 goes back to Hestenes [20] and Powell [34], whereas the update (7.24) for inequality-constrained problems is due to Rockafellar [38]. For more about Lagrangian methods we refer the reader to the book of Bertsekas [9].

Exercises

1. Show that from (3.3) on page 19 it follows that with $A_i \in \mathcal{A},\ i = 1, 2, \cdots,$

$$\bigcap_{i=1}^{\infty} A_i \in \mathcal{A} \text{ and } A_i - A_j \in \mathcal{A} \ \ \forall i, j.$$

2. Find an example of a two-dimensional discrete probability distribution that is not quasi-concave.

3. Show that $A = \{(x, y) \in \mathbb{R}^2 \mid x \geq 1,\ 0 \leq y \leq 1/x\}$ is measurable with respect to the natural measure μ in $\mathbb{R}^2$ and that $\mu(A) = \infty$ (see Section 1.3.1, page 16). [Hint: Show first that for

$$I_n := \{(x, y) \mid n \leq x \leq n + 1,\ 0 \leq y < 2\},\ n \in \mathbb{N},$$

the sets $A \cap I_n$ are measurable. For $n \in \mathbb{N}$ the interval

$$C_n := \{(x, y) \mid n \leq x < n + 1,\ 0 \leq y < 1/(n + 1)\}$$

is a packing of $A \cap I_n$ with $\mu(C_n) = 1/(n + 1)$. Hence $\mu(A) = \sum_{n=1}^{\infty} \mu(A \cap I_n) \geq \sum_{n=1}^{\infty} \mu(C_n)$ implies $\mu(A) = \infty$.]

4. Show that $A := \{(x, y) \in \mathbb{R}^2 \mid x \geq 0,\ 0 \leq y \leq e^{-x}\}$ is measurable and that $\mu(A) = 1$. [Hint: Consider $A_\alpha := \{(x, y) \mid 0 \leq x \leq \alpha,\ 0 \leq y \leq e^{-x}\}$ for arbitrary $\alpha > 0$. Observe that $\mu(A_\alpha)$, according to its definition in Section 1.3.1, page 16, coincides with the Riemann integral $J(\alpha) = \int_0^\alpha e^{-x}\, dx$. Hence $\mu(A) = \lim_{\alpha \to \infty} \mu(A_\alpha) = \lim_{\alpha \to \infty} J(\alpha)$.]

5. Consider the line segment $B := \{(x, y) \in \mathbb{R}^2 \mid 3 \leq x \leq 7,\ y = 5\}$. Show that for the natural measure μ in $\mathbb{R}^2$, $\mu(B) = 0$ (see Section 1.3.1, page 16).

6. Assume that the linear program $\gamma(b) := \min\{c^{\mathrm{T}} x \mid Ax = b,\ x \geq 0\}$ is solvable for all $b \in \mathbb{R}^m$. Show that the optimal value function $\gamma(\cdot)$ is piecewise linear and convex in b.

7. In Section 1.7 we discussed various regularity conditions for nonlinear programs. Let $\hat{x}$ be a local solution of problem (7.1), page 75. Show that if $\underline{\mathcal{RC}\_1}$ is satisfied in $\hat{x}$ then $\underline{\mathcal{RC}\_0}$ also holds true in $\hat{x}$. (See (7.8) on page 80.)

8. Assume that $(\hat{x}, \hat{u})$ is a saddle point of

$$L(x, u) := f(x) + \sum_{i=1}^{m} u_i g_i(x).$$

Show that $\hat{x}$ is a global solution of

$$\begin{aligned}&\min f(x)\\&\text{s.t. } g_i(x) \leq 0, \quad i = 1, \cdots, m.\end{aligned}$$

(See Proposition 1.25 for the definition of a saddle point.)

References

[1] Abadie J. and Carpentier J. (1969) Generalization of the Wolfe reduced gradient method to the case of nonlinear constraints. In Fletcher R. (ed) *Optimization*, pages 37–47. Academic Press, London.

[2] Attouch H. and Wets R. J.-B. (1981) Approximation and convergence in nonlinear optimization. In Mangasarian O. L., Meyer R. M., and Robinson S. M. (eds) *NLP 4*, pages 367–394. Academic Press, New York.

[3] Avriel M. and Williams A. (1970) The value of information and stochastic programming. *Oper. Res.* 18: 947–954.

[4] Bazaraa M. S. and Shetty C. M. (1979) *Nonlinear Programming—Theory and Algorithms.* John Wiley & Sons, New York.

[5] Beale E. M. L. (1955) On minimizing a convex function subject to linear inequalities. *J. R. Stat. Soc.* B17: 173–184.

[6] Beale E. M. L. (1961) The use of quadratic programming in stochastic linear programming. Rand Report P-2404, The RAND Corporation.

[7] Benders J. F. (1962) Partitioning procedures for solving mixed-variables programming problems. *Numer. Math.* 4: 238–252.

[8] Bereanu B. (1967) On stochastic linear programming distribution problems, stochastic technology matrix. *Z. Wahrsch. theorie u. verw. Geb.* 8: 148–152.

[9] Bertsekas D. P. (1982) *Constrained Optimization and Lagrange Multiplier Methods.* Academic Press, New York.

[10] Borell C. (1975) Convex set functions in *d*-space. *Period. Math. Hungar.* 6: 111–136.

[11] Dantzig G. B. (1955) Linear programming under uncertainty. *Management Sci.* 1: 197–206.

[12] Dantzig G. B. (1963) *Linear Programming and Extensions.* Princeton University Press, Princeton, New Jersey.

[13] Dantzig G. B. and Madansky A. (1961) On the solution of two-stage linear programs under uncertainty. In Neyman I. J. (ed) *Proc. 4th Berkeley Symp. Math. Stat. Prob.*, pages 165–176. Berkeley.

[14] Dempster M. A. H. (ed) (1980) *Stochastic Programming.* Academic Press, London.

[15] Dupačová J., Gaivoronski A., Kos Z., and Szántai T. (1991) Stochastic programming in water resources system planning: A case study and a

comparison of solution techniques. *Eur. J. Oper. Res.* 52: 28–44.
[16] Ermoliev Y. and Wets R. J.-B. (eds) (1988) *Numerical Techniques for Stochastic Optimization.* Springer-Verlag, Berlin.
[17] Fiacco A. V. and McCormick G. P. (1968) *Nonlinear Programming: Sequential Unconstrained Minimization Techniques.* John Wiley & Sons, New York.
[18] Frauendorfer K. (1992) *Stochastic Two-Stage Programming*, volume 392 of *Lecture Notes in Econ. Math. Syst.* Springer-Verlag, Berlin.
[19] Halmos P. R. (1950) *Measure Theory.* D. van Nostrand, Princeton, New Jersey.
[20] Hestenes M. R. (1969) Multiplier and gradient methods. In Zadeh L. A., Neustadt L. W., and Balakrishnan A. V. (eds) *Computing Methods in Optimization Problems—2*, pages 143–163. Academic Press, New York.
[21] Kall P. (1966) Qualitative Aussagen zu einigen Problemen der stochastischen Programmierung. *Z. Wahrsch. theorie u. verw. Geb.* 6: 246–272.
[22] Kall P. (1976) *Stochastic Linear Programming.* Springer-Verlag, Berlin.
[23] Kall P. (1986) Approximation to optimization problems: An elementary review. *Math. Oper. Res.* 11: 9–18.
[24] Kall P. and Prékopa A. (eds) (1980) *Recent Results in Stochastic Programming*, volume 179 of *Lecture Notes in Econ. Math. Syst.* Springer-Verlag, Berlin.
[25] Kelley J. E. (1960) The cutting plane method for solving convex programs. *SIAM J. Appl. Math.* 11: 703–712.
[26] King A. J. (1986) *Asymptotic Behaviour of Solutions in Stochastic Optimization: Nonsmooth Analysis and the Derivation of Non-Normal Limit Distributions.* PhD thesis, University of Washington, Seattle.
[27] Kleibohm K. (1966) *Ein Verfahren zur approximativen Lösung von konvexen Programmen.* PhD thesis, Universität Zürich. Mentioned in C.R. Acad. Sci. Paris 261:306–307 (1965).
[28] Kolbin V. V. (1977) *Stochastic Programming.* D. Reidel, Dordrecht.
[29] Luenberger D. G. (1973) *Introduction to Linear and Nonlinear Programming.* Addison-Wesley, Reading, Massachusetts.
[30] Mayer J. (1988) Probabilistic constrained programming: A reduced gradient algorithm implemented on pc. Working Paper WP-88-39, IIASA, Laxenburg.
[31] Olsen P. (1976) Multistage stochastic programming with recourse: The equivalent deterministic problem. *SIAM J. Contr. Opt.* 14: 495–517.
[32] Olsen P. (1976) When is a multistage stochastic programming problem well defined? *SIAM J. Contr. Opt.* 14: 518–527.
[33] Olsen P. (1976) Discretizations of multistage stochastic programming problems. *Math. Prog. Study* 6: 111–124.
[34] Powell M. J. D. (1969) A method for nonlinear constraints in

minimization problems. In Fletcher R. (ed) *Optimization*, pages 283–298. Academic Press, London.

[35] Prékopa A. (1971) Logarithmic concave measures with applications to stochastic programming. *Acta Sci. Math. (Szeged)* 32: 301–316.

[36] Prékopa A., Ganczer S., Deák I., and Patyi K. (1980) The STABIL stochastic programming model and its experimental application to the electricity production in Hungary. In Dempster M. A. H. (ed) *Stochastic Programming*, pages 369–385. Academic Press, London.

[37] Rinott Y. (1976) On convexity of measures. *Ann. Prob.* 4: 1020–1026.

[38] Rockafellar R. T. (1973) The multiplier method of Hestenes and Powell applied to convex programming. *J. Opt. Theory Appl.* 12: 555–562.

[39] Rockafellar R. T. and Wets R. J.-B. (1976) Stochastic convex programming: Relatively complete recourse and induced feasibility. *SIAM J. Contr. Opt.* 14: 574–589.

[40] Rockafellar R. T. and Wets R. J.-B. (1976) Nonanticipativity and L^1-martingales in stochastic optimization problems. *Math. Prog. Study* 6: 170–187.

[41] Ruszczyński A. (1986) A regularized decomposition method for minimizing a sum of polyhedral functions. *Math. Prog.* 35: 309–333.

[42] Sengupta J. K. (1972) *Stochastic programming. Methods and applications.* North-Holland, Amsterdam.

[43] Tintner G. (1955) Stochastic linear programming with applications to agricultural economics. In Antosiewicz H. (ed) *Proc. 2nd Symp. Linear Programming*, volume 2, pages 197–228. National Bureau of Standards, Washington D.C.

[44] Topkis D. M. and Veinott A. F. (1967) On the convergence of some feasible direction algorithms for nonlinear programming. *SIAM J. Contr. Opt.* 5: 268–279.

[45] Vajda S. (1972) *Probabilistic Programming.* Academic Press, New York.

[46] Walkup D. W. and Wets R. J. B. (1967) Stochastic programs with recourse. *SIAM J. Appl. Math.* 15: 1299–1314.

[47] Wets R. (1974) Stochastic programs with fixed recourse: The equivalent deterministic program. *SIAM Rev.* 16: 309–339.

[48] Wets R. J.-B. (1989) Stochastic programming. In Nemhauser G. *et al.* (eds) *Handbooks in OR & MS*, volume 1, pages 573–629. Elsevier, Amsterdam.

[49] Williams A. (1966) Approximation formulas for stochastic linear programming. *SIAM J. Appl. Math.* 14: 668–877.

[50] Wolfe P. (1963) Methods of nonlinear programming. In Graves R. L. and Wolfe P. (eds) *Recent Advances in Mathematical Programming*, pages 67–86. McGraw-Hill, New York.

[51] Zenios S. A. (1992) Progress on the massively parallel solution of network "mega"-problems. *COAL* 20: 13–19.

[52] Zoutendijk G. (1960) *Methods of Feasible Directions.* Elsevier, Amsterdam/D. Van Nostrand, Princeton, New Jersey.
[53] Zoutendijk G. (1966) Nonlinear programming: A numerical survey. *SIAM J. Contr. Opt.* 4: 194–210.

2

Dynamic Systems

2.1 The Bellman Principle

As discussed in Chapter 1, optimization problems can be of various types. The differences may be found in the goal, i.e. minimization or maximization, in the constraints, i.e. inequalities or equalities and free or nonnegative variables, and in the mathematical properties of the functions involved in the objective or the constraints. We have met linear functions in Section 1.6, nonlinear functions in Section 1.7 and even integral functions in Section 1.3. Despite their differences, all these problems may be presented in the unified form

$$\max\{F(x_1,\cdots,x_n) \mid x \in X\}.$$

Here X is the prescribed feasible set of decisions over which we try to maximize, or sometimes minimize, the given objective function F.

This general setting also covers a class of somewhat special decision problems. They are illustrated in Figure 1.

Consider a system that is inspected at finitely many *stages*. Often stages are just points in time—the reason for using the term "dynamic". The example in Figure 1 has four stages. This is seen from the fact that there are four columns. Assume that at any stage the system can be in one out of finitely many *states*. In Figure 1 there are four possible states in each stage, represented by the four dots in each column. Also, at any stage (except maybe the last one) a decision has to be made which possibly will have an influence on the system's state at the subsequent stage. Attached to the decision is an *immediate return* (or else an immediate cost). In Figure 1 the three arrows in the right part of the figure indicate that in this example there are three possible decisions: one bringing us to a lower state in the next stage, one keeping us in the same state, and one bringing us to a higher state. (We must assume that if we are at the highest or lowest possible state then only two decisions are possible). Given the initial state of the system, the overall objective is to maximize (or minimize) some given function of the immediate returns for all stages and states the system goes through as a result of our decisions. Formally the problem is described

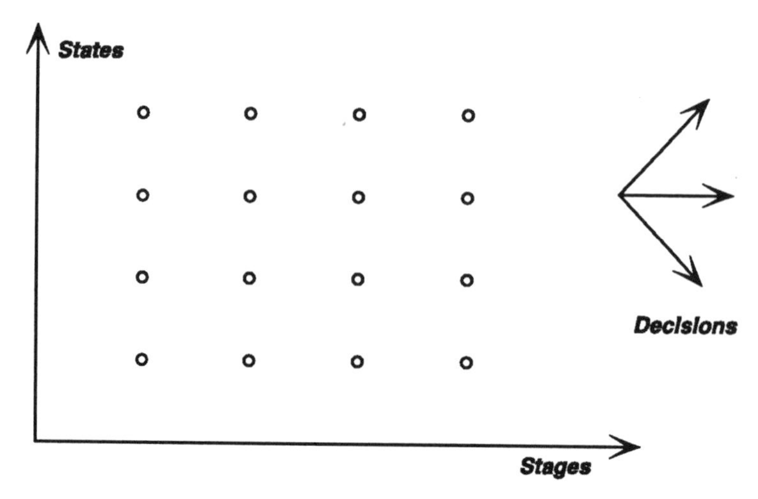

Figure 1 Basic set-up for a dynamic program with four states, four stages and three possible decisions.

as follows. With

t	the stages, $t = 1, \cdots, T$,
z_t	the state at stage t,
x_t	the decision taken at stage t, (in general depending on the state z_t),
$G_t(z_t, x_t)$	the transformation (or transition) of the system from the state z_t and the decision taken at stage t into the state z_{t+1} at the next stage, i.e. $z_{t+1} = G_t(z_t, x_t)$,
$r_t(z_t, x_t)$	the immediate return if at stage t the system is in state z_t and the decision x_t is taken,
F	the overall objective, given by $F(r_1(z_1, x_1), \cdots, r_T(z_T, x_T))$, and
$X_t(z_t)$	the set of feasible decisions at stage t, (which may depend on the state z_t),

our problem can be stated as

$$\max\{F(r_1(z_1, x_1), \cdots, r_T(z_T, x_T)) \mid x_t \in X_t, t = 1, \cdots, T\}.$$

Observe that owing to the relation $z_{t+1} = G_t(z_t, x_t)$, the objective function can be rewritten in the form $\Phi(z_1, x_1, x_2, \cdots, x_T)$. To get an idea of the possible structures we can face, let us revisit the example in Figure 1. The purpose of the example is not to be realistic, but to illustrate a few points. A more realistic problem will be discussed in the next section.

Example 2.1 Assume that stages are years, and that the system is inspected annually, so that the three stages correspond to 1 January of the first, second and third years, and 31 December of the third year. Assume further that four different levels are distinguished as states for the system, i.e. at any stage one may observe the state $z_t = 1, 2, 3$ or 4. Finally, depending on the state of the system in stages $1, 2$ and 3, one of the following decisions may be made:

$$x_t = \begin{cases} 1, & \text{leading to the immediate return } r_t = 2, \\ 0, & \text{leading to the immediate return } r_t = 1, \\ -1, & \text{leading to the immediate return } r_t = -1. \end{cases}$$

The transition from one stage to the next is given by

$$z_{t+1} = z_t + x_t.$$

Note that, since $z_t \in \{1, 2, 3, 4\}$ for all t, we have that $x_t = 1$ in state $z_t = 4$ and $x_t = -1$ in state $z_t = 1$ are not feasible, and are therefore excluded. Finally, assume that there are no decisions in the final stage $T = 4$. There are immediate returns, however, given as

$$r_T = \begin{cases} -2 & \text{if } z_T = 4, \\ -1 & \text{if } z_T = 3, \\ 1 & \text{if } z_T = 2, \\ 2 & \text{if } z_T = 1. \end{cases}$$

To solve $\max F(r_1, \cdots, r_4)$, we have to fix the overall objective F as a function of the immediate returns r_1, r_2, r_3, r_4. To demonstrate possible effects of properties of F on the solution procedure, we choose two variants.

(a) Let

$$F(r_1, \cdots, r_4) := r_1 + r_2 + r_3 + r_4$$

and assume that the initial state is $z_1 = 4$. This is illustrated in Figure 2, which has the same structure as Figure 1. Using the figure, we can check that an optimal policy (i.e. sequence of decisions), is $x_1 = x_2 = x_3 = 0$ keeping us in $z_t = 4$ for all t with the optimal value $F(r_1, \cdots, r_4) = 1 + 1 + 1 - 2 = 1$.

We may determine this optimal policy iteratively as follows. First, we determine the decision for each of the states in stage 3 by determining

$$f_3^*(z_3) := \max_{x_3}[r_3(z_3, x_3) + r_4(z_4)]$$

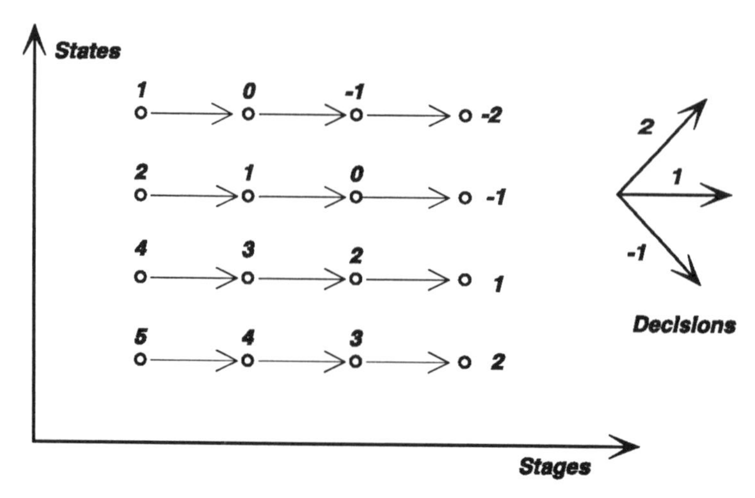

Figure 2 Dynamic program: additive composition. The solid lines show the result of the backward recursion.

for $z_3 = 1, \cdots, 4$, and $z_4 = G_3(z_3, x_3) = z_3 + x_3$. For example, if we are in state 2, i.e. $z_3 = 2$, we have three options, namely -1, 0 and 1. If $x_3 = 1$, we receive an immediate income of 2, and a final value of -1, since this decision will result in $z_4 = 2 + 1 = 3$. The second option is to let $x_3 = 0$, yielding an immediate income of 1 and a final value of 1. The third possibility is to let $x_3 = -1$, yielding incomes of -1 and 1. The total incomes are therefore 1, 2 and 0 respectively, so the best option is to let $z_3 = 0$. This is illustrated in the figure by putting an arrow from state 2 in stage 3 to state 2 in stage 4. Letting "$(z_3 = i) \rightarrow (x_3, f_3^*)$" indicate that in state $z_3 = i$ the optimal decision is x_3 and the sum of the immediate and final income is f_3^*, we can repeat the above procedure for each state in stage 3 to obtain $(z_3 = 1) \rightarrow (0, 3)$, $(z_3 = 2) \rightarrow (0, 2)$, $(z_3 = 3) \rightarrow (0, 0)$, $(z_3 = 4) \rightarrow (0, -1)$. This is all illustrated in Figure 2, by adding the f_3^*-values above the state nodes in stage 3.

Once $f_3^*(z_3)$ is known for all values of z_3, we can turn to stage 2 and similarly determine

$$f_2^*(z_2) := \max_{x_2}[r_2(z_2, x_2) + f_3^*(z_3)],$$

where $z_3 = G_2(z_2, x_2) = z_2 + x_2$. This yields $f_2^*(1) = 4$, $f_2^*(2) = 3$, $f_2^*(3) = 1$ *and* $f_2^*(4) = 0$. This is again illustrated in Figure 2, together with the corresponding optimal decisions.

Finally, given that $z_1 = 4$, the problem can be rephrased as

$$f_1^*(z_1) := \max_{x_1}[r_1(z_1, x_1) + f_2^*(z_2)],$$

where $z_2 = G_1(z_1, x_1) = z_1 + x_1$. This immediately yields $f_1^*(z_1) = 1$ for $x_1 = 0$.

In this simple example it is easy to see that $f_1^*(z_1)$ coincides with the optimal value of $F(r_1, \cdots, r_4)$, given the initial state z_1, so that the problem can be solved with the above backward recursion. (The recursion is called *backward* because we start in the last period and move backwards in time, ending up in period 1.)

Note that an alternative way to solve this problem would be to enumerate all possible sequences of decisions. For this small problem, that would have been a rather simple task. But for larger problems, both in terms of states and stages (especially when both are multi-dimensional), it is easy to see that this will become an impossible task. This reduction from a full enumeration of all possible sequences of decisions to that of finding the optimal decision in all states for all stages is the major reason for being interested in the backward recursion, and more generally, for being interested in dynamic programming.

(b) As an alternative, let us use multiplication to obtain

$$F(r_1, \cdots, r_4) := r_1 r_2 r_3 r_4$$

and perform the backward recursion as above, yielding Figure 3. With

$$f_t^*(z_t) := \max_{x_t}[r_t(z_t, x_t) f_{t+1}^*(z_{t+1})]$$

for $t = 3, 2, 1$, where $z_{t+1} = G_t(z_t, x_t) = z_t + x_t$ and $f_4^*(z_4) = r_4(z_4)$, we should get $f_1^*(z_1 = 4) = 1$ with an "optimal" policy $(0, 0, -1)$. However, the policy $(-1, 1, 0)$ yields $F(r_1, \cdots, r_4) = 4$. Hence the backward recursion does not yield the optimal solution when the returns are calculated in a multiplicative fashion.

□

In this example we had

$$F(r_1(z_1, x_1), \cdots, r_T(z_T, x_T)) = r_1(z_1, x_1) \oplus r_2(z_2, x_2) \oplus \cdots \oplus r_T(z_T, x_T),$$

where the composition operation "$\oplus$" was chosen as addition in case (a) and multiplication in case (b). For the backward recursion we have made use of the so-called *separability* of F. That is, there exist two functions φ_1, ψ_2 such that

$$F(r_1(z_1, x_1), \cdots, r_T(z_T, x_T)) \\ = \varphi_1(r_1(z_1, x_1), \psi_2(r_2(z_2, x_2), \cdots, r_T(z_T, x_T))). \tag{1.1}$$

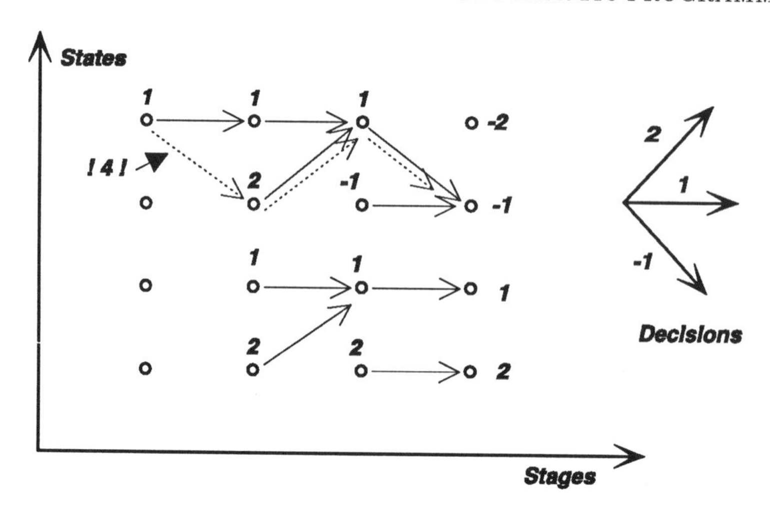

Figure 3 Dynamic program: multiplicative composition. Solid lines show the result of the backward recursion (with $z_1 = 4$), whereas the dotted line shows the optimal sequence of decisions.

Furthermore, we proceeded "as if" the following relation held:

$$\begin{aligned}&\max\{F(r_1(z_1,x_1),\cdots,r_T(z_T,x_T)) \mid x_t \in X_t, t=1,\cdots,T\} \\ &\quad = \max_{x_1\in X_1}[\varphi_1(r_1(z_1,x_1), \max_{x_2\in X_2,\cdots,x_T\in X_T} \psi_2(r_2(z_2,x_2),\cdots,r_T(z_T,x_T)))].\end{aligned} \tag{1.2}$$

This relation is the formal equivalent of the well-known *optimality principle*, which was expressed by Bellman as follows (quote).

Proposition 2.1 *"An optimal policy has the property that whatever the initial state and initial decision are, the remaining decisions must constitute an optimal policy with regard to the state resulting from the first decision."*

As we have seen in Example 2.1, this principle, applied repeatedly in the backward recursion, gave the optimal solution for case (a) but not for case (b). The reason for this is that, although the composition operation "$\oplus$" is separable in the sense of (1.1), this is not enough to guarantee that the repeated application of the optimality principle (i.e. through backward recursion) will yield an optimal policy. A sufficient condition under which the optimality principle holds involves a certain monotonicity of our composition operation "$\oplus$". More precisely, we have the following.

Proposition 2.2 *If F satisfies the separability condition (1.1) and if φ_1 is monotonically nondecreasing in ψ_2 for every r_1 then the optimality principle (1.2) holds.*

Proof The unique meaning of "max" implies that we have that

$$\max_{\{x_t \in X_t,\ t \geq 1\}} \varphi_1(r_1(z_1, x_1), \psi_2(r_2(z_2, x_2), \cdots, r_T(z_T, x_T)))$$
$$\geq \varphi_1(r_1(z_1, x_1), \max_{\{x_t \in X_t,\ t \geq 2\}} [\psi_2(r_2(z_2, x_2), \cdots, r_T(z_T, x_T))]),$$

for all x_1. Therefore this also holds when the right-hand side of this inequality is maximized with respect to x_1.

On the other hand, it is also obvious that

$$\max_{\{x_t \in X_t,\ t \geq 2\}} \psi_2(r_2(z_2, x_2), \cdots, r_T(z_T, x_T))$$
$$\geq \psi_2(r_2(z_2, x_2), \cdots, r_T(z_T, x_T)) \ \ \forall x_t \in X_t, t \geq 2.$$

Hence, by the assumed monotonicity of ψ_1 with respect to ψ_2, we have that

$$\varphi_1(r_1(z_1, x_1), \max_{\{x_t \in X_t, t \geq 2\}} \psi_2(r_2(z_2, x_2), \cdots, r_T(z_T, x_T)))$$
$$\geq \varphi_1(r_1(z_1, x_1), \psi_2(r_2(z_2, x_2), \cdots, r_T(z_T, x_T))) \ \ \forall x_t \in X_t, t \geq 1.$$

Taking the maximum with respect to $x_t, t \geq 2$, on the right-hand side of this inequality and maximizing afterwards both sides with respect to $x_1 \in X_1$ shows that the optimality principle (1.2) holds. □

Needless to say, all problems considered by Bellman in his first book on dynamic programming satisfied this proposition. In case (b) of our example, however, the monotonicity does not hold. The reason is that when "$\oplus$" involves multiplication of possibly negative factors (i.e. negative immediate returns), the required monotonicity is lost. On the other hand, when "$\oplus$" is summation, the required monotonicity is always satisfied.

Let us add that the optimality principle applies to a much wider class of problems than might seem to be the case from this brief sketch. For instance, if for finitely many states we denote by ρ_t the vector having as ith component the immediate return $r_t(z_t = i)$, and if we define the composition operation "$\oplus$" such that, with a positive matrix S (i.e. all elements of S nonnegative),

$$\rho_t \oplus \rho_{t+1} = \rho_t + S\rho_{t+1}, \ t = 1, \cdots, T-1,$$

then the monotonicity assumed for Proposition 2.2 follows immediately. This case is quite common in applications. Then S is the so-called transition matrix, which means that an element s_{ij} represents the probability of entering state j at stage $t+1$, given that the system is in state i at stage t. Iterating the above composition for $T-1, T-2, \cdots, 1$ we get that $F(\rho_1, \cdots, \rho_T)$ is the vector of the expected total returns. The ith component gives the expected overall return if the system starts from state i at stage 1. Putting it this way we see that multistage stochastic programs with recourse (formula (3.13) in Chapter 1) belong to this class.

2.2 Dynamic Programming

The purpose of this section is to look at certain aspects of the field of dynamic programming. The example we looked at in the previous section is an example of a dynamic programming problem. It will not represent a fair description of the field as a whole, but we shall concentrate on aspects that are useful in our context. This section will not consider randomness. That will be discussed later.

We shall be interested in dynamic programming as a means of solving problems that evolve over time. Typical examples are production planning under varying demand, capacity expansion to meet an increasing demand and investment planning in forestry. Dynamic programming can also be used to solve problems that are not sequential in nature. Such problems will not be treated in this text.

Important concepts in dynamic programming are the time horizon, state variables, decision variables, return functions, accumulated return functions, optimal accumulated returns and transition functions. The *time horizon* refers to the number of *stages* (time periods) in the problem. *State variables* describe the state of the system, for example the present production capacity, the present age and species distribution in a forest or the amount of money one has in different accounts in a bank. *Decision variables* are the variables under one's control. They can represent decisions to build new plants, to cut a certain amount of timber, or to move money from one bank account to another. The *transition function* shows how the state variables change as a function of decisions. That is, the transition function dictates the state that will result from the combination of the present state and the present decisions. For example, the transition function may show how the forest changes over the next period as a result of its present state and of cutting decisions, how the amount of money in the bank increases, or how the production capacity will change as a result of its present size, investments (and detoriation). A *return function* shows the immediate returns (costs or profits) as a result of making a specific decision in a specific state. *Accumulated return functions* show the accumulated effect, from now until the end of the time horizon, associated with a specific decision in a specific state. Finally, *optimal accumulated returns* show the value of making the optimal decision based on an accumulated return function, or in other words, the best return that can be achieved from the present state until the end of the time horizon.

Example 2.2 Consider the following simple investment problem, where it is clear that the Bellman principle holds. We have some money S_0 in a bank account, called account B. We shall need the money two years from now, and today is the first of January. If we leave the money in the account we will face an interest rate of 7% in the first year and 5% in the second. You also have

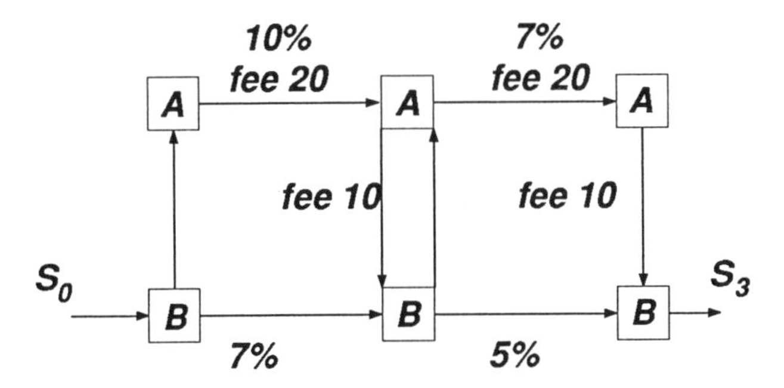

Figure 4 Graphical description of a simple investment problem.

the option of moving the money to account A. You will there face an interest rate of 10% the first year and 7% the second year. However, there is a fixed charge of 20 per year and a charge of 10 each time we withdraw money from account A. The fixed charge is deducted from the account at the end of a year, whereas the charges on withdrawals are deducted immediately. The question is: Should we move our money to account A for the first year, the second year or both years? In any case, money left in account A at the end of the second year will be transferred to account B. The goal is to solve the problem for all initial $S_0 > 1000$. Figure 4 illustrates the example.

Note that all investments will result in a case where the wealth increases, and that it will never be profitable to split the money between the accounts (why?).

Let us first define the two-dimensional state variables $z_t = (z_t^1, z_t^2)$. The first state variable, z_t^1, refers to the account name (A or B); the second state variable, z_t^2, refers to the amount of money S_t in that account. So $z_t = (B, S_t)$ refers to a state where there is an amount S_t in account B in stage t. Decisions are where to put the money for the next time period. If x_t is our decision variable then $x_t \in \{A, B\}$. The transition function will be denoted by $G_t(z_t, x_t)$, and is defined via interest rates and charges. It shows what will happen to the money over one year, based on where the money is now, how much there is, and where it is put next. Since the state space has two elements, the function G_t is two-valued. For example

$$z_{t+1}^1 = G_t^1\left(\begin{pmatrix} A \\ S_t \end{pmatrix}, A\right) = A, \qquad z_{t+1}^2 = G_t^2\left(\begin{pmatrix} A \\ S_t \end{pmatrix}, A\right) = S_t \times 1.07 - 20.$$

Accumulated return functions will be denoted by $f_t(z_t^1, z_t^2, x_t)$. They describe how the amount z_t^2 in account z_t^1 will grow, up to the end of the time horizon, if the money is put into account x_t in the next period, and optimal decisions are made thereafter. So if $f_1(A, S_1, B) = \overline{S}$, we know that in stage 1 (i.e. at the end of period 1), if we have S_1 in account A and then move it to account B, we shall be left with $S_3 = \overline{S}$ in account B at the end of the time horizon, *given that we make optimal decisions at all stages after stage 1*. By maximizing over all possible decisions, we find the optimal accumulated returns $f_t^*(z_t^1, z_t^2)$ for a given state. For example,

$$f_1^*(A, S_1) = \max_{x_1 \in \{A,B\}} f_1(A, S_1, x_1).$$

The calculations for our example are as follows. Note that we have three stages, which we shall denote Stage 0, Stage 1 and Stage 2. Stage 2 represents the point in time (after two years) when all funds must be transferred to account B. Stage 1 is one year from now, where we, if we so wish, may move the money from one account to another. Stage 0 is now, where we must decide if we wish to keep the money in account B or move it to account A.

Stage 2 At Stage 2, all we can do is to transfer whatever money we have in account A to account B:

$$\begin{aligned} f_2^*(A, S_2) &= S_2 - 10, \\ f_2^*(B, S_2) &= S_2, \end{aligned}$$

indicating that a cost of 10 is incurred if the money is in account A and needs to be transferred to account B.

Stage 1 Let us first consider account A, and assume that the account contains S_1. We can keep the money in account A, making $S_2 = S_1 \times 1.07 - 20$ (this is the transition function), or move it to B, making $S_2 = (S_1 - 10) \times 1.05$. This generates the following two evaluations of the accumulated return function:

$$\begin{aligned} f_1(A, S_1, A) &= f_2^*(A, S_1 \times 1.07 - 20) = S_1 \times 1.07 - 30, \\ f_1(A, S_1, B) &= f_2^*(B, (S_1 - 10) \times 1.05) = S_1 \times 1.05 - 10.5. \end{aligned}$$

By comparing these two, we find that, as long as $S_1 \geq 975$ (which is always the case since we have assumed that $S_0 > 1000$), account A is best, making

$$f_1^*(A, S_1) = S_1 \times 1.07 - 30.$$

Next, consider account B. If we transfer the amount S_1 to account A, we get $S_2 = S_1 \times 1.07 - 20$. If it stays in B, we get $S_2 = S_1 \times 1.05$. This gives us

$$\begin{aligned} f_1(B, S_1, A) &= f_2^*(A, S_1 \times 1.07 - 20) = S_1 \times 1.07 - 30, \\ f_1(B, S_1, B) &= f_2^*(B, S_1 \times 1.05) = S_1 \times 1.05. \end{aligned}$$

By comparing these two, we find that

$$f_1^*(B, S_1) = \begin{cases} S_1 \times 1.07 - 30 & \text{if } S_1 \geq 1500, \\ S_1 \times 1.05 & \text{if } S_1 \leq 1500. \end{cases}$$

Stage 0 Since we start out with all our money in account B, we only need to check that account. Initially we have S_0. If we transfer to A, we get $S_1 = S_0 \times 1.1 - 20$, and if we keep it in B, $S_1 = S_0 \times 1.07$. The accumulated returns are

$$\begin{aligned} f_0(B, S_0, A) &= f_1^*(A, S_1) = f_1^*(A, S_0 \times 1.1 - 20) \\ &= (S_0 \times 1.1 - 20) \times 1.07 - 30 = 1.177 \times S_0 - 51.4, \\ f_0(B, S_0, B) &= f_1^*(B, S_1) = f_1^*(B, S_0 \times 1.07) \\ &= \begin{cases} S_0 \times 1.1449 - 30 & \text{if } S_0 \geq 1402, \\ S_0 \times 1.1235 & \text{if } S_0 \leq 1402. \end{cases} \end{aligned}$$

Comparing the two options, we see that account A is always best, yielding

$$f_0^*(B, S_0) = 1.177 \times S_0 - 51.4.$$

So we should move our money to account A and keep it there until the end of the second period. Then we move it to B as required. We shall be left with a total interest of 17.7% and fixed charges of 51.4 (including lost interest on charges).

□

As we can see, the main idea behind dynamic programming is to take one stage at a time, starting with the last stage. For each stage, find the optimal decision for all possible states, thereby calculating the optimal accumulated return from then until the end of the time horizon for all possible states. Then move one step towards the present, and calculate the returns from that stage until the end of the time horizon by adding together the immediate returns, and the returns for all later periods based on the calculations made at the previous stage. In the example we found that $f_1^*(A, S_1) = S_1 \times 1.07 - 30$. This shows us that if we end up in stage 1 with S_1 in account A, we shall (if we behave optimally) end up with $S_1 \times 1.07 - 30$ in account B at the end of the time horizon. However, f_1^* does not tell us *what* to do, since that is not needed to calculate optimal decisions at stage 0.

Formally speaking, we are trying to solve the following problem, where $x = (x_0, \ldots, x_T)^{\mathrm{T}}$:

$$\begin{array}{lll} \max_x & F(r_0(z_0, x_0), \ldots, r_T(z_T, x_T), Q(z_{T+1})) & \\ \text{s.t.} & z_{t+1} = G_t(z_t, x_t) & \text{for } t = 0, \ldots, T, \\ & A_t(z_t) \leq x_t \leq B_t(z_t) & \text{for } t = 0, \ldots, T, \end{array}$$

where F satisfies the requirements of Proposition 2.2. This is to be solved for one or more values of the initial state z_0. In this set-up, r_t is the return function for all but the last stage, Q the return function for the last stage, G_t the transition function, T the time horizon, z_t the (possibly multi-dimensional) state variable in stage t and x_t the (possibly multi-dimensional) decision variable in stage t. The accumulated return function $f_t(z_t, x_t)$ and optimal accumulated returns $f_t^*(z_t)$ are not part of the problem formulation, but rather part of the solution procedure. The solution procedure, justified by the Bellman principle, runs as follows.

$$\text{Find } f_0^*(z_0)$$

by solving recursively

$$\begin{aligned} f_t^*(z_t) &= \max_{A_t(z_t) \le x_t \le B_t(z_t)} f_t(z_t, x_t) \\ &= \max_{A_t(z_t) \le x_t \le B_t(z_t)} \varphi_t(r_t(z_t, x_t), f_{t+1}^*(z_{t+1})) \text{ for } t = T, \ldots, 0 \end{aligned}$$

with

$$\begin{aligned} z_{t+1} &= G_t(z_t, x_t) \text{ for } t = T, \ldots, 0, \\ f_{T+1}^*(z_{T+1}) &= Q(z_{T+1}). \end{aligned}$$

In each case the problem must be solved for all possible values of the state variable z_t, which might be multi-dimensional.

Problems that are not dynamic programming problems (unless rewritten with a large expansion of the state space) would be problems where, for example,

$$z_{t+1} = G_t(z_0, \ldots, z_t, x_0, \ldots, x_t),$$

or where the objective function depends in an arbitrary way on the whole history up til stage t, represented by

$$r_t(z_0, \ldots, z_t, x_0, \ldots, x_t).$$

Such problems may more easily be solved using other approaches, such as decision trees, where these complicated functions cause little concern.

2.3 Deterministic Decision Trees

We shall not be overly interested in decision trees in a deterministic setting. However, since they might be used to analyse sequential decision problems, we shall mention them. Let us consider our simple investment problem in Example 2.2. A decision tree for that problem is given in Figure 5. A decision tree consists of nodes and arcs. The nodes represent states, the arcs decisions. For each possible state in each stage we must create one arc for each possible decision. Therefore the number of possible decisions must be very limited for

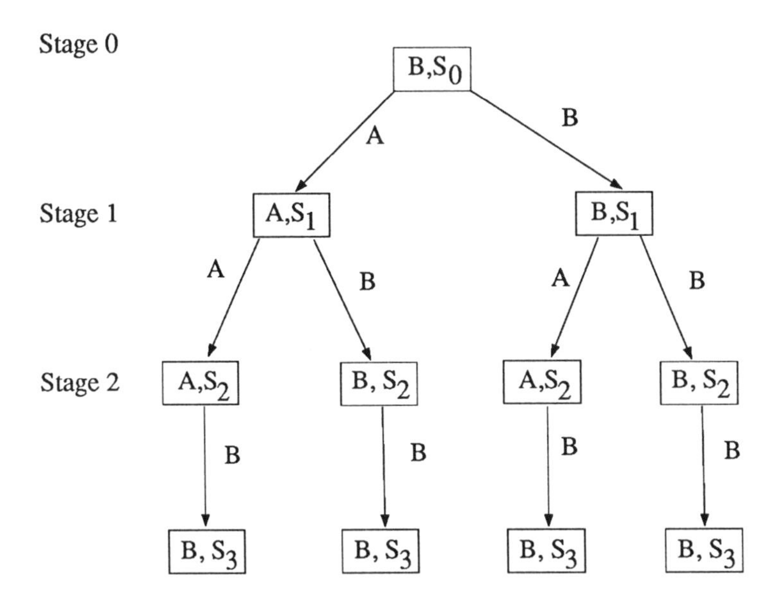

Figure 5 Deterministic decision tree for the small investment problem.

this method to be useful, since there is one leaf in the tree for each possible *sequence* of decisions.

The tree indicates that at stage 0 we have S_0 in account B. We can then decide to put them into A (go left) or keep them in B (go right). Then at stage 1 we have the same possible decisions. At stage 2 we have to put them into B, getting S_3, the final amount of money. As before we could have skipped the last step. To be able to solve this problem, we shall first have to follow each path in the tree from the root to the bottom (the leaves) to find S_3 in all cases. In this way, we enumerate all possible sequences of decisions that we can possibly make. (Remember that this is exactly what we avoid in dynamic programming). The optimal sequence, must, of course, be one of these sequences. Let (AAB) refer to the path in the tree with the corresponding indices on the arcs. We then get

$$\begin{aligned}
(ABB) &: \quad S_3 = ((S_0 \times 1.1 - 20) - 10) \times 1.05 = S_0 \times 1.155 - 31.5, \\
(AAB) &: \quad S_3 = ((S_0 \times 1.1 - 20) \times 1.07 - 20) - 10 = S_0 \times 1.177 - 51.4, \\
(BAB) &: \quad S_3 = ((S_0 \times 1.07) \times 1.07 - 20) - 10 = S_0 \times 1.1449 - 30, \\
(BBB) &: \quad S_3 = S_0 \times 1.07 \times 1.05 = S_0 \times 1.1235.
\end{aligned}$$

We have now achieved numbers in all leaves of the tree (for some reason decision trees always grow with the root up). We are now going to move back towards the root, using a process called *folding back*. This implies moving one step up the tree at a time, finding for each node in the tree the best decision for that node.

This first step is not really interesting in this case (since we must move the money to account B), but, even so, let us go through it. We find that the best we can achieve after two decisions is as follows.

$$\begin{aligned}
(AB) &: \quad S_3 = S_0 \times 1.155 - 31.5, \\
(AA) &: \quad S_3 = S_0 \times 1.177 - 51.4, \\
(BA) &: \quad S_3 = S_0 \times 1.1449 - 30, \\
(BB) &: \quad S_3 = S_0 \times 1.1235.
\end{aligned}$$

Then we move up to stage 1 to see what is the best we can achieve if we presently have S_1 in account A. The answer is

$$\max\{S_0 \times 1.155 - 31.5, S_0 \times 1.177 - 51.4\} = S_0 \times 1.177 - 51.4$$

so long as $S_0 > 1000$, the given assumption. If we have S_1 in account B (the right node in stage 1), we get

$$\max\{S_0 \times 1.1449 - 30, S_0 \times 1.1235\} = \begin{cases} S_0 \times 1.1449 - 30 & \text{if } S_0 \geq 1402, \\ S_0 \times 1.1235 & \text{if } S_0 \leq 1402. \end{cases}$$

We can then fold back to the top, finding that it is best going left, obtaining the given $S_3 = S_0 \times 1.177 - 51.4$. Of course, we recognize most of these computations from Section 2.2 on dynamic programming.

You might feel that these computations are not very different from those in the dynamic programming approach. However, they are. For example, assume that we had 10 periods, rather than just 2. In dynamic programming we would then have to calculate the optimal accumulated return as a function of S_t for both accounts in 10 periods, a total of 210 = 20 calculations, each involving a maximization over the two possible decisions. In the decision tree case the number of such calculations will be $2^{10} + 2^9 + \ldots + 1 = 2^{11} - 1 = 2047$. (The counting depends a little on how we treat the last period.) This shows the strength of dynamic programming. It investigates many fewer cases. It should be easy to imagine situations where the use of decision trees is absolutely impossible due to the mere size of the tree.

On the other hand, the decision tree approach certainly has advantages. Assume, for example, that we were not to find the optimal investments for all $S_0 > 1000$, but just for $S_0 = 1000$. That would not help us much in the dynamic programming approach, except that $f_1^*(B, S_1) = S_1 \times 1.05$, since $S_1 < 1500$. But that is a minor help. The decision tree case, on the other

Table 1 Distribution of the interest rate on account A. All outcomes have probability 0.5, and the outcomes in period 2 are independent of the outcomes in period 1.

Period	Outcome 1	Outcome 2
1	8	12
2	5	9

hand, would produce *numbers* in the leaves, not functions of S_0, as shown above. Then folding back will of course be *very* simple.

2.4 Stochastic Decision Trees

We shall now see how decision trees can be used to solve certain classes of stochastic problems. We shall initiate this with a look at our standard investment problem in Example 2.2. In addition, let us now assume that the interest on account B is unchanged, but that the interest rate on account A is random, with the previously given rates as expected values. Charges on account A are unchanged. The distribution for the interest rate is given in Table 1. We assume that the interest rates in the two periods are described by independent random variables.

Based on this information, we can give an update of Figure 4, where we show the deterministic and stochastic parameters of the problem. The update is shown in Figure 6.

Consider the decision tree in Figure 7. As in the deterministic case, square nodes are decision nodes, from which we have to choose between account A and B. Circular nodes are called *chance nodes*, and represent points at which something happens, in this case that the interest rates become known.

Start at the top. In stage 0, we have to decide whether to put the money into account A or into B. If we choose A, we shall experience an interest rate of 8% or 12% for the first year. After that we shall have to make a new decision for the second year. That decision will be allowed to depend on what interest rate we experienced in the first period. If we choose A, we shall again face an uncertain interest rate. Whenever we choose B, we shall know the interest rate with certainty.

Having entered a world of randomness, we need to specify what our decisions will be based on. In the deterministic setting we maximized the final amount in account B. That does not make sense in a stochastic setting. A given series

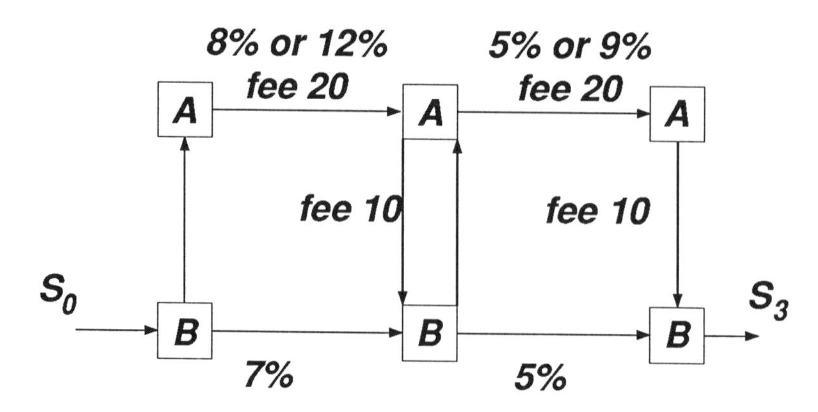

Figure 6 Simple investment problem with uncertain interest rates.

of decisions does not produce a certain amount in account B, but rather an uncertain amount. In other words, we have to compare distributions. For example, keeping the money in account A for both periods will result in one out of four sequences of interest rates, namely (8,5), (8,9), (12,5) or (12,9). Hence, if we start out with, say 1000, we can end up with (remember the fees) 1083, 1125, 1125 or 1169 (rounded numbers).

An obvious possibility is to look for the decision that produces the highest expected amount in account B after two periods. However—and this is a very important point—this does not mean that we are looking for the *sequence* of decisions that has the highest expected value. We are only looking for the best possible *first* decision. If we decide to put the money in account A in the first period, we can wait and observe the actual interest rate on the account before we decide what to do in the next period. (Of course, if we decide to use B in the first period, we can as well decide what to do in the second period immediately, since no new information is made available during the first year!) Let us see how this works. First we do as we have done before: we follow each path down the tree to see what amount we end up with in account B. We have assumed $S_0 = 1000$. That is shown in the leaves of the tree in Figure 8.

We then fold back. Since the next node is a chance node, we take the expected value of the two square nodes below. Then for stage 1 we check which of the two possible decisions has the largest expectation. In the far left of Figure 8 it is to put the money into account A. We therefore *cross out* the other alternative. This process is repeated until we reach the top level. In stage 0 we see that it is optimal to use account A in the first period and regardless of the interest rate in the first period, we shall also use account A in the second period. In general, the second-stage decision depends on the outcome in the first stage, as we shall see in a moment.

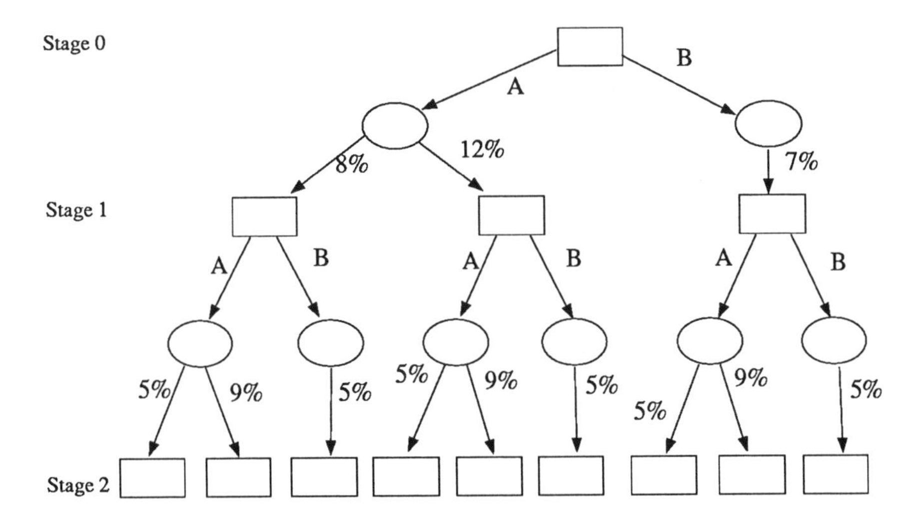

Figure 7 Stochastic decision tree for the simple investment problem.

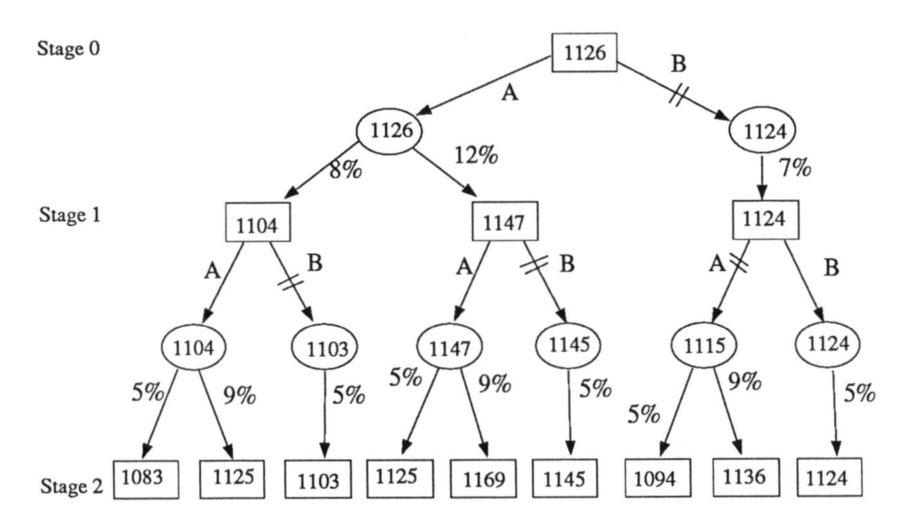

Figure 8 Stochastic decision tree for the investment problem when we maximize the expected amount in account B at the end of stage 2.

You might have observed that the solution derived here is exactly the same as we found in the deterministic case. This is caused by two facts. First, the interest rate in the deterministic case equals the expected interest rate in the stochastic case, and, secondly, the objective function is linear. In other words, if $\tilde{\xi}$ is a random variable and a and b are constants then

$$E_{\tilde{\xi}}(a\tilde{\xi} + b) = aE\tilde{\xi} + b.$$

For the stochastic case we calculated the left-hand side of this expression, and for the deterministic case the right-hand side.

In many cases it is natural to maximize expected profits, but not always. One common situation for decision problems under uncertainty is that the decision is repeated many times, often, in principle, infinitely many. Investments in shares and bonds, for example, are usually of this kind. The situation is characterized by long time series of data, and by many minor decisions. Should we, or should we not, maximize expected profits in such a case?

Economics provide us with a tool to answer that question, called a *utility function*. Although it is not going to be a major point in this book, we should like to give a brief look into the area of utility functions. It is certainly an area very relevant to decision making under uncertainty. If you find the topic interesting, consult the references listed at the end of this chapter. The area is full of pitfalls and controversies, something you will probably not discover from our little glimpse into the field. More than anything, we simply want to give a small taste, and, perhaps, something to think about.

We may think of a utility function as a function that measures our happiness (utility) from a certain wealth (let us stick to money). It does not measure utility in any fixed unit, but is only used to *compare* situations. So we can say that one situation is *preferred* to another, but *not* that one situation is twice as good as another. An example of a utility function is found in Figure 9.

Note that the utility function is concave. Let us see what that means. Assume that our wealth is w_0, and we are offered a game. With 50% probability we shall win δw; with 50% probability we shall lose the same amount. It costs nothing to take part. We shall therefore, after the game, either have a wealth of $w_0 + \delta w$ or a wealth of $w_0 - \delta w$. If the function in Figure 9 is our utility function, and we calculate the utility of these two possible future situations, we find that the decrease in utility caused by losing δw is larger than the increase in utility caused by winning δw. What has happened is that we do not think that the advantage of possibly increasing our wealth by δw is good enough to offset our worry about losing the same amount. In other words, our expected utility after having taken part in the game is smaller than our certain utility of not taking part. We prefer w_0 with certainty to a distribution of possible wealths having expected value w_0. We are *risk-averse*. If we found the two situations equally good, we are *risk-neutral*. If we prefer

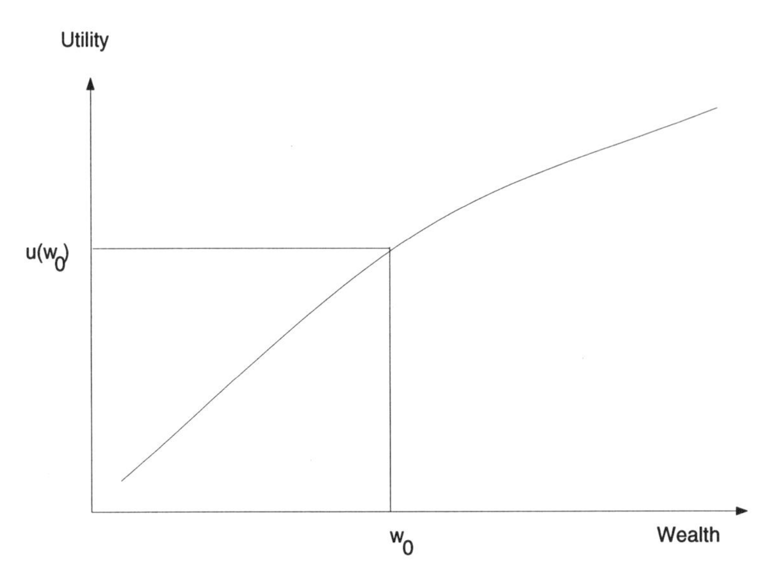

Figure 9 Example of a typical concave utility function representing risk aversion.

the game to the certain wealth, we are risk seekers or *gamblers*.

It is generally believed that people are risk-averse, and that they need a premium to take part in a gamble like the one above. Empirical investigations of financial markets confirm this idea. The premium must be high enough to make the expected utility of taking part in the game (including the premium) equal to the utility of the wealth w_0.

Now, finally, we are coming close to the question we started out with. Should we maximize expected profit? We have seen above that maximizing expected profit can be interpreted as maximizing expected utility with a risk-neutral attitude. In other words, it puts us in a situation where a fair gamble (i.e. one with expected value zero) is acceptable. When can that be the case?

One reason can be that the project under consideration is very small compared with the overall wealth of the decision maker, so that risk aversion is not much of an issue. As an example, consider the purchase of a lottery ticket. Despite the fact that the expected value of taking part in a lottery is negative, people buy lottery tickets. This fact can create some theoretical problems in utility theory, problems that we shall not discuss here. A reference is given at the end of the chapter.

A more important case arises in public investments. One can argue that the government should not trade expected values for decreased risks, since the overall risk facing a government is very small, even if the risk in one single project is large. The reason behind this argument is that, with a very large number of projects at hand (which certainly the government has), some will win, some will lose. Over all, owing to offsetting effects, the government will face very little risk. (It is like in a life insurance company, where the death of costumers is not considered a random event. With a large number of costumers, they "know" how many will die the next year.)

In all, as we see, we must argue in each case whether or not a linear or concave utility function is appropriate. Clearly, in most cases a linear utility function creates easier problems to solve. But in some cases risk should indeed be taken into account.

Let us now continue with our example, and assume we are faced with a concave utility function

$$u(s) = \ln(s - 1000)$$

and that we wish to maximize the expected utility of the final wealth s. In the deterministic case we found that it would never be profitable to split the money between the two accounts. The argument is the same when we simply maximized the expected value of S_3 as outlined above. However, when maximizing expected utility, that might no longer be the case. On the other hand, the whole set-up used in this chapter assumes implicitly that we do not split the funding. Hence in what follows we shall *assume* that all the money must be in one and only one account. The idea in the decision tree is to determine which decisions to make, not how to combine them. Figure 10 shows how we fold back with expected utilities. The numbers in the leaves represent the utility of the numbers in the leaves of Figure 8. For example $u(1083) = \ln(1083 - 1000) = \ln 83 = 4.419$.

We observe that, with this utility function, it is optimal to use account B. The reason is that we fear the possibility of getting only 8% in the first period combined with the charges. The result is that we choose to use B, getting the certain amount $S_3 = 1124$. Note that if we had used account A in the first period (which is not optimal), the optimal second-stage decision would depend on the actual outcome of the interest on account A in the first period. With 8%, we pick B in the second period; with 12%, we pick A.

2.5 Stochastic Dynamic Programming

Looking back at Section 2.2 on dynamic programming, we observe two major properties of the solution and solution procedure. First, the procedure (i.e. dynamic programming) produces one solution per possible state in each stage. These solutions are not stored, since they are not needed in the procedure, but the extra cost incurred by doing so would be minimal. Secondly, if there is only one given value for the initial state z_0, we can use these decisions to produce a series of optimal solutions—one for each stage. In other words, given an

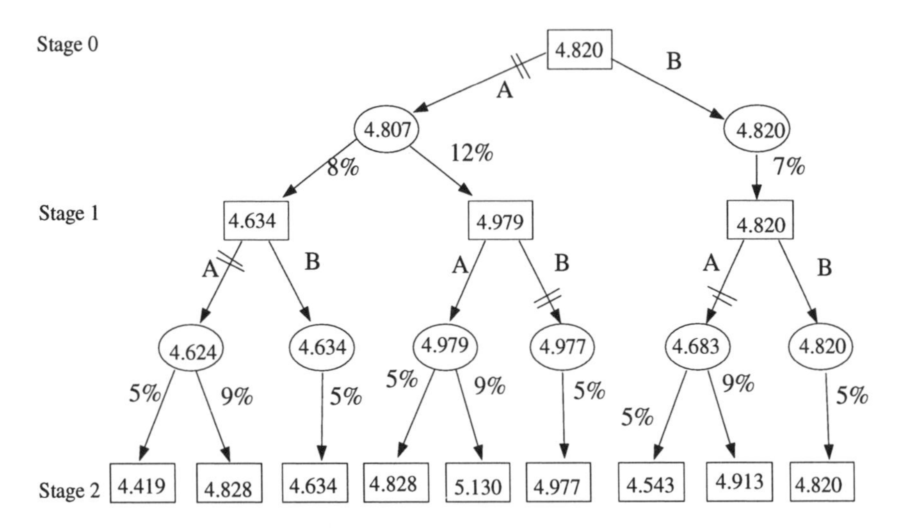

Figure 10 Stochastic decision tree for the investment problem when we maximize the expected utility of the amount in account B at the end of period 2.

initial state, we can make plans for all later periods. In our small investment Example 2.2 (to which we added randomness in the interest rates in Section 2.4 we found, in the deterministic case, that with $S_0 > 1000$, $x_0 = x_1 = A$ and $x_2 = B$ was the optimal solution. That is, we put the money in account A for the two periods, before we send the money to account B as required at the end of the time horizon.

When we now move into the area of stochastic dynamic programming, we shall keep one property of the dynamic programming algorithm, namely that there will be one decision for each state in each stage, but it will no longer be possible to plan for the whole period ahead of time. Decisions for all but the first period will depend on what happens in the mean time. This is the same as we observed for stochastic decision trees.

Let us turn to the small investment example, keeping the extra requirement that the money must stay in one account, and using the utility function $u(s) = \ln(s - 1000)$.

Stage 2 As for the deterministic case, we find that

$$f_2^*(A, S_2) = \ln(S_2 - 1010),$$
$$f_2^*(B, S_2) = \ln(S_2 - 1000),$$

since we must move the money into account B at the end of the second year.

Stage 1 We have to consider the two accounts separately.

Account A If we keep the money in account A, we get the following expected return:

$$\begin{aligned} f_1(A, S_1, A) &= 0.5[f_2^*(A, S_1 \times 1.05 - 20) + f_2^*(A, S_1 \times 1.09 - 20)] \\ &= 0.5 \ln[(S_1 \times 1.05 - 1030)(S_1 \times 1.09 - 1030)]. \end{aligned}$$

If we move the money to account B, we get

$$\begin{aligned} f_1(A, S_1, B) &= f_2^*(B, (S_1 - 10) \times 1.05) \\ &= \ln(S_1 \times 1.05 - 1010.5). \end{aligned}$$

To find the best possible solution, we compare these two possibilities by calculating

$$\begin{aligned} f_1^*(A, S_1) &= \max\{f_1(A, S_1, A), f_1(A, S_1, B)\} \\ &= \max\{0.5 \ln[(S_1 \times 1.05 - 1030)(S_1 \times 1.09 - 1030)], \\ &\qquad \ln(S_1 \times 1.05 - 1010.5)\}, \end{aligned}$$

from which we find, (remembering that $S_1 > 1000$)

$$f_1^*(A, S_1) = \begin{cases} \ln(S_1 \times 1.05 - 1010.5) & \text{if } S_1 < 1077, \\ 0.5 \ln[(S_1 \times 1.05 - 1030)(S_1 \times 1.09 - 1030)] & \\ & \text{if } S_1 > 1077. \end{cases}$$

Account B. For account B we can either move the money to account A to get

$$\begin{aligned} & f_1(B, S_1, A) \\ &\quad = 0.5[f_2^*(A, S_1 \times 1.05 - 20) + f_2^*(A, S_1 \times 1.09 - 20)] \\ &\quad = 0.5 \ln[(S_1 \times 1.05 - 1030)(S_1 \times 1.09 - 1030)], \end{aligned}$$

or we can keep the money in B to obtain

$$\begin{aligned} f_1(B, S_1, B) &= f_2^*(B, S_1 \times 1.05) \\ &= \ln(S_1 \times 1.05 - 1000). \end{aligned}$$

To find the best possible solution, we calculate

$$\begin{aligned} f_1^*(B, S_1) &= \max\{f_1(B, S_1, A), f_1(B, S_1, B)\} \\ &= \max\{0.5 \ln[(S_1 \times 1.05 - 1030)(S_1 \times 1.09 - 1030)], \\ &\qquad \ln(S_1 \times 1.05 - 1000)\}. \end{aligned}$$

From this, we find that (remembering that $S_1 > 1000$)

$$f_1^*(B, S_1) = \begin{cases} \ln(S_1 \times 1.05 - 1000) & \text{if } S_1 < 1538 \\ 0.5 \ln[(S_1 \times 1.05 - 1030)(S_1 \times 1.09 - 1030)] & \\ & \text{if } S_1 > 1538. \end{cases}$$

Stage 0 We here have to consider only the case when the amount $S_0 > 1000$ sits in account B. The basis for these calculations will be the following two expressions. The first calculates the expected result of using account A, the second the certain result of using account B.

$$f_0(B, S_0, A) = 0.5[f_1^*(A, S_0 \times 1.08 - 20) + f_1^*(A, S_0 \times 1.12 - 20)],$$
$$f_0(B, S_0, B) = f_1^*(B, S_0 \times 1.07).$$

Using these two expressions, we then calculate

$$f_0^*(B, S_0) = \max\{f_0(B, S_0, A), f_0(B, S_0, B)\}.$$

To find the value of this expression for $f_0^*(B, S_0)$, we must make sure that we use the correct expressions for f_1^* from stage 1. To do that, we must know how conditions on S_1 relate to conditions on S_0. There are three different ways S_0 and S_1 can be connected (see e.g. the top part of Figure 10):

$$S_1 = S_0 \times 1.08 - 20 \Rightarrow (S_1 = 1077 \iff S_0 = 1016),$$
$$S_1 = S_0 \times 1.12 - 20 \Rightarrow (S_1 = 1077 \iff S_0 = 979),$$
$$S_1 = S_0 \times 1.07 \Rightarrow (S_1 = 1538 \iff S_0 = 1437).$$

From this, we see that three different cases must be discussed, namely $1000 < S_0 < 1016$, $1016 < S_0 < 1437$ and $1437 < S_0$.

Case 1 Here $1000 < S_0 < 1016$. In this case

$$f_0^*(B, S_0) = \ln(S_0 \times 1.1235 - 1000),$$

which means that we always put the money into account B. (Make sure you understand this by actually performing the calculations.)

Case 2 Here $1016 < S_0 < 1437$. In this case

$$f_0^*(B, S_0) = \begin{cases} \ln(S_0 \times 1.1235 - 1000) & \text{if } S_0 < 1022, \\ 0.25 \times \ln[(S_0 \times 1.134 - 1051) & \\ \quad \times(S_0 \times 1.1772 - 1051.8) \times (S_0 \times 1.176 - 1051) & \\ \quad \times(S_0 \times 1.2208 - 1051.8)] & \text{if } S_0 > 1022, \end{cases}$$

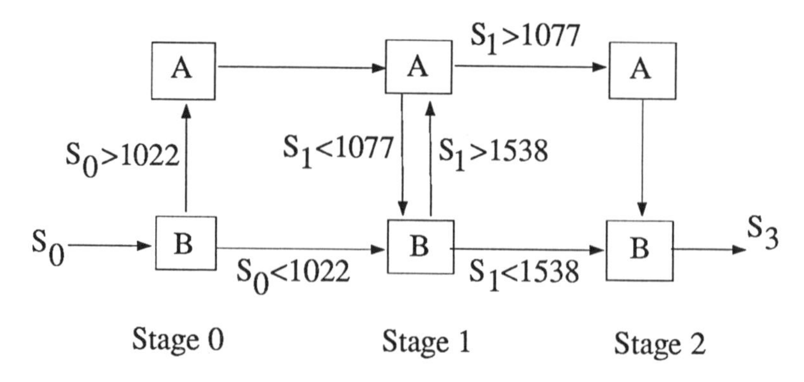

Figure 11 Description of the solution to the stochastic investment problem using stochastic dynamic programming.

which means that we use account B for small amounts and account A for large amounts within the given interval.

Case 3 Here we have $S_0 > 1437$. In this case

$$\begin{aligned} f_0^*(B, S_0) = \tfrac{1}{4}\ln[&(S_0 \times 1.134 - 1051) \times (S_0 \times 1.1772 - 1051.8) \\ &\times (S_0 \times 1.176 - 1051) \times (S_0 \times 1.2208 - 1051.8)], \end{aligned}$$

which means that we should use account A.

Summing up all cases, for stage 0, we get

$$f_0^*(B, S_0) = \begin{cases} \ln(S_0 \times 1.1235 - 1000) & \text{if } S_0 < 1022, \\ 0.25 \times \ln[(S_0 \times 1.134 - 1051) & \\ \quad \times (S_0 \times 1.1772 - 1051.8) \times (S_0 \times 1.176 - 1051) & \\ \quad \times (S_0 \times 1.2208 - 1051.8)] & \text{if } S_0 > 1022. \end{cases}$$

If we put these results into Figure 4, we obtain Figure 11. From the latter, we can easily construct a solution similar to the one in Figure 10 for any $S_0 > 1000$. Verify that we do indeed get the solution shown in Figure 10 if $S_0 = 1000$.

But we see more than that from Figure 11. We see that if we choose account B in the first period, we shall *always* do the same in the second period. There is no way we can start out with $S_0 < 1022$ and get $S_1 > 1538$.

Formally, what we are doing is as follows. We use the vocabulary of Section 2.2. Let the random vector for stage t be given by $\tilde{\xi}_t$ and let the return and transition functions become $r_t(z_t, x_t, \xi_t)$ and $z_{t+1} = G_t(z_t, x_t, \xi_t)$. Given this, the procedure becomes

$$\text{find } f_0^*(z_0)$$

by recursively calculating

$$\begin{aligned} f_t^*(z_t) &= \min_{A_t(z_t)\le x_t\le B_t(z_t)} f_t(z_t, x_t) \\ &= \min_{A_t(z_t)\le x_t\le B_t(z_t)} E_{\tilde{\xi}_t}\{\varphi_t(r_t(z_t, x_t, \tilde{\xi}_t), f_{t+1}^*(z_{t+1}))\},\ t = T,\ldots,0, \end{aligned}$$

with

$$\begin{aligned} z_{t+1} &= G_t(z_t, x_t, \xi_t) \text{ for } t = 0,\ldots,T, \\ f_{T+1}^*(z_{T+1}) &= Q(z_{T+1}), \end{aligned}$$

where the functions satisfy the requirements of Proposition 2.2. In each stage the problem must be solved for all possible values of the state z_t. It is possible to replace expectations (represented by E above) by other operators with respect to $\tilde{\xi}_t$, such as max or min. In such a case, of course, probability distributions are uninteresting—only the support matters.

2.6 Scenario Aggregation

So far we have looked at two different methods for formulating and solving multistage stochastic problems. The first, stochastic decision trees, requires a tree that branches off for each possible decision x_t and each possible realization of $\tilde{\xi}_t$. Therefore these must both have finitely many possible values. The state z_t is not part of the tree, and can therefore safely be continuous. A stochastic decision tree easily grows out of hand.

The second approach was stochastic dynamic programming. Here we must make a decision for each possible state z_t in each stage t. Therefore, it is clearly an advantage if there are finitely many possible states. However, the theory is also developed for a continuous state space. Furthermore, a continuous set of decisions x_t is acceptable, and so is a continuous distribution of $\tilde{\xi}_t$, provided we are able to perform the expectation with respect to $\tilde{\xi}_t$.

The method we shall look at in this section is different from those mentioned above with respect to where the complications occur. We shall now operate on an *event tree* (see Figure 12 for an example). This is a tree that branches off for each possible value of the random variable $\tilde{\xi}_t$ in each stage t. Therefore, compared with the stochastic decision tree approach, the new method has similar requirements in terms of limitations on the number of possible values of $\tilde{\xi}_t$. Both need finite discrete distributions. In terms of x_t we must have finitely many values in the decision tree, the new method prefers continuous variables. Neither of them has any special requirements on z_t.

The second approach we have discussed so far for stochastic problems is stochastic dynamic programming. The new method we are about to outline is called *scenario aggregation*. We shall see that stochastic dynamic programming is more flexible than scenario aggregation in terms of distributions of $\tilde{\xi}_t$, is similar with respect to x_t, but is much more restrictive with respect to the state variable z_t, in the sense that the state space is hardly of any concern in scenario aggregation.

If we have T time periods and ξ_t is a vector describing what happens in time period t (i.e. a realization of $\tilde{\xi}_t$) then we call

$$s = (\xi_0^s, \xi_1^s, \ldots, \xi_T^s)$$

a *scenario*. It represents one possible future. So assume we have a set of scenarios S describing all (or at least the most interesting) possible futures. What do we do? Assume our "world" can be described by state variables z_t and decision variables x_t and that the cost (i.e. the return function) in time period t is given by $r_t(z_t, x_t, \xi_t)$. Furthermore, as before, the state variables can be calculated from

$$z_{t+1} = G_t(z_t, x_t, \xi_t),$$

with z_0 given. Let α be a discount factor. What is often done in this case is to solve for each $s \in S$ the following problem

$$\left.\begin{array}{ll} \min \displaystyle\sum_{t=0}^{T} \alpha^t r_t(z_t, x_t, \xi_t^s) + \alpha^{T+1} Q(z_{T+1}) & \\ \text{s.t.} \quad z_{t+1} = G_t(z_t, x_t, \xi_t^s) \text{ for } t = 0, \ldots, T \text{ with } z_0 \text{ given}, & \\ \qquad A_t(z_t) \le x_t \le B_t(z_t) \text{ for } t = 0, \ldots, T, & \end{array}\right\} \qquad (6.1)$$

where $Q(z)$ represents the value of ending the problem in state z, yielding an optimal solution $x^s = (x_0^s, x_1^s, \ldots, x_T^s)$. Now what? We have a number of different solutions—one for each $s \in S$. Shall we take the average and calculate for each t

$$\underline{x}_t = \sum_{s \in S} p^s x_t^s,$$

where p^s is the probability that we end up on scenario s? This is very often done, either by explicit probabilities or by more subjective methods based on "looking at the solutions". However, several things can go wrong. First, if $\underline{x}$ is chosen as our policy, there might be cases (values of s) for which it is not even feasible. We should not like to suggest to our superiors a solution that might be infeasible (infeasible probably means "going broke", "breaking down" or something like that). But even if feasibility is no problem, is using $\underline{x}$ a good idea?

In an attempt to answer this, let us again turn to event trees. In Figure 12 we have $T = 1$. The top node represents "today". Then one out of three things can happen, or, in other words, we have a random variable with three outcomes. The second row of nodes represents "tomorrow", and after tomorrow a varying

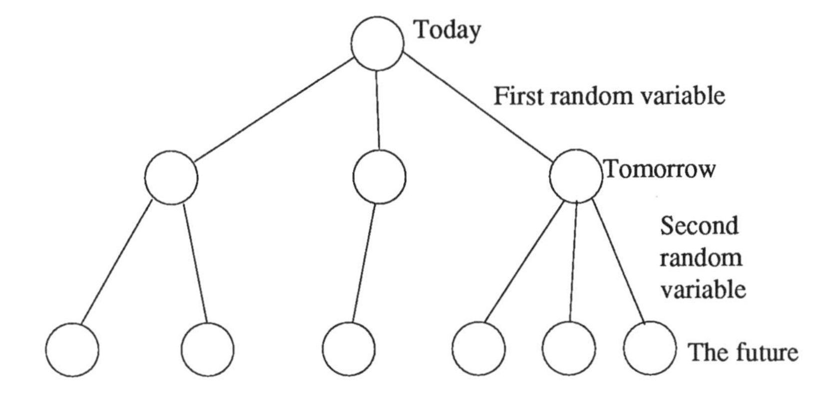

Figure 12 Example of an event tree for $T = 1$.

number of things can happen, depending on what happens today. The bottom row of nodes takes care of the rest of the time—the future.

This tree represents six scenarios, since the tree has six leaves. In the setting of optimization that we have discussed, there will be two decisions to be made, namely one "today" and one "tomorrow". However, note that what we do tomorrow will depend on what happens today, so there is not *one* decision for tomorrow, but rather one for each of the three nodes in the second row. Hence $\underline{x}_0$ works as a suggested first decision, but $\underline{x}_1$ isn't very interesting. However, if we are in the leftmost node representing tomorrow, we can talk about an $\underline{x}_1$ for the two scenarios going through that node. We can therefore calculate, for each version of "tomorrow", an average x_1, where the expectation is conditional upon being on one of the scenarios that goes through the node.

Hence we see that the nodes in the event tree are decision points and the arcs are realizations of random variables. From our scenario solutions x^s we can therefore calculate decisions for each node in the tree, and these will all make sense, because they are all possible decisions, or what are called *implementable decisions.*

For each time period t let $\{s\}_t$ be the set of all scenarios having $\xi_0^s, \ldots, \xi_{t-1}^s$ in common with scenario s. In Figure 12, $\{s\}_0 = S$, whereas each $\{s\}_2$ contains only one scenario. There are three sets $\{s\}_1$. Let $p(\{s\}_t)$ be the sum of the probabilities of all scenarios in $\{s\}_t$. Hence, after solving (6.1) for all s, we calculate for all $\{s\}_t$

$$\underline{x}(\{s\}_t) = \sum_{s' \in \{s\}_t} \frac{p^{s'} x_t^{s'}}{p(\{s\}_t)}.$$

So what does this solution mean? It has the advantage that it is implementable, but is it the optimal solution to any problem we might want to solve? Let us now turn to a formal mathematical description of a multistage problem that lives on an event tree, to see how $\underline{x}(\{s\}_t)$ may be used. In this description we are assuming that we have finite discrete distributions.

$$\min \sum_{s \in S} p^s \left(\sum_{t=0}^{T} \alpha^t r_t(z_t^s, x_t^s, \xi_t^s) + \alpha^{T+1} Q(z_{T+1}^s) \right)$$

subject to

$$\begin{aligned} z_{t+1}^s &= G_t(z_t^s, x_t^s, \xi_t^s) \text{ for } t = 0, \ldots, T \text{ with } z_0^s = z_0 \text{ given}, \\ A_t(z_t^s) &\leq x_t^s \leq B_t(z_t^s) \text{ for } t = 0, \ldots, T, \\ x_t^s &= \sum_{s' \in \{s\}_t} \frac{p^{s'} x_t^{s'}}{p(\{s\}_t)} \text{ for } t = 0, \ldots, T \text{ and all } s. \end{aligned} \tag{6.2}$$

Note that only (6.2), the implementability constraints, connect the scenarios. As discussed in Section 1.7, a common approach in nonlinear optimization is to move constraints that are seen as complicated into the objective function, and penalize deviations. We outlined a number of different approaches. For scenario aggregation, the appropriate one is the augmented Lagrangian method. Its properties, when used with equality constraints such as (6.2), were given in Propositions 1.27 and 1.28. Note that if we move the implementability constraints into the objective, the remaining constraints are separable in the scenarios (meaning that there are no constraints containing information from more than one scenario). Our objective then becomes

$$\begin{aligned} \sum_{s \in S} p(s) \Bigg\{ \sum_{t=0}^{T} \alpha^t \Big[& r_t(z_t^s, x_t^s, \xi_t^s) \\ & + w_t^s \Big(x_t^s - \sum_{s' \in \{s\}_t} \frac{p^{s'} x_t^{s'}}{p(\{s\}_t)}\Big)\Big] + \alpha^{T+1} Q(z_{T+1}^s) \Bigg\} \end{aligned} \tag{6.3}$$

where w_t^s is the multiplier for implementability for scenario s in period t.

If we add an augmented Lagrangian term, this problem can, in principle, be solved by an approach where we first fix w, then solve the overall problem, then update w and so on until convergence, as outlined in Section 1.7.2.4. However, a practical problem (and a severe one as well) results from the fact that the augmented Lagrangian term will change the objective function from one where the different variables are separate, to one where products between variables occur. Hence, although this approach is acceptable in principle, it does not work well numerically, since we have one large problem instead of many scenario problems that can be solved separately. What we then do is to replace

procedure scenario$(s, \underline{x}, x^s, z^s)$;
begin
Solve the problem

$$\min\left\{\sum_{t=0}^{T}\alpha^t[r_t(z_t,x_t,\xi_t^s)+w_t^s x_t+\tfrac{1}{2}\rho(x_t-\underline{x})^2]+\alpha^{T+1}Q(z_{T+1})\right\}$$

$$\begin{aligned}\text{s.t.}\quad z_{t+1} &= G_t(z_t,x_t,\xi_t^s)\ \text{ for } t=0,\ldots,T,\ \text{with } z_0 \text{ given},\\ A_t(z_t) &\le x_t \le B_t(z_t)\ \text{for } t=0,\ldots,T,\end{aligned}$$

to obtain $x^s = (x_0^s, \ldots, x_T^s)$ and $z^s = (z_0^s, \ldots, z_{T+1}^s)$;
end;

Figure 13 Procedure for solving individual scenario problems.

$$\sum_{s'\in\{s\}_t}\frac{p^{s'}x_t^{s'}}{p(\{s\}_t)}$$

with

$$\underline{x}(\{s\}_t) = \sum_{s'\in\{s\}_t}\frac{p^{s'}x_t^{s'}}{p(\{s\}_t)}$$

from the *previous* iteration. Hence, we get

$$\sum_{s\in S}p(s)\left\{\sum_{t=0}^{T}\alpha^t\Big[r_t(z_t^s,x_t^s,\xi_t^s)+w_t^s[x_t^s-\underline{x}(\{s\}_t)]\Big]+\alpha^{T+1}Q(z_{T+1}^s)\right\}.$$

But since, for a fixed w, the terms $w_t^s\underline{x}(\{s\}_t)$ are fixed, we can as well drop them. If we then add an augmented Lagrangian term, we are left with

$$\sum_{s\in S}p(s)\left\{\sum_{t=0}^{T}\alpha^t\Big[r_t(z_t^s,x_t^s,\xi_t^s)+w_t^s x_t^s+\tfrac{1}{2}\rho[x_t^s-\underline{x}(\{s\}_t)]^2\Big]+\alpha^{T+1}Q(z_{T+1}^s)\right\}.$$

Our problem is now totally separable in the scenarios. That is what we need to define the scenario aggregation method. See the algorithms in Figures 13 and 14 for details. A few comments are in place. First, to find an initial $\underline{x}(\{s\}_t)$, we can solve (6.1) using expected values for all random variables.

procedure scen-agg;
begin
  for all s **and** t **do** $w_t^s := 0$;
  Find initial $\underline{x}(\{s\}_t)$;
  Initiate $\rho > 0$;
  repeat
    for all $s \in S$ **do** scenario$(s, \underline{x}(\{s\}_t), x^s, z^s)$;
    for all $\underline{x}(\{s\}_t)$ **do**

$$\underline{x}(\{s\}_t) = \sum_{s' \in \{s\}_t} \frac{p^{s'} x_t^{s'}}{p(\{s\}_t)};$$

    Update ρ if needed;
    for all s **and** t **do**

$$w_t^s := w_t^s + \rho\,[x_t^s - \underline{x}(\{s\}_t)];$$

  until result good enough;
end;

Figure 14 Principal set-up of the scenario aggregation method.

Finding the correct value of ρ, and knowing how to update it, is very hard. We discussed that to some extent in Chapter 1: see in particular (7.17). This is a general problem for augmented Lagrange methods, and will not be discussed here. Also, we shall not go into the discussion of stopping criteria, since the details are beyond the scope of the book. Roughly speaking, though, the goal is to have the scenario problems produce implementable solutions, so that x^s equals $\underline{x}(\{s\}_t)$.

Example 2.3 This small example concerns a very simple fisheries management model. For each time period we have one state variable, one decision variable, and one random variable. Let z_t be the state variable, representing the biomass of a fish stock in time period t, and assume that z_0 is known. Furthermore, let x_t be a decision variable, describing the portion of the fish stock caught in a given year. The implicit assumption made here is that it requires a fixed effort (measured, for example, in the number of participating vessels) to catch a fixed portion of the stock. This seems to be a fairly correct description of demersal fisheries, such as for example the cod fisheries. The catch in a given year is hence $z_t x_t$.

During a year, fish grow, some die, and there is a certain recruitment. A common model for the total effect of these factors is the so called Schaefer model, where the total change in the stock, due to *natural* effects listed above, is given by

$$sz_t\left(1-\frac{z_t}{K}\right),$$

where s is a growth ratio and K is the carrying capacity of the environment. Note that if $z_t = K$ there is no net change in the stock size. Also note that if $z_t > K$, then there is a negative net effect, decreasing the size of the stock, and if $z_t < K$, then there is a positive net effect. Hence $z_t = K$ is a stable situation (as $z_t = 0$ is), and the fish stock will, according to the model, stabilize at $z = K$ if no fishing takes place.

If fish are caught, the catch has to be subtracted from the existing stock, giving us the following transition function:

$$z_{t+1} = z_t - x_t z_t + sz_t\left(1-\frac{z_t}{K}\right).$$

This transition function is clearly nonlinear, with both a $z_t x_t$ term and a z_t^2 term. If the goal is to catch as much as possible, we might choose to maximize

$$\sum_{t=0}^{\infty} \alpha^t z_t x_t,$$

where α is a discount factor. (For infinite horizons we need $0 \le \alpha < 1$, but for finite problems we can choose to let $\alpha = 1$.) In addition to this, we have the natural constraint

$$0 \le x_t \le 1.$$

So far, this is a deterministic control problem. It is known, however, that predicting the net effects of growth, natural mortality and recruitment is very difficult. In particular, the recruitment is not well understood. Therefore, it seems unreasonable to use a deterministic model to describe recruitment, as we have in fact done above. Let us therefore assume that the growth ratio s is not known, but rather given by a random vector $\tilde{\xi}_t$ in time period t.

To fit into the framework of scenario aggregation, let us assume that we are able to cut the problem after T periods, giving it a finite horizon. Furthermore, assume that we have found a reasonable finite discretization of $\tilde{\xi}_t$ for all $t \le T$. It can be hard to do that, but we shall offer some discussion in Section 3.4.

A final issue when making an infinite horizon problem finite is to construct a function $Q(z_{T+1})$ that, in a reasonable way, approximates the value of ending up in state z_{T+1} at time $T+1$. Finding Q can be difficult. However, let us briefly show how one approximation can be found for our problem.

Let us assume that all $\tilde{\xi}_t$ are independent and identically distributed with expected value $\overline{\xi}$. Furthermore, let us simply replace all random variables with their means, and assume that each year we catch exactly the net recruitment, i.e. we let

$$x_t = \bar{\xi}\left(1 - \frac{z_t}{K}\right).$$

But since this leaves $z_t = z_{T+1}$ for all $t \geq T+1$, and therefore all x_t for $t \geq T+1$ equal, we can let

$$Q(z_{T+1}) = \sum_{t=T+1}^{\infty} \alpha^{t-T-1} x_t z_t = \frac{\bar{\xi} z_{T+1}(1 - z_{T+1}/K)}{1-\alpha}.$$

With these assumptions on the horizon, the existence of $Q(z_{T+1})$ and a finite discretization of the random variables, we arrive at the following optimization problem, (the objective function amounts to the expected catch, discounted over the horizon of the problem; of course, it is easy to bring this into monetary terms):

$$\begin{aligned}
&\min \textstyle\sum_{s\in S} p(s) \left[\sum_{t=0}^{T} \alpha^t z_t^s x_t^s + \alpha^{T+1} Q(z_{T+1}^s)\right] \\
&\text{s.t. } z_{t+1}^s = z_t^s\left[1 - x_t^\xi + \xi_t^s\left(1 - \tfrac{z_t^s}{K}\right)\right], \quad \text{with } z_0^s = z_0 \text{ given,} \\
&\qquad 0 \leq x_t^s \leq 1, \\
&\qquad x_t^s = \textstyle\sum_{s'\in\{s\}_t} \frac{p^{s'} x_t^{s'}}{p(\{s\}_t)}.
\end{aligned}$$

We can then apply scenario aggregation as outlined before.

□

2.6.1 *Approximate Scenario Solutions*

Consider the algorithm just presented. If the problem being solved is genuinely a stochastic problem (in the sense that the optimal decisions change compared with the optimal decisions in the deterministic—or expected value—setting), we should expect scenario solutions x^s to be very different initially, before the dual variables w^s obtain their correct values. Therefore, particularly in early iterations, it seems a waste of energy to solve scenario problems to optimality. What will typically happen is that we see a sort of "fight" between the scenario solutions x^s and the implementable solution $\underline{x}(\{s\}_t)$. The scenario solutions try to pull away from the implementable solutions, and only when the penalty (in terms of w_t^s) becomes properly adjusted will the scenario solutions agree with the implementable solutions. In fact, the convergence criterion, vaguely stated, is exactly that the scenario solutions and the implementable solutions agree.

From this observation, it seems reasonable to solve scenario problems only approximately, but precisely enough to capture the direction in which the scenario problem moves relative to the implementable solution. Of course, as the iterations progress, and the dual variables w_t^s adjust to their correct values, the scenario solutions and the implementable solutions agree more and more. In the end, if things are properly organized, the overall set-up converges. It must be noted that the convergence proof for the scenario aggregation method does indeed allow for approximate scenario solutions. From an algorithmic point of view, this would mean that we replaced the solution procedure in Figure 13 by one that found only an approximate solution.

It has been observed that by solving scenario problems only very approximately, instead of solving them to optimality, one obtains a method that converges *much faster*, also in terms of the number of outer iterations. It simply is not wise to solve scenario problems to optimality. Not only *can* one solve scenario problems approximately, one *should* solve them approximately.

2.7 The Value of Using a Stochastic Model

We have so far embarked on formulating and solving stochastic programming models, without much concern about whether or not that is a worthwhile thing to do. Most decision problems are certainly affected by randomness, but that is not the same as saying that the randomness should be introduced into a model. We all know that the art of modelling amounts to describing the important aspects of a problem, and dropping the unimportant ones. We must remember that randomness, although present in the situation, may turn out to be one of the unimportant issues.

We shall now, briefly, outline a few approaches for evaluating the importance of randomness. We shall see that randomness can be (un)important in several different ways.

2.7.1 Comparing the Deterministic and Stochastic Objective Values

The most straightforward way to check if randomness is unimportant is to compare the optimal objective value of the stochastic model with the corresponding optimal value of the deterministic model (probably produced by replacing all random variables by their means).

When we compare the optimal objective values (and also the solutions) in these two cases, we must be aware that what we are observing is composed of several elements. First, while the deterministic solution has one decision for each time period, the stochastic solution "lives" on a tree, as we have discussed in this chapter. The major point here is that the deterministic model has lost all elements of dynamics (it has several time periods, but all decisions are made *here and now*). Therefore decisions that have elements of options in them will never be of any use. In a deterministic world there is never a need to do something *just in case.*

Secondly, replacing random variables by their means will in itself have an effect, as we shall discuss in much more detail in the next chapter.

Therefore, even if these two models come out with about the same optimal objective value, one does not really know much about whether or not it is wise to work with a stochastic model. These models are simply too different to say much in most situations.

From this short discussion, you may have observed that there are really two major issues when solving a model. One is the optimal objective value, the other the optimal solution. It depends on the situation which of these is more important. Sometimes one's major concern is *if* one should do something or not; in other cases the question is not if one should do something, but *what* one should do.

When we continue, we shall be careful, and try to distinguish these cases.

2.7.2 Deterministic Solutions in the Event Tree

To illustrate this idea we shall use the following example.

Example 2.4 Assume that we have a container that can take up to 10 units, and that we have two possible items that can be put into the container. The items are called A and B, and some of their properties are given in Table 2.

Table 2 Properties of the two items A and B.

Item	Value	Minimum size	Maximum size
A	6	5	8
B	4	3	6

The goal is to fill the container with as valuable items as possible. However, the size of an item is uncertain. For simplicity, we assume that each item can have two different sizes, as given in Table 2. All sizes occur with the same probability of 0.5. As is always the case with a stochastic model, we must decide on how the stages are defined. We shall assume that we must pick an item before we learn its size, and that once it is picked, it must be put into the container. If the container becomes overfull, we obtain a penalty of 2 per unit in excess of 10. We have the choice of picking only one item, and they can be picked in any order.

A stochastic decision tree for the problem is given in Figure 15, where we have already folded back and crossed out nonoptimal decisions. We see that the expected value is 7.5. That is obtained by first picking item A, and then, if item A turns out to be small, also pick item B. If item A turns out to be large, we choose not to pick item B.

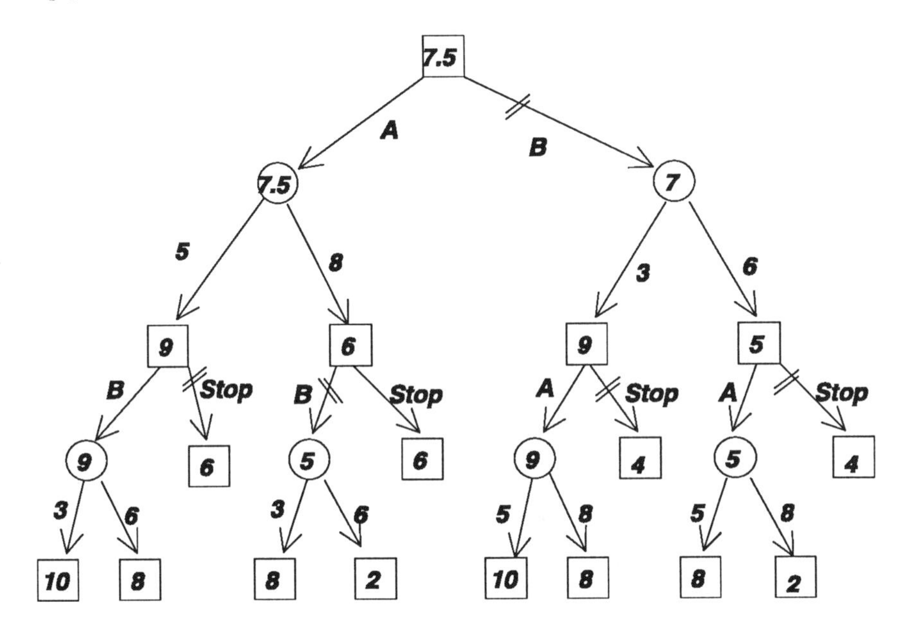

Figure 15 Stochastic decision tree for the container problem.

□

If we assume that the event tree (or the stochastic part of the stochastic decision tree) is a fair description of the randomness of a model, the following simple approach gives a reasonable measure of how good the deterministic model really is. Start in the root of the event tree, and solve the deterministic model. (Probably this means replacing random variables by their means. However, this approach can be used for any competing deterministic model.) Take that part of the deterministic solution that corresponds to the first stage of the stochastic model, and let it represent an implementable solution in the root of the event tree. Then go to each node at level two of the event tree and repeat the process. Taking into consideration what has happened in stage 1 (which is different for each node), solve the deterministic model from stage 2 onwards, and use that part of the solution that corresponds to stage 2 as an implementable solution. Continue until you have reached the leaves of the event tree.

This is a fair comparison, since even people who prefer deterministic models resolve them as new information becomes available (represented by the event tree). In this setting we can compare both decisions and (expected) optimal objective values. What we may observe is that although the solutions are different, the optimal values are almost the same. If that is the case, we are observing flat objective functions with many (almost) optimal solutions. If we observe large differences in objective values, we have a clear indication that solving a stochastic model is important.

Let us return to Example 2.4. Let the following simple deterministic algorithm be an alternative to the stochastic programming approach in Figure 15. Consider all items not put into the container so far. For each item, calculate the value of adding it to the container, *given that it has its expected size*. If at least one item adds a positive value to the content of the container, pick the one with the highest added value. Then put it in, and repeat.

This is not meant to be a specially efficient algorithm—it is only presented for its simplicity to help us make a few points. If we apply this algorithm to our case, we see that with an empty container, item A will add 6 to the value of the container and item B will add 4. Hence we pick item A. The algorithm will next determine if B should be picked or not. However, for the comparison between the deterministic and stochastic approach, it suffices to observe that item A is picked first. This coincides with the solution in Figure 15.

Next we observe the size of A. If it is small, there is still room for 5 units in the container. Since B has an expected size of 4.5, it will add 4 to the value of the container, and will therefore be picked. On the other hand, if A turns out to be large, there is only room for 2 more units, and B will add $4 - 2.5 \times 2 = -1$ to the value, and it will therefore not be picked. Again, we get exactly the same solution as in Figure 15.

So what have we found out? We have seen that for this problem, with its structure *and* data, the deterministic approach was as good as the stochastic approach. However, it is not possible to draw any general conclusions from this. In fact, it illustrates a very important point: *it is extremely difficult to know if randomness is important before we have solved the problem and checked the results.* But, in this special case, anyone claiming that using stochastic decision trees on this problem was like shooting sparrows with cannons will be proved correct.

2.7.3 Expected Value of Perfect Information

For simplicity, assume that we have a two-stage model. Now compare the optimal objective value of the stochastic model with the expected value of the *wait-and-see* solutions. The latter is calculated by finding the optimal solution for each possible realization of the random variables. Clearly, it is better to know the value of the random variable before making a decision than having to make the decision before knowing. The difference between these two expected objective values is called the *expected value of perfect information* (EVPI), since it shows how much one could expect to win if one were told what would happen before making one's decisions. Another interpretation is that this difference is what one would be willing to pay for that information.

What does it mean to have a large EVPI? Does it mean that it is important to solve a stochastic model? The answer is no! It shows that randomness plays an important role in the problem, but it does not necessarily show that a deterministic model cannot function well. By resorting to the set-up of the previous subsection, we may be able to find that out. We can be quite sure, however, that a small EVPI means that randomness plays a minor role in the model.

In the multistage case the situation is basically the same. It is, however, possible to have a very low EVPI, but at the same time have a node far down in the tree with a very high EVPI (but low probability.)

Let us again turn to Example 2.4. Table 3 shows the optimal solutions for the four cases that can occur, if we make the decisions *after* the true values have become known. Please check that you agree with the numbers.

Table 3 The four possible wait-and-see solutions for the container problem in Example 2.4.

Size of A	Size of B	Solution	Value
5	3	A, B	10
5	6	A, B	8
8	3	A, B	8
8	6	A	6

With each case in Table 3 equally probable, the expected value of the wait-and-see solution is 8, which is 0.5 more than what we found in Figure 15. Hence EVPI equals 0.5; The value of knowing the true sizes of the items *before* making decisions is 0.5. This is therefore also the maximal price one would pay to know this.

What if we were offered to pay for knowing the value of A or B before making our first pick? In other words, does it help to know the size of for example item B before choosing what to do? This is illustrated in Figure 16.

We see that the EVPI for knowing the size of item B is 0.5, which is the same as that for knowing both A and B. The calculation for item A is left as

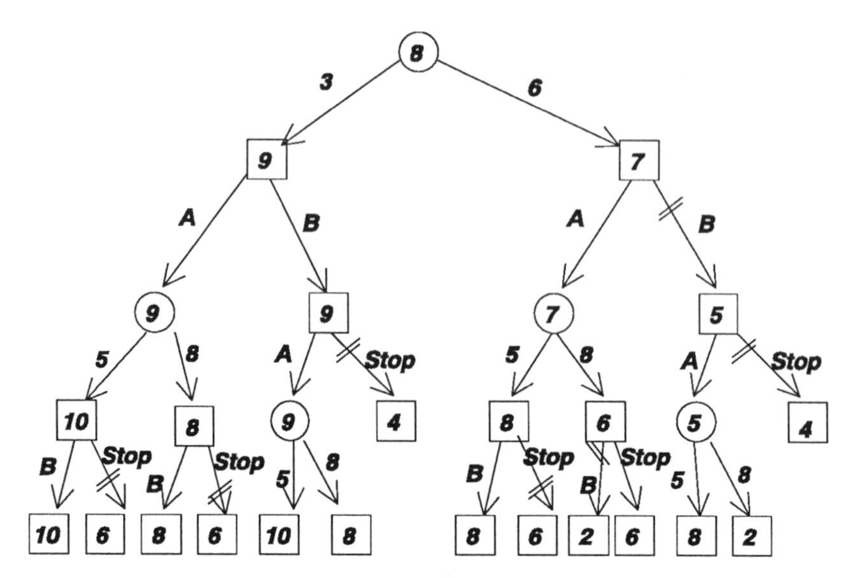

Figure 16 Stochastic decision tree for the container problem when we know the size of B before making decisions.

an exercise.

Example 2.5 Let us conclude this section with another similar example. You are to throw a die twice, and you will win 1 if you can guess the total number of eyes from these two throws. The optimal guess is 7 (if you did not know that already, check it out!), and that gives you a chance of winning of $\frac{1}{6}$. So the expected win is also $\frac{1}{6}$.

Now, you are offered to pay for knowing the result of the first throw. How much will you pay (or alternatively, what is the EVPI for the first throw)? A close examination shows that knowing the result of the first throw does not help at all. Even if you knew, guessing a total of 7 is still optimal (but that is no longer a unique optimal solution), and the probability that that will happen is still $\frac{1}{6}$. Hence, the EVPI for the first stage is zero.

Alternatively, you are offered to pay for learning the value of both throws before "guessing". In that case you will of course make a correct guess, and be certain of winning one. Therefore the expected gain has increased from $\frac{1}{6}$ to 1, so the EVPI for knowing the value of both random variables is $\frac{5}{6}$. □

As you see, EVPI is not *one* number for a stochastic program, but can be calculated for any combination of random variables. If only one number is given, it usually means the value of learning everything, in contrast to knowing nothing.

2.8 Bibliographical Notes

For more insight into the area of dynamic programming we refer to Bellman [1], Howard [3] and Nemhauser [4].

Scenario aggregation was developed by Rockafellar and Wets [5]. A more simplified version can be found in Wets [9]. A discussion of data structures can be found in Wallace and Helgason [7]. Approximate scenario solutions, using Hamiltonians, are discussed in Helgason and Wallace [2]. They also show that by such an approach the scenario aggregation method and the solution procedure for the individual problems melt together into what they call the *Lagrangian scenario aggregation method.*

Example 2.3 originally comes from Schaefer [6], but the stochastic programming setting is discussed by Helgason and Wallace [2].

For an introduction to utility theory, and many of the questions posed abouts its applicability, we refer to Watson and Buede [8].

Exercises

1. Consider some problem where the Bellman principle is appropriate, so that stochastic decision trees, stochastic dynamic programming and scenario aggregation all can be applied. Compare the three methods with respect to the following characteristics of the problem:
   - discrete or continuous probability distributions;
   - discrete or continuous state space;
   - discrete or continuous decision variables;
   - dependent or independent random variables within the stages (time periods);
   - dependent or independent random variables between the stages;
   - utilization of problem structure.
2. Look back at Example 2.3. Assume that $T = 1$. Use this fisheries example to write down all necessary functions needed in the scenario aggregation method, detailed in Figure 14.
3. Look back at Example 2.3. Assume that we change the model slightly to take into account that there are young and adult fish, and that the characteristics of catch and recruitment depend on the age composition of the stock. We now need a two-dimensional state space:
   - z_t^1: the number of young fish in period t;

- z_t^2: the number of adult fish in period t.

First, we need to know the growth and recruitment characteristics of the stock. With w the average weight of an adult fish, these are given by

$$\begin{aligned} z_{t+1}^1 &= 0.4(z_t^1 - 0.5x_t z_t^1) + swz_t^2\left(1 - \tfrac{wz_t^2}{K}\right) \\ z_{t+1}^2 &= 0.8(z_t^2 - x_t z_t^2) + 0.5(z_t^1 - 0.5x_t z_t^1). \end{aligned}$$

As before, we assume z_0 to be known.

(a) Give a verbal interpretation of the two transition functions given above, including the decision variable x_t.

(b) What is now a natural objective function, in your view?

(c) Assume s is random, and formalize the use of scenario aggregation for solving the new version of the Schaefer model.

4. Look back at Example 2.4

 (a) In Figure 16 we calculated the expected value of perfect information (EVPI) for knowing the size of item B. It was found to be 0.5. Carry out a similar calculation for knowing the size of item A!

 (b) Was this result predictable from what we already knew from the text? [Hint: Look at the discussions about Table 3].

References

[1] Bellman R. (1957) *Dynamic Programming.* Princeton University Press, Princeton, New Jersey.

[2] Helgason T. and Wallace S. W. (1991) Approximate scenario solutions in the progressive hedging algorithm. *Ann. Oper. Res.* 31: 425–444.

[3] Howard R. A. (1960) *Dynamic Programming and Markov Processes.* MIT Press, Cambridge, Massachusetts.

[4] Nemhauser G. L. (1966) *Dynamic Programming.* John Wiley & Sons, New York.

[5] Rockafellar R. T. and Wets R. J.-B. (1991) Scenarios and policy aggregation in optimization under uncertainty. *Math. Oper. Res.* 16: 119–147.

[6] Schaefer M. B. (1954) Some aspects of the dynamics of populations important to the management of the commercial marine fisheries. *Inter-Am. Trop. Tuna Comm. Bull.* 1: 27–56.

[7] Wallace S. W. and Helgason T. (1991) Structural properties of the progressive hedging algorithm. *Ann. Oper. Res.* 31: 445–456.

[8] Watson S. R. and Buede D. M. (1987) *Decision Synthesis. The Principles and Practice of Decision Analysis.* Cambridge University Press, Cambridge, UK.

[9] Wets R. J.-B. (1989) The aggregation principle in scenario analysis and stochastic optimization. In Wallace S. W. (ed) *Algorithms and Model Formulations in Mathematical Programming*, pages 91–113. Springer-Verlag, Berlin.

3

Recourse Problems

The purpose of this chapter is to discuss principal questions of linear recourse problems. We shall cover general formulations, solution procedures and bounds and approximations.

Figure 1 shows a simple example from the fisheries area. The assumption is that we know the position of the fishing grounds, and potential locations for plants. The cost of building a plant is known, and so are the distances between grounds and potential plants. The fleet capacity is also known, but quotas, and therefore catches, are only known in terms of distributions. Where should the plants be built, and how large should they be?

This is a typical two-stage problem. In the first stage we determine which plants to build (and how big they should be), and in the second stage we catch and transport the fish when the quotas for a given year are known. Typically, quotas can vary as much as 50% from one year to the next.

3.1 Outline of Structure

Let us formulate a *two-stage stochastic linear program*. This formulation differs from (3.16) of Chapter 1 only in the randomness in the objective of the recourse problem.

$$\begin{aligned}&\min c^{\mathrm{T}}x + \mathcal{Q}(x)\\ &\text{s.t. } Ax = b,\ x \geq 0,\end{aligned}$$

where

$$\mathcal{Q}(x) = \sum_j p^j Q(x, \xi^j)$$

and

$$Q(x, \xi) = \min\{q(\xi)^{\mathrm{T}}y \mid W(\xi)y = h(\xi) - T(\xi)x,\ y \geq 0\},$$

where p^j is the probability that $\tilde{\xi} = \xi^j$, the jth realization of $\tilde{\xi}$, $h(\xi) = h_0 + H\xi = h_0 + \sum_i h_i\xi_i$, $T(\xi) = T_0 + \sum_i T_i\xi_i$ and $q(\xi) = q_0 + \sum_i q_i\xi_i$.

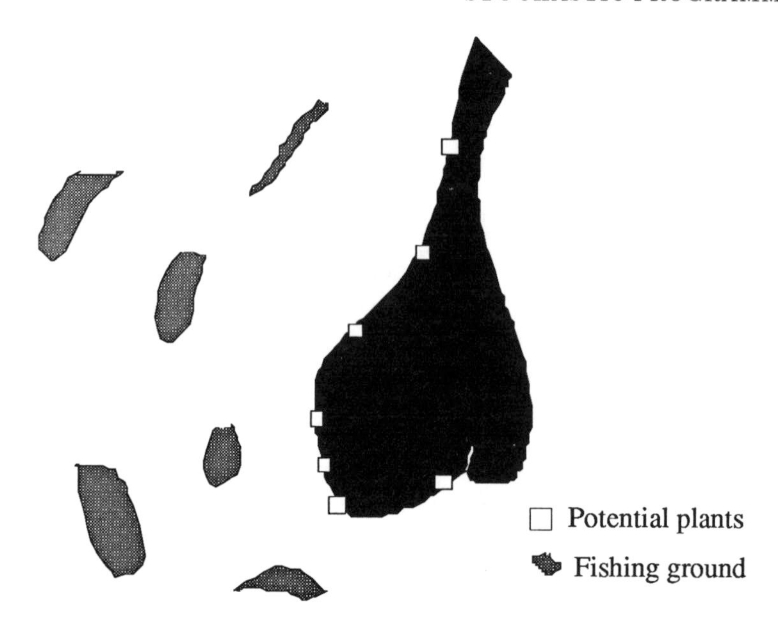

Figure 1 A map showing potential plant sites and actual fishing grounds for Southern Norway and the North Sea.

The function $Q(x,\xi)$ is called the *recourse function*, and $\mathcal{Q}(x)$ therefore the *expected recourse function.*

In this chapter we shall look at only the case with fixed recourse, i.e. the case where $W(\xi) \equiv W$. Let us repeat a few terms from Section 1.3, in order to prepare for the next section. The cone pos W, mentioned in (3.17) of Chapter 1, is defined by

$$\text{pos } W = \{t \mid t = Wy,\ y \geq 0\}.$$

The cone pos W is illustrated in Figure 2. Note that

$$Wy = h,\ y \geq 0 \text{ is feasible} \iff h \in \text{pos } W.$$

Recall that a problem has *complete recourse* if

$$\text{pos } W = R^m.$$

Among other things, this implies that

$$h(\xi) - T(\xi)x \in \text{pos } W \text{ for all } \xi \text{ and all } x.$$

But that is definitely more than we need in most cases. Usually, it is more than enough to know that

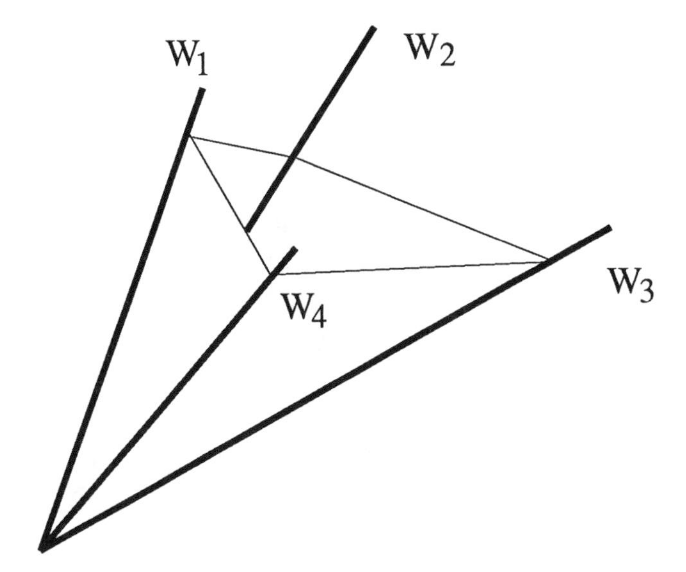

Figure 2 The cone pos W for a case where W has three rows and four columns.

$$h(\xi) - T(\xi)x \in \text{pos } W \text{ for all } \xi \text{ and all } x \geq 0 \text{ satisfying } Ax = b.$$

If this is true, we have *relatively complete recourse*. Of course, complete recourse implies relatively complete recourse.

3.2 The L-shaped Decomposition Method

This section contains a much more detailed version of the material found in Section 1.6.4. In addition to adding more details, we have now added randomness more explicitly, and have also chosen to view some of the aspects from a different perspective. It is our hope that a new perspective will increase the understanding.

3.2.1 Feasibility

The material treated here coincides with step 2(a) in the dual decomposition method of Section 1.6.4. Let the second-stage problem be given by

$$Q(x, \xi) = \min\{q(\xi)^{\mathrm{T}} y \mid Wy = h(\xi) - T(\xi)x,\ y \geq 0\},$$

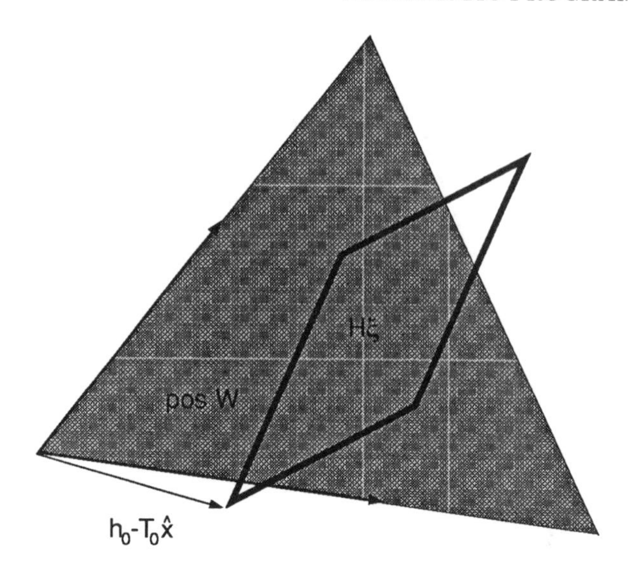

Figure 3 Illustration showing that if infeasibility is to occur for a fixed $\hat{x}$, it must occur for an extreme point of the support of $H\tilde{\xi}$, and hence of $\tilde{\xi}$. In this example $T(\xi)$ is assumed to be equal to T_0.

where W is fixed. Assume we are given an $\hat{x}$ and should like to know if that $\hat{x}$ yields a feasible second-stage problem for all possible values of $\tilde{\xi}$. We assume that $\tilde{\xi}$ has a rectangular and bounded support. Consider Figure 3. We have there drawn pos W plus a parallelogram that represents all possible values of $h_0 + H\tilde{\xi} - T_0\hat{x}$. We have assumed that $T(\xi) \equiv T_0$, only to make the illustration simpler.

Figure 3 should be interpreted as representing a case where H is a 2×2 matrix, so that the extreme points of the parallelogram correspond to the extreme points of the support Ξ of $\tilde{\xi}$. This is a known result from linear algebra, namely that if one polyhedron is a linear transformation of another polyhedron, then the extreme points of the latter are maps of extreme points in the first.

What is important to note from Figure 3 is that if the second-stage problem is to be infeasible for some realizations of $\tilde{\xi}$ then at least one of these realizations will correspond to an extreme point of the support. The figure shows such a case. And conversely, if all extreme points of the support produce feasible problems, all other possible realizations of $\tilde{\xi}$ will also produce feasible problems. Therefore, to check feasibility, we shall in the worst case have to check all extreme points of the support. With k random variables, and Ξ a k-dimensional rectangle, we get 2^k points. Let us define $\mathcal{A}$ to be a set containing these points. In Chapter 5 we shall discuss how we can often reduce the number of points in $\mathcal{A}$ without removing the property that if all points in $\mathcal{A}$ yield a feasible second-stage problem, so will *all* other points in the support.

We shall next turn to another aspect of feasibility, namely the question of how to decide if a given $x = \hat{x}$ will yield feasible second-stage problems for all possible values of $\tilde{\xi}$ in a setting where we are not aware of relatively complete recourse. What we shall outline now corresponds to Farkas' lemma (Proposition 1.19, page 69). Farkas' lemma states that

$$\{y \mid Wy = h,\ y \geq 0\} \neq \emptyset$$

if and only if

$$W^{\mathrm{T}}u \geq 0 \ \text{ implies that } \ h^{\mathrm{T}}u \geq 0.$$

The first of these equivalent statements is just an alternative way of saying that $h \in \text{pos}\, W$, which we now know means that h represents a feasible problem.

By changing the sign of u, the second of the equivalent statements can be rewritten as

$$W^{\mathrm{T}}u \leq 0 \ \text{ implies that } \ h^{\mathrm{T}}u \leq 0.$$

or equivalently

$$h^{\mathrm{T}}t \leq 0 \ \text{ whenever } \ t \in \{u \mid W^{\mathrm{T}}u \leq 0\}.$$

However, this may be reformulated as

$$\begin{aligned}\{u \mid W^{\mathrm{T}}u \leq 0\} &= \{u \mid u^{\mathrm{T}}Wy \leq 0 \text{ for all } y \geq 0\}\\ &= \{u \mid u^{\mathrm{T}}h \leq 0 \text{ for all } h \in \text{pos}\, W\}.\end{aligned}$$

The last expression defines the *polar cone* of pos W as

$$\text{pol pos}\, W = \{u \mid u^{\mathrm{T}}h \leq 0 \text{ for all } h \in \text{pos}\, W\}.$$

Using Figure 4, we can now restate Farkas' lemma the following way. The system $Wy = h$, $y \geq 0$, is feasible if and only if the right-hand side h has a non-positive inner product with all vectors in the cone pol pos W, in particular with its *generators*. Generators were discussed in Chapter 1 (see e.g. Remark 1.6, page 63). The matrix W^*, containing as columns all generators of pol pos W, is denoted the *polar matrix* of W.

We shall see in Chapter 5 how this understanding can be used to generate relatively complete recourse in a problem that does not possess that property. For now, we are satisfied by understanding that if we knew all the generators of pol pos W, that is the polar matrix W^*, then we could check feasibility of a second-stage problem by performing a number of inner products (one for each

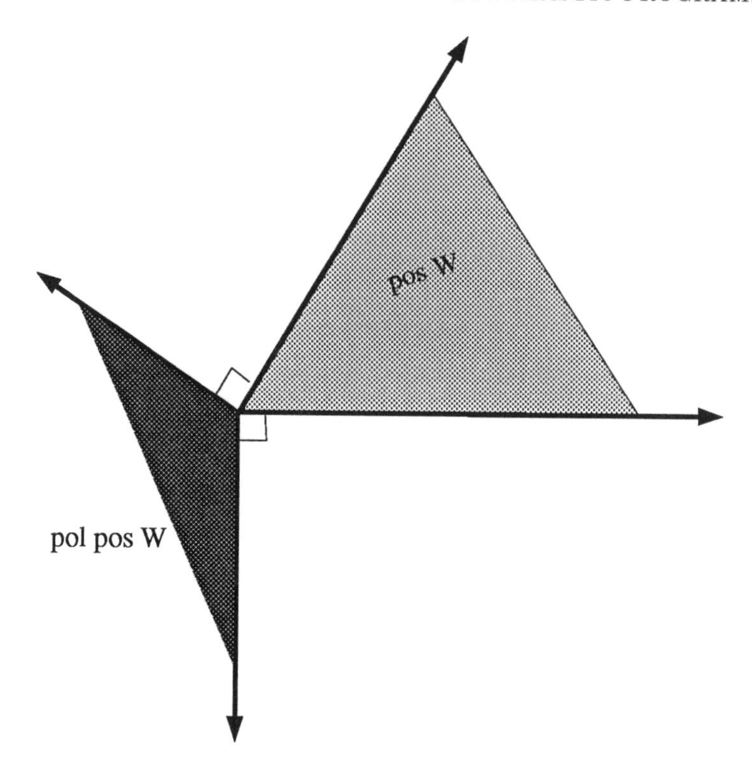

Figure 4 The polar of a cone.

generator), and if at least one of them gave a positive value then we could conclude that the problem was indeed infeasible.

If we do not know all the generators of pol pos W, and we are not aware of relatively complete recourse, for a given $\hat{x}$ and all $\xi \in \mathcal{A}$ we must check for feasibility. We should like to check for feasibility in such a way that if the given problem is not feasible, we automatically come up with a generator of pol pos W. For the discussion, we shall use Figure 5.

We should like to find a σ such that

$$\sigma^{\mathrm{T}} t \leq 0 \text{ for all } t \in \text{pos } W.$$

This is equivalent to requiring that $\sigma^{\mathrm{T}} W \leq 0$. In other words, σ should be in the cone pol pos W. But, assuming that the right-hand side $h(\xi) - T(\xi)\hat{x}$ produces an infeasible problem, we should at the same time require that

$$\sigma^{\mathrm{T}}[h(\xi) - T(\xi)\hat{x}] > 0,$$

because if we later add the constraint $\sigma^{\mathrm{T}}[h(\xi) - T(\xi)x] \leq 0$ to our problem,

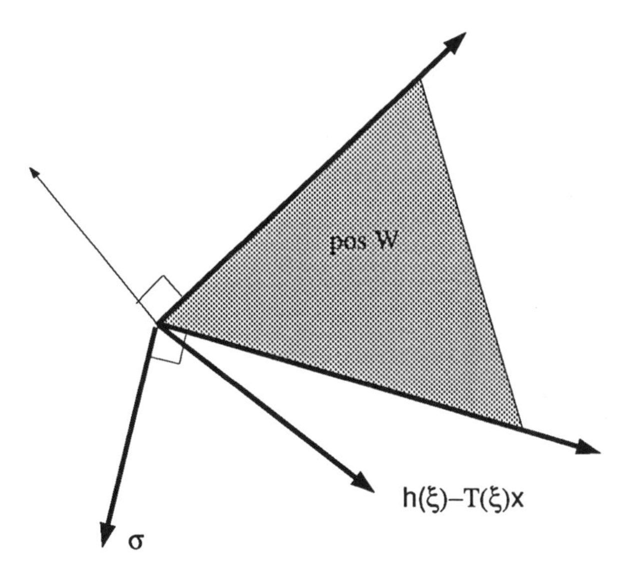

Figure 5 Generation of feasibility cuts.

we shall exclude the infeasible right-hand side $h(\xi) - T(\xi)\hat{x}$ without leaving out any feasible solutions. Hence we should like to solve

$$\max_{\sigma}\{\sigma^{\mathrm{T}}(h(\xi) - T(\xi)\hat{x}) \mid \sigma^{\mathrm{T}}W \leq 0,\ \|\sigma\| \leq 1\},$$

where the last constraint has been added to bound σ. We can do that, because otherwise the maximal value will be $+\infty$, and that does not interest us since we are looking for the *direction* defined by σ. If we had chosen the ℓ_2 norm, the maximization would have made sure that σ came as close to $h(\xi) - T(\xi)\hat{x}$ as possible (see Figure 5). Computationally, however, we should not like to work with quadratic constraints. Let us therefore see what happens if we choose the ℓ_1 norm. Let us write our problem differently to see the details better. To do that, we need to let the unconstrained σ be replaced by $\sigma^1 - \sigma^2$, where $\sigma^1, \sigma^2 \geq 0$. We then get the following:

$$\max\{(\sigma^1-\sigma^2)^{\mathrm{T}}(h(\xi)-T(\xi)\hat{x}) \mid (\sigma^1-\sigma^2)^{\mathrm{T}}W \leq 0,\ e^{\mathrm{T}}(\sigma^1+\sigma^2) \leq 1,\ \sigma^1, \sigma^2 \geq 0\},$$

where e is a vector of ones. To more easily find the dual of this problem, let us write it down in a more standard format:

$$\begin{array}{ll|c}
\max(\sigma^1 - \sigma^2)^{\mathrm{T}}(h(\xi) - T(\xi)\hat{x}) & & \text{dual variables} \\
W^{\mathrm{T}}\sigma^1 - W^{\mathrm{T}}\sigma^2 \leq 0 & & y \\
\ \ e^{\mathrm{T}}\sigma^1 + \ \ e^{\mathrm{T}}\sigma^2 \leq 1 & & t \\
\qquad\quad \sigma^1, \sigma^2 \geq 0 & &
\end{array}$$

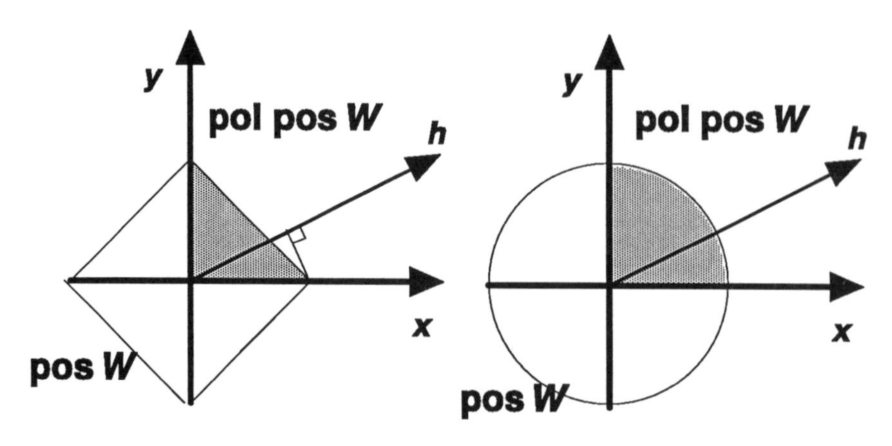

Figure 6 Illustration of the difference between the ℓ_1 and ℓ_2 norms when generating feasibility cuts.

From this, we find the dual linear program to be

$$\min\{t \mid Wy + et \geq (h(\xi) - T(\xi)\hat{x}),\ -Wy + et \geq -(h(\xi) - T(\xi)\hat{x}),\ y, t \geq 0\}.$$

Note that if the optimal value in this problem is zero, we have $Wy = h(\xi) - T(\xi)\hat{x}$, so that we do indeed have $h(\xi) - T(\xi)\hat{x} \in \text{pos } W$, contrary to our assumption. We also see that if t gets large enough, the problem is always feasible. This is what we solve for all $\xi \in \mathcal{A}$. If for some ξ we find a positive optimal value, we have found a ξ for which $h(\xi) - T(\xi)\hat{x} \notin \text{pos } W$, and we create the cut

$$\sigma^{\mathrm{T}}(h(\xi) - T(\xi)x) \leq 0 \iff \sigma^{\mathrm{T}}T(\xi)x \geq \sigma^{\mathrm{T}}h(\xi). \tag{2.1}$$

The σ used here is a generator of pol pos W, but it is not in general as close to $h(\xi) - T(\xi)\hat{x}$ as possible. This is in contrast to what would have happened had we used the ℓ_2 norm. (See Example 3.1 below for an illustration of this point.)

Note that if $T(\xi) \equiv T_0$, the expression $\sigma^{\mathrm{T}}T_0x$ in (2.1) does not depend on ξ. Since at the same time (2.1) must be true for *all* ξ, we can for this special case strengthen the inequality by calculating

$$\sigma^{\mathrm{T}}T_0x \geq \sigma^{\mathrm{T}}h_0 + \max_{t\in\Xi}\left(\sigma^{\mathrm{T}}H\right)t.$$

Since $\sigma^{\mathrm{T}}T_0$ is a vector and the right-hand side is a scalar, this can conveniently be written as $-\gamma^{\mathrm{T}}x \geq \delta$. The $\hat{x}$ we started out with will not satisfy this constraint.

Example 3.1 We present this little example to indicate why the ℓ_1 and ℓ_2 norms give different results when we generate feasibility cuts. The important point is how the two norms limit the possible σ values. The ℓ_1 norm is given in the left part of Figure 6, the ℓ_2 norm in the right part.

For simplicity, we have assumed that pol pos W equals the positive quadrant, so that the constraints $\sigma^{\mathrm{T}}W \leq 0$ reduce to $\sigma \geq 0$. Since at the same time $||\sigma|| \leq 1$, we get that σ must be within the shaded part of the two figures.

For convenience, let us denote the right-hand side by h, and let $\sigma = (\sigma^x, \sigma^y)^{\mathrm{T}}$, to reflect the x and y parts of the vector. In this example $h = (4, 2)^{\mathrm{T}}$. For the ℓ_1 norm the problem now becomes.

$$\max_{\sigma}\{4\sigma^x + 2\sigma^y || \sigma^x + \sigma^y \leq 1,\ \sigma \geq 0\}.$$

The optimal solution here is $\sigma = (1, 0)^{\mathrm{T}}$. Graphically this can be seen from the figure from the fact that an inner product equals the length of one vector multiplied by the length of the projection of the second vector on the first. If we take the h vector as the fixed first vector, the feasible σ vector with the largest projection on h is $\sigma = (1, 0)^{\mathrm{T}}$.

For the ℓ_2 norm the problem becomes

$$\max_{\sigma}\{4\sigma^x + 2\sigma^y || (\sigma^x)^2 + (\sigma^y)^2 \leq 1,\ \sigma \geq 0\}.$$

The optimal solution here is $\sigma = \sqrt{\frac{1}{5}}\,(2, 1)^{\mathrm{T}}$, which is a vector in the same direction as h.

In this example we see that if σ is found using the ℓ_1 norm, it becomes a generator of pol pos W, but it is not as close to h as possible. With the ℓ_2 norm, we did not get a generator, but we got a vector as close to h as possible.

□

3.2.2 Optimality

The material discussed here concerns step 1(b) of the dual decomposition method in Section 1.6.4. Let us first note that if we have relatively complete recourse, or if we have checked that $h(\xi) - T(\xi)x \in \text{pos } W$ for all $\xi \in \mathcal{A}$, then the second-stage problem

$$\min\{q(\xi)^{\mathrm{T}}y \mid Wy = h(\xi) - T(\xi)x,\ y \geq 0\}$$

is feasible. Its dual formulation is given by

$$\max\{\pi^{\mathrm{T}}(h(\xi) - T(\xi)x) \mid \pi^{\mathrm{T}}W \leq q(\xi)^{\mathrm{T}}\}.$$

```
procedure LP(W:matrix; b, q, y:vectors; feasible:boolean);
begin
  if min{q^T y' | W y' = b, y' >= 0} is feasible then begin
    feasible := true;
    y is the optimal y';
  end
  else feasible := false;
end;
```

Figure 7 LP solver.

As long as $q(\xi) \equiv q_0$, the dual is either feasible or infeasible for *all* x and ξ, since x and ξ do not enter the constraints. We see that this is more complicated if q is also affected by randomness. But even when ξ enters the objective function, we can at least say that if the dual is feasible for one x and a given ξ then it is feasible for all x for that value of ξ, since x enters only the objective function. Therefore, from standard linear programming duality, since the primal is feasible, the primal must be unbounded if and only if the dual is infeasible, and that would happen for all x for a given ξ, if randomness affects the objective function. If $q(\xi) \equiv q_0$ then it would happen for all x and ξ. Therefore we can check in advance for unboundedness, and this is particularly easy if randomness does not affect the objective function. Note that this discussion relates to Proposition 1.18. Assume we know that our problem is bounded.

Now consider

$$\mathcal{Q}(x) = \sum_j p^j Q(x, \xi^j),$$

with

$$Q(x, \xi) = \min\{q(\xi)^{\mathrm{T}} y \mid W y = h(\xi) - T(\xi)x,\ y \geq 0\}.$$

It is clear from standard linear programming theory that $Q(x, \xi)$ is piecewise linear and convex in x (for fixed ξ). Provided that $q(\xi) \equiv q_0$, $Q(x, \xi)$ is also piecewise linear and convex in ξ (for fixed x). (Remember that $T(\xi) = T_0 + \sum T_i \xi_i$.) Similarly, if $h(\xi) - T(\xi)x \equiv h_0 - T_0 x$, while $q(\xi) = q_0 + \sum_i q_i \xi_i$, then, from duality, $Q(x, \xi)$ is piecewise linear and concave in ξ. Each linear piece corresponds to a basis (possibly several in the case of degeneracy). Therefore $\mathcal{Q}(x)$, being a finite sum of such functions, will also be convex and piecewise linear in x. If, instead of minimizing, we were maximizing, convexity and concavity would change places in the statements.

procedure master(K, L:integer;x, θ:real;feasible:boolean);
begin
  if $L > 0$ **then begin**

$$\text{LP}\left(\begin{pmatrix} A & 0 & 0 & 0 & 0 \\ -\Gamma & 0 & 0 & -I & 0 \\ -\beta & e & -e & 0 & -I \end{pmatrix}, \begin{pmatrix} b \\ \Delta \\ \alpha \end{pmatrix}, \begin{pmatrix} c \\ 1 \\ -1 \\ 0 \\ 0 \end{pmatrix}, \begin{pmatrix} \hat{x} \\ \theta^+ \\ \theta^- \\ s_1 \\ s_2 \end{pmatrix}, \text{feasible}\right);$$

    if (feasible) **then** $\hat{\theta} = \theta^+ - \theta^-$;
  end
  else begin

$$\text{LP}\left(\begin{pmatrix} A & 0 \\ -\Gamma & -I \end{pmatrix}, \begin{pmatrix} b \\ \Delta \end{pmatrix}, \begin{pmatrix} c \\ 0 \end{pmatrix}, \begin{pmatrix} \hat{x} \\ s \end{pmatrix}, \text{feasible}\right);$$

    if feasible **then** $\hat{\theta} := -\infty$;
  end;
end;

Figure 8 Master problem solver for the L-shaped decomposition method.

procedure feascut($\mathcal{A}$:set; $\hat{x}$:real; newcut:boolean);
begin
  $\mathcal{A}' := \mathcal{A}$; newcut := false;
  while $\mathcal{A}' \neq \emptyset$ **and not** (newcut) **do begin**
    pick$\xi(\mathcal{A}', \xi)$; $\mathcal{A}' := \mathcal{A}' \setminus \{\xi\}$;

$$\text{LP}\left(\begin{pmatrix} W & e & -I & 0 \\ -W & e & 0 & -I \end{pmatrix}, \begin{pmatrix} h(\xi) - T(\xi)\hat{x} \\ -h(\xi) + T(\xi)\hat{x} \end{pmatrix}, \begin{pmatrix} 0 \\ 1 \\ 0 \\ 0 \end{pmatrix}, \begin{pmatrix} \hat{y} \\ \hat{t} \\ s_1 \\ s_2 \end{pmatrix}, \text{feasible}\right);$$

    newcut := $(\hat{t} > 0)$;
    if newcut **then begin**
      (* Create a feasibility cut—see page 149. *)
      $K := K + 1$;
      Construct the cut $-\gamma_K^{\mathrm{T}} x \geq \delta_K$;
    end;
  end;
end;

Figure 9 Procedure used to find feasibility cuts.

In order to arrive at an algorithm for our problem, let us now reformulate the latter by introducing a new variable θ:

$$\begin{array}{ll} \min c^{\mathrm{T}}x + \theta & \\ \text{s.t.} \quad Ax = b, & \\ \quad \theta \geq \mathcal{Q}(x), & \\ \quad -\gamma_k^{\mathrm{T}}x \geq \delta_k & \text{for } k = 1, \ldots, K, \\ \quad x \geq 0, & \end{array}$$

where, as before,

$$\mathcal{Q}(x) = \sum_j p^j Q(x, \xi^j)$$

and

$$Q(x, \xi) = \min\{q(\xi)^{\mathrm{T}}y \mid Wy = h(\xi) + T(\xi)x,\ y \geq 0\}.$$

Of course, computationally we cannot use $\theta \geq \mathcal{Q}(x)$ as a constraint since $\mathcal{Q}(x)$ is only defined implicitly by a large number of optimization problems. Instead, let us for the moment drop it, and solve the above problem without it, simply hoping it will be satisfied (assuming so far that all feasibility cuts $-\gamma_k^{\mathrm{T}}x \geq \delta_k$ are there, or that we have relatively complete recourse). We then get some $\hat{x}$ and $\hat{\theta}$ (the first time $\hat{\theta} = -\infty$). Now we calculate $\mathcal{Q}(\hat{x})$, and then check if $\hat{\theta} \geq \mathcal{Q}(\hat{x})$. If it is, we are done. If not, our $\hat{x}$ is not optimal—dropping $\theta \geq \mathcal{Q}(x)$ was not acceptable.

Now

$$\mathcal{Q}(\hat{x}) = \sum_j p^j Q(\hat{x}, \xi^j) = \sum_j p^j q(\xi^j)^{\mathrm{T}} y^j$$

where y^j is the optimal second-stage solution yielding $Q(\hat{x}, \xi^j)$. But, owing to linear programming duality, we also have

$$\sum_j p^j q(\xi^j)^{\mathrm{T}} y^j = \sum_j p^j (\hat{\pi}^j)^{\mathrm{T}} [h(\xi^j) - T(\xi^j)\hat{x}],$$

where $\hat{\pi}^j$ is the optimal dual solution yielding $Q(\hat{x}, \xi^j)$. The constraints in the dual problem are, as mentioned before, $\pi W \leq q(\xi^j)$, which are independent of x. Therefore, for a general x, and corresponding optimal vectors $\pi^j(x)$, we have

$$\mathcal{Q}(x) = \sum_j p^j (\pi^j(x))^{\mathrm{T}} [h(\xi^j) - T(\xi^j)x] \geq \sum_j p^j (\hat{\pi}^j)^{\mathrm{T}} [h(\xi^j) - T(\xi^j)x],$$

since $\hat{\pi}$ is feasible but not necessarily optimal, and the dual problem is a maximization problem. Since what we dropped from the constraint set was $\theta \geq \mathcal{Q}(x)$, we now add in its place

```
procedure L-shaped;
begin
   K := 0, L := 0;
   θ̂ := −∞
   LP(A, b, c, x̂, feasible);
   stop := not (feasible);
   while not (stop) do begin
      feascut(𝒜, x̂, newcut);
      if not (newcut) then begin
         Find 𝒬(x̂);
         stop := (θ̂ ≥ 𝒬(x̂));
         if not (stop) then begin
            (* Create an optimality cut—see page 155. *)
            L := L + 1;
            Construct the cut −β_L^T x + θ ≥ α_L;
         end;
      end;
      if not (stop) then begin
         master(K, L, x̂, θ̂, feasible);
         stop := not (feasible);
      end;
   end;
end;
```

Figure 10 The L-shaped decomposition algorithm.

$$\theta \geq \sum_j p^j (\hat{\pi}^j)^{\mathrm{T}} [h(\xi^j) - T(\xi^j)x] = \alpha + \beta^{\mathrm{T}} x,$$

or

$$-\beta^{\mathrm{T}} x + \theta \geq \alpha.$$

Since there are finitely many feasible bases coming from the matrix W, there are only finitely many such cuts.

We are now ready to present the basic setting of the L-shaped decomposition algorithm. It is shown in Figure 10. To use it, we shall need a procedure that solves LPs. It can be found in Figure 7. Also, to avoid too complicated expressions, we shall define a special procedure for solving the master problem; see Figure 8. Furthermore, we refer to **procedure** pick$\xi(\mathcal{A}, \xi)$, which simply

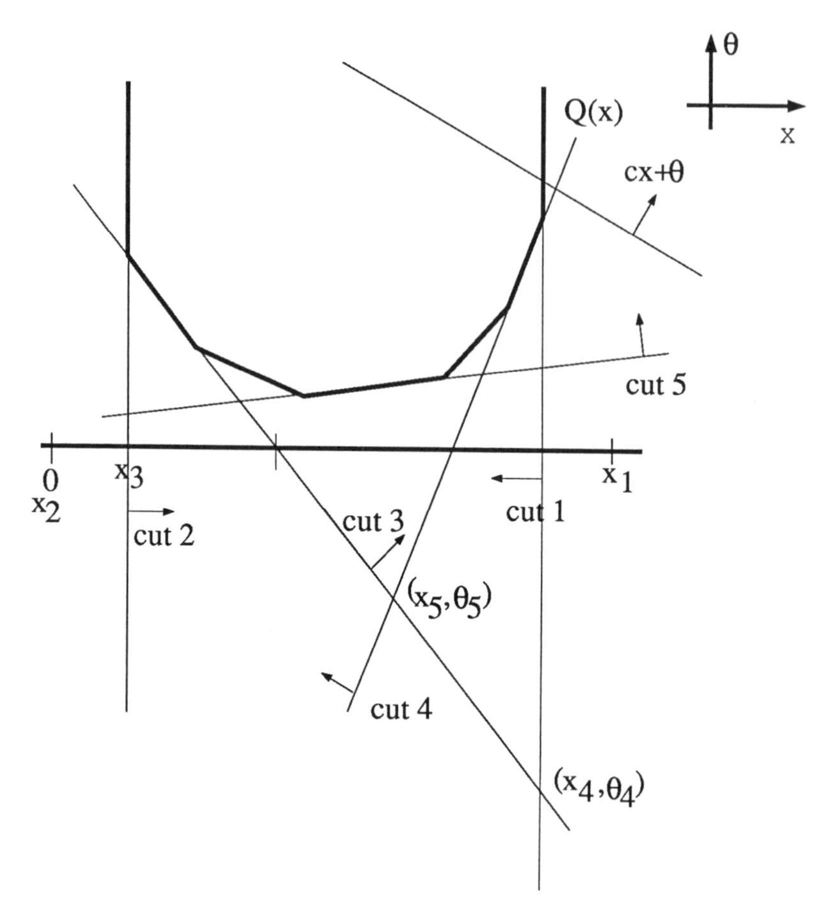

Figure 11 Example of the progress of the L-shaped decomposition algorithm.

picks an element ξ from the set $\mathcal{A}$, and, finally, we use **procedure** feascut which is given in Figure 9. The set $\mathcal{A}$ was defined on page 150.

In the algorithms to follow, let $-\Gamma x \geq \Delta$ represent the K feasibility cuts $-\gamma_k^{\mathrm{T}} x \geq \delta_k$, and let $-\beta x + I\theta \geq \alpha$ represent the L optimality cuts $-\beta_l^{\mathrm{T}} x + \theta \geq \alpha_l$. Furthermore, let e be a column of 1s of appropriate size.

The example in Figure 11 can be useful in understanding the L-shaped decomposition algorithm. The five first solutions and cuts are shown. The initial $\hat{x}_1$ was chosen arbitrarily. Cuts 1 and 2 are feasibility cuts, and the rest optimality cuts. $\hat{\theta}_1 = \hat{\theta}_2 = \hat{\theta}_3 = -\infty$. To see if you understand this, try to find $(\hat{x}_6, \hat{\theta}_6)$, cut 6 and then the final optimal solution.

3.3 Regularized Decomposition

As mentioned at the end of Section 1.6.4, the recourse problem (for a discrete distribution) looks like

$$\left.\begin{array}{ll} \min\{c^{\mathrm{T}}x + \sum_{i=1}^{K} p^i (q^i)^{\mathrm{T}} y^i\} & \\ \text{s.t. } Ax \qquad\qquad = b & \\ \quad T^i x + W y^i = h^i, & i = 1, \cdots, K \\ \qquad\qquad x \geq 0, & \\ \qquad\qquad y^i \geq 0, & i = 1, \cdots, K. \end{array}\right\} \tag{3.1}$$

To use the multicut method mentioned in Section 1.6.4, we simply have to introduce feasibility and optimality cuts for all the recourse functions $f_i(x) := \min\{(q^i)^{\mathrm{T}} y^i \mid W y^i = h^i - T^i x,\ y^i \geq 0\}$, $i = 1, \cdots, K$, until the overall procedure has converged.

In general, with the notation of the previous section, these cuts have the form

$$\gamma^{\mathrm{T}} x + \delta \leq 0, \text{ where } \gamma = -T^{\mathrm{T}} \overline{\sigma},\ \delta = h^{\mathrm{T}} \overline{\sigma}, \tag{3.2}$$

$$\gamma^{\mathrm{T}} x + \delta \leq \theta, \text{ where } \gamma = -T^{\mathrm{T}} \hat{\pi},\ \delta = f(\hat{x}) - \gamma^{\mathrm{T}} \hat{x}, \tag{3.3}$$

where (3.2) denotes a feasibility cut and (3.3) denotes an optimality cut, the $\overline{\sigma}$ and $\hat{\pi}$ resulting from step 2 of the dual decomposition method of Section 1.6.4, as explained further in Section 3.2. Of course, the matrix T and the right-hand side vector h will vary, depending on the block i for which the cut is derived. One cycle of a multicut solution procedure for problem (3.1) looks as follows:

Let $\mathcal{B}_{1i} = \{(x, \theta_1, \cdots, \theta_K) \mid \cdots\}$, $i = 1, \cdots, K$, be feasible for the cuts generated so far for block i (obviously for block i restricting only (x, θ_i)). Given $\mathcal{B}_0 = \{(x, \theta) \mid Ax = b,\ x \geq 0, \theta \in \mathbb{R}^K\}$ and the sets $\mathcal{B}_{1i}$, solve the *master program*

$$\min\left\{ c^{\mathrm{T}} x + \sum_{i=1}^{K} p^i \theta_i \,\middle|\, (x, \theta_1, \cdots, \theta_K) \in \mathcal{B}_0 \cap \left(\bigcap_{i=1}^{K} \mathcal{B}_{1i} \right) \right\}, \tag{3.4}$$

yielding $(\hat{x}, \hat{\theta}_1, \cdots, \hat{\theta}_K)$ as a solution. With this solution try to construct further cuts for the blocks.

- If there are no further cuts to generate, then stop (optimal solution);
- otherwise repeat the cycle.

The advantage of a method like this lies in the fact that we obviously make use of the particular structure of problem (3.1) in that we have to deal in the master program only with $n + K$ variables instead of $n + \sum_i n_i$, if $y^i \in \mathbb{R}^{n_i}$.

The drawback is easy to see as well: we may have to add very many cuts, and so far we have no reliable criterion to drop cuts that are obsolete for further iterations. Moreover, initial iterations are often inefficient. This is not surprising, since in the master (3.4) we deal only with

$$\theta_i \geq \max_{j \in J_i}[(\gamma^{ij})^{\mathrm{T}} x + \delta_{ij}]$$

for J_i denoting the set of optimality cuts generated so far for block i with the related dual basic solutions $\hat{\pi}^{ij}$ according to (3.3), and not, as we intend to, with

$$\theta_i \geq f_i(x) = \max_{j \in \hat{J}_i}[(\gamma^{ij})^{\mathrm{T}} x + \delta_{ij}]$$

where $\hat{J}_i$ enumerates *all* dual feasible basic solutions for block i. Hence we are working in the beginning with a piecewise linear convex function ($\max_{j \in J_i}[(\gamma^{ij})^{\mathrm{T}} x + \delta_{ij}]$) supporting $f_i(x)$ that does not sufficiently reflect the shape of f_i (see e.g. Figure 25 of Chapter 1, page 73). The effect may be—and often is—that even if we start a cycle with an (almost) optimal first-stage solution $x^\star$ of (3.1), the first-stage solution $\hat{x}$ of the master (3.4) may be far away from $x^\star$, and it may take many further cycles to come back towards $x^\star$. The reason for this is now obvious: if the set of available optimality cuts, J_i, is a small subset of the collection $\hat{J}_i$ then the piecewise linear approximation of $f_i(x)$ may be inadequate near $x^\star$. Therefore it seems desirable to modify the master program in such a way that, when starting with some overall feasible first-stage iterate z^k, its solution x^k does not move too far away from z^k. Thereby we can expect to improve the approximation of $f_i(x)$ by an optimality cut for block i at x^k. This can be achieved by introducing into the objective of the master the term $\|x - z^k\|^2$, yielding a so-called *regularized* master program

$$\min \left\{ \frac{1}{2\rho}\|x - z^k\|^2 + c^{\mathrm{T}} x + \sum_{i=1}^{K} p^i \theta_i \,\middle|\, (x, \theta_1, \cdots, \theta_K) \in \mathcal{B}_0 \cap \left(\bigcap_{i=1}^{K} \mathcal{B}_{1i} \right) \right\}, \tag{3.5}$$

with a control parameter $\rho > 0$. To avoid too many constraints in (3.5), let us start with some $z^0 \in \mathcal{B}_0$ such that $f_i(z^0) < \infty \ \forall i$ and $\mathcal{G}_0$ being the feasible set defined by the first-stage equations $Ax = b$ and all optimality cuts at z^0. Hence we start (for $k = 0$) with the reduced regularized master program

$$\min \left\{ \frac{1}{2\rho}\|x - z^k\|^2 + c^{\mathrm{T}} x + \sum_{i=1}^{K} p^i \theta_i \,\middle|\, (x, \theta_1, \cdots, \theta_K) \in \mathcal{G}_k \right\}. \tag{3.6}$$

Observe that the objective of (3.6) implicitly contains the function[1]

$$\hat{F}(x) = c^{\mathrm{T}} x + \min_{\theta}\{p^{\mathrm{T}} \theta \mid (x, \theta) \in \mathcal{G}_k\},$$

[1] With $p = (p^1, \cdots, p^K)^{\mathrm{T}}$.

which, according to the above discussion, is a piecewise linear convex function supporting from below our original piecewise linear objective

$$\begin{aligned} F(x) &= c^{\mathrm{T}}x + p^{\mathrm{T}}f(x) \\ &= c^{\mathrm{T}}x + \sum_i p^i f_i(x). \end{aligned}$$

Excluding by assumption degeneracy in the constraints defining $\mathcal{G}_k$, a point $(x, \theta) \in \mathbb{R}^{n+K}$ is a vertex, i.e. a basic solution, of $\mathcal{G}_k$ iff (including the first-stage equations $Ax = b$) exactly $n + K$ constraints are active (i.e. satisfied as equalities), owing to the simple fact that a point in $\mathbb{R}^{n+K}$ is uniquely determined by the intersection of $n + K$ independent hyperplanes.[2] In the following we sometimes want to check whether at a certain overall feasible $\hat{x} \in \mathbb{R}^n$ the support function $\hat{F}$ has a kink, which in turn implies that for $\hat{\theta} \in \arg\min_\theta\{p^{\mathrm{T}}\theta \mid (\hat{x}, \theta) \in \mathcal{G}_k\}$ at $(\hat{x}, \hat{\theta})$ we have a vertex of $\mathcal{G}_k$. Hence we have to check whether at $(\hat{x}, \hat{\theta})$ exactly $n + K$ constraints are active.

Having solved (3.6) with a solution x^k, and x^k not being overall feasible, we just add the violated constraints (either $x_i \geq 0$ from the first-stage or the necessary feasibility cuts from the second stage) and resolve (3.6). If x^k is overall feasible, we have to decide whether we maintain the candidate solution z^k or whether we replace it by x^k. As shown in Figure 12, there are essentially three possibilities:

- $F(x^k) = \hat{F}(x^k)$, i.e. the supporting function coincides at x^k with the true objective function (see x^1 in Figure 12);
- $\hat{F}(x^k) < F(x^k)$, but at x^k there is a kink of $\hat{F}$ and the decrease of the true objective from z^k to x^k is 'substantial' as compared with the decrease $\hat{F}(x^k) - \hat{F}(z^k) = \hat{F}(x^k) - F(z^k)$ (< 0) (we have $\hat{F}(z^k) = F(z^k)$ in view of the overall feasibility of z^k); more precisely, for some fixed $\mu \in (0, 1)$, $F(x^k) - F(z^k) \leq (1 - \mu)[\hat{F}(x^k) - F(z^k)]$ (see x^2 in Figure 12 with $\mu = 0.75$);
- neither of the two above situations arises (see x^3 in Figure 12).

In these cases we should decide respectively

- $z^2 := x^1$, observing that no cut was added, and therefore keeping z^1 unchanged would block the pocedure;
- $z^3 := x^2$, realizing that x^2 is "substantially" better than z^2—in terms of the original objective—and that at the same time $\hat{F}$ has a kink at x^2 such that we might intuitively expect—thus clearly making use of a heuristic argument—to make a good step forward towards the optimal kink of the true objective;

[2] Recall that in $\mathbb{R}^{n+K}$ never more than $n + K$ independent hyperplanes intersect at one point.

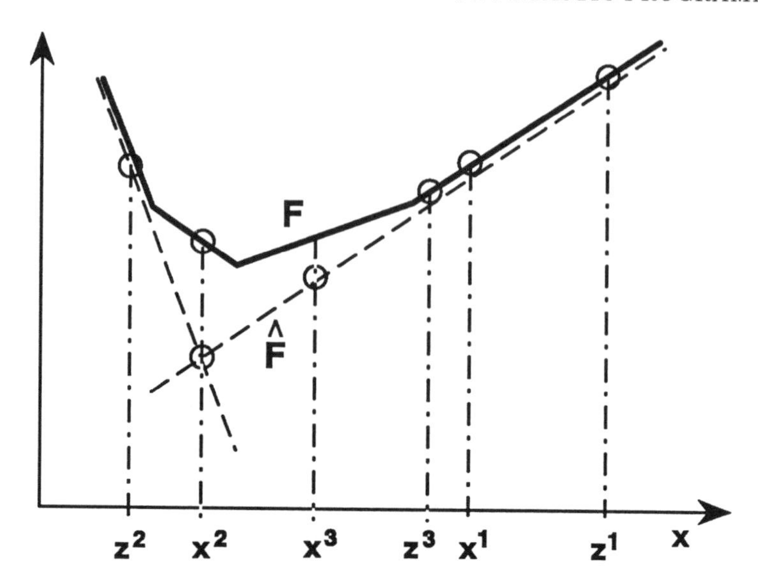

Figure 12 Keeping or changing the candidate solutions in QDECOM.

- $z^4 := z^3$, since—neither rationally nor heuristically—can we see any convincing reason to change the candidate solution. Hence it seems preferable to first improve the approximation of $\hat{F}$ to F by introducing the necessary optimality cuts.

After these considerations, motivating the measures to be taken in the various steps, we want to formulate precisely one cycle of the *regularized decomposition method* (RD), which with

$$F(x) := c^{\mathrm{T}}x + \sum_{i=1}^{K} p^i f_i(x)$$

for $\mu \in (0,1)$, is described as follows.

Step 1 Solve (3.6) at z^k, getting x^k as first-stage solution and $\theta^k = (\theta_1^k, \cdots, \theta_K^k)^{\mathrm{T}}$ as recourse approximates. If, for $\hat{F}_k := c^{\mathrm{T}}x^k + p^{\mathrm{T}}\theta^k$, $\hat{F}_k = F(z^k)$ then stop (z^k is an optimal solution of (3.1)). Otherwise, go to step 2.

Step 2 Delete from (3.6) some constraints that are inactive at (x^k, θ^k) such that no more than $n + K$ constraints remain.

Step 3 If x^k satisfies the first-stage constraints (i.e. $x^k \geq 0$) then go to step 4; otherwise add to (3.6) no more than K violated (first-stage) constraints, yielding the feasible set $\mathcal{G}_{k+1}$, put $z^{k+1} := z^k$, $k := k+1$, and go to step 1.

Step 4 For $i = 1, \cdots, K$ solve the second-stage problems at x^k and

(a) if $f_i(x^k) = \infty$ then add to (3.6) a feasibility cut;

(b) otherwise, if $f_i(x^k) > \theta_i^k$ then add to (3.6) an optimality cut.

Step 5 If $f_i(x^k) = \infty$ for at least one i then put $z^{k+1} := z^k$ and go to step 7. Otherwise, go to step 6.

Step 6 If $F(x^k) = \hat{F}_k$, or else if $F(x^k) \leq \mu F(z^k) + (1-\mu)\hat{F}_k$ and if exactly $n + K$ constraints were active at (x^k, θ^k), then put $z^{k+1} := x^k$; otherwise, put $z^{k+1} := z^k$.

Step 7 Determine $\mathcal{G}_{k+1}$ as resulting from $\mathcal{G}_k$ after deleting and adding constraints due to step 2 and step 4 respectively. With $k := k+1$, go to step 1.

It can be shown that this algorithm converges in finitely many steps. The parameter ρ can be controlled during the procedure so as to increase it whenever steps (i.e. $||x^k - z^k||$) seem too short, and decrease it when $F(x^k) > F(z^k)$.

3.4 Bounds

Section 3.2 was devoted to the L-shaped decomposition method. We note that the deterministic methods very quickly run into dimensionality problems with respect to the number of random variables. With much more than 10 random variables,we are in trouble.

This section discusses bounds on stochastic problems. These bounds can be useful and interesting in their own right, or they can be used as subproblems in larger settings. An example of where we might need to bound a problem, and where this problem is not a subproblem, is the following. Assume that a company is facing a decision problem. The decision itself will be made next year, and at that time all parameters describing the problem will be known. However, today a large number of relevant parameters are unknown, so it is difficult to predict how profitable the operation described by the decision problem will actually be. It is desired to know the expected profitability of the operation. The reason is that, for planning purposes, the firm needs to know the expected activities and profits for the next year. Given the large number of uncertain parameters, it is not possible to calculate the exact expected value. However, using bounding techniques it may be possible to identify an interval that contains the expected value. Technically speaking, one needs to find the expected value of the "wait-and-see" solution discussed in Chapter 1, and also in Example 2.4. Another example, which we shall see later in Section 6.6, is that of calculating the expected project duration time in a project consisting of activities with random durations.

Bounding methods are also useful if we wish to use deterministic decomposition methods (such as the L-shaped decomposition method or scenario aggregation), on problems with a large number of random variables. That will be discussed later in Section 3.5.2. One alternative to bounding involves the development of approximations using stochastic methods. We shall outline two of them later, they are called stochastic decomposition (Section 3.8) and stochastic quasi-gradient methods (Section 3.9).

As discussed above, bounds can be used either to approximate the expected value of some linear program or to bound the second-stage problem in a two-stage problem. These two settings are principally the same, and we shall therefore consider the problem of finding the expected value of a linear program. We shall discuss this in terms of a function $\phi(\xi)$, which in the two-stage case represents $Q(\hat{x}, \xi)$ for a fixed $\hat{x}$. To illustrate, we shall look at the refinery example of Section 1.2. The problem is repeated here for convenience:

$$\left.\begin{array}{rl} \phi(\xi) = \text{“min”} & \{2x_{raw1} + 3x_{raw2}\} \\ \text{s.t.} & x_{raw1} + x_{raw2} \leq 100, \\ & 2x_{raw1} + 6x_{raw2} \geq 180 + \xi_1, \\ & 3x_{raw1} + 3x_{raw2} \geq 162 + \xi_2, \\ & x_{raw1} \geq 0, \\ & x_{raw2} \geq 0. \end{array}\right\} \quad (4.1)$$

where both ξ_1 and ξ_2 are normally distributed with mean 0. As discussed in Section 1.2, we shall look at the 99% intervals for both (as if that was the support). This gives us

$$\xi_1 \in [-30.91, 30.91], \quad \xi_2 \in [-23.18, 23.18].$$

The interpretation is that 100 is the production limit of a refinery, which refines crude oil from two countries. The variable x_{raw1} represents the amount of crude oil from Country 1 and x_{raw2} the amount from Country 2. The qualities of the crude oils are different, so one unit of crude oil from Country 1 gives two units of Product 1 and three units of Product 2, whereas the crude oil from the second country gives 6 and 3 units of the same products. Company 1 wants at least $180 + \xi_1$ units of Product 1 and Company 2 at least $162 + \xi_2$ units of Product 2. The goal now is to find the expected value of $\phi(\xi)$; in other words, we seek the expected value of the “wait-and-see” solution. Note that this interpretation is not the one we adopted in Section 1.2.

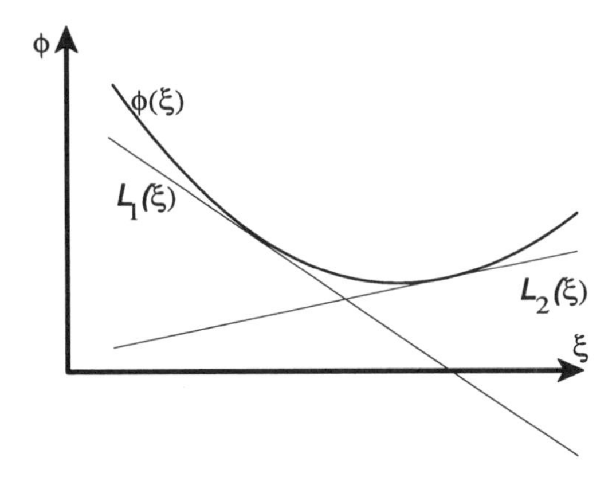

Figure 13 Two possible lower bounding functions.

3.4.1 The Jensen Lower Bound

Assume that $q(\xi) \equiv q_0$, so that randomness affects only the right-hand side. The purpose of this section is to find a lower bound on $Q(\hat{x}, \xi)$, for fixed $\hat{x}$, and for that purpose we shall, as just mentioned, use $\phi(\xi) \equiv Q(\hat{x}, \xi)$ for a fixed $\hat{x}$.

Since $\phi(\xi)$ is a convex function, we can bound it from below by a linear function $L(\xi) = c\xi + d$. Since the goal will always be to find a lower bound that is as large as possible, we shall require that the linear lower bound be tangent to $\phi(\xi)$ at some point $\hat{\xi}$. Figure 13 shows two examples of such lower-bounding functions. But the question is which one should we pick. Is $L_1(\xi)$ or $L_2(\xi)$ the better?

If we let the lower bounding function $L(\xi)$ be tangent to $\phi(\xi)$ at $\hat{\xi}$, the slope must be $\phi'(\hat{\xi})$, and we must have

$$\phi(\hat{\xi}) = \phi'(\hat{\xi})\hat{\xi} + b,$$

since $\phi(\hat{\xi}) = L(\hat{\xi})$. Hence, in total, the lower-bounding function is given by

$$L(\xi) = \phi(\hat{\xi}) + \phi'(\hat{\xi})(\xi - \hat{\xi}).$$

Since this is a linear function, we easily calculate the expected value of the lower-bounding function:

$$EL(\tilde{\xi}) = \phi(\hat{\xi}) + \phi'(\hat{\xi})(E\tilde{\xi} - \hat{\xi}) = L(E\tilde{\xi}).$$

In other words, we find the expected lower bound by evaluating the lower bounding function in $E\tilde{\xi}$. From this, is is easy to see that we obtain the best (largest) lower bound by letting $\hat{\xi} = E\tilde{\xi}$. This can be seen not only from the fact that no linear function that supports $\phi(\xi)$ can have a value larger than $\phi(E\tilde{\xi})$ in $E\tilde{\xi}$, but also from the following simple differentiation:

$$\frac{d}{d\hat{\xi}}L(E\tilde{\xi}) = \phi'(\hat{\xi}) - \phi'(\hat{\xi}) + \phi''(\hat{\xi})(E\tilde{\xi} - \hat{\xi}).$$

If we set this equal to zero we find that $\hat{\xi} = E\tilde{\xi}$. What we have developed is the so-called *Jensen lower bound*, or the *Jensen inequality*.

Proposition 3.1 *If $\phi(\xi)$ is convex over the support of $\tilde{\xi}$ then*

$$E\phi(\tilde{\xi}) \geq \phi(E\tilde{\xi})$$

This best lower bound is illustrated in Figure 14. We can see that the Jensen lower bound *can be viewed* two different ways. First, it can be seen as a bound where a distribution is replaced by its mean and the problem itself is unchanged. This is when we calculate $\phi(E\tilde{\xi})$. Secondly, it can be viewed as a bound where the distribution is left unchanged and the function is replaced by a linear affine function, represented by a straight line. This is when we integrate $L(\xi)$ over the support of $\tilde{\xi}$. Depending on the given situation, both these views can be useful.

There is even a third interpretation. We shall see it used later in the stochastic decomposition method. Assume we first solve the dual of $\phi(E\tilde{\xi})$ to obtain an optimal basis B. This basis, since ξ does not enter the constraints of the dual of ϕ, is dual feasible for all possible values of ξ. Assume now that we solve the dual version of $\phi(\xi)$ for *all* ξ, but constrain our optimization so that we are allowed to use only the given basis B. In such a setting, we might claim that we use the correct function, the correct distribution, but optimize only in an approximate way. (In stochastic decomposition we use not one, but a finite number of bases.) The Jensen lower bound can in this setting be interpreted as representing approximate optimization using the correct problem and correct distribution, but only one dual feasible basis.

It is worth pointing out that these interpretations of the Jensen lower bound are put forward to help you see how a bound can be interpreted in different ways, and that these interpretations can lead you in different directions when trying to strengthen the bound. An interpretation is not necessarily motivated by computational efficiency.

Looking back at our example in (4.1), we find the Jensen lower bound by calculating $\phi(E\tilde{\xi}) = \phi(0)$. That has been solved already in Section 1.2, where we found that $\phi(0) = 126$.

3.4.2 Edmundson–Madansky Upper Bound

Again let $\tilde{\xi}$ be a random variable. Let the support $\Xi = [a, b]$, and assume that $q(\xi) \equiv q_0$. As in the previous section, we define $\phi(\xi) = Q(\hat{x}, \xi)$. (Remember that x is fixed at $\hat{x}$.) Consider Figure 14, where we have drawn a linear function $U(\xi)$ between the two points $(a, \phi(a))$ and $(b, \phi(b))$. The line is clearly above $\phi(\xi)$ for all $\xi \in \Xi$. Also this straight line has the formula $c\xi + d$, and since we know two points, we can calculate

$$c = \frac{\phi(b) - \phi(a)}{b - a}, \qquad d = \frac{b}{b-a}\phi(a) - \frac{a}{b-a}\phi(b).$$

We can now integrate, and find (using the linearity of $U(\xi)$)

$$\begin{aligned} EU(\tilde{\xi}) &= \frac{\phi(b) - \phi(a)}{b - a} E\tilde{\xi} + \frac{b}{b-a}\phi(a) - \frac{a}{b-a}\phi(b) \\ &= \phi(a)\frac{b - E\tilde{\xi}}{b - a} + \phi(b)\frac{E\tilde{\xi} - a}{b - a}. \end{aligned}$$

In other words, if we have a function that is convex in ξ over a bounded support $\Xi = [a, b]$, it is possible to replace an arbitrary distribution by a two point distribution, such that we obtain an upper bound. The important parameter is

$$p = \frac{E\tilde{\xi} - a}{b - a},$$

so that we can replace the original distribution with

$$P\{\tilde{\xi} = a\} = 1 - p, \qquad P\{\tilde{\xi} = b\} = p. \tag{4.2}$$

As for the Jensen lower bound, we have now shown that the Edmundson–Madansky upper bound can be seen as either changing the distribution and keeping the problem, or changing the problem and keeping the distribution.

Looking back at our example in (4.1), we have two independent random variables. Hence we have $2^2 = 4$ LPs to solve to find the Edmundson–Madansky upper bound. Since both distributions are symmetric, the probabilities attached to these four points will all be 0.25. Calculating this we find an upper bound of

$$\tfrac{1}{4}(106.6825 + 129.8625 + 122.1375 + 145.3175) = 126.$$

This is exactly the same as the lower bound, and hence it is the true value of $E\phi(\tilde{\xi})$. We shall shortly comment on this situation where the bounds turn out to be equal.

In higher dimensions, the Jensen lower bound corresponds to a hyperplane, while the Edmundson–Madansky bound corresponds to a more general polynomial. A two-dimensional illustration of the Edmundson–Madansky bound is given in Figure 15. Note that if we fix the value of all but one of the

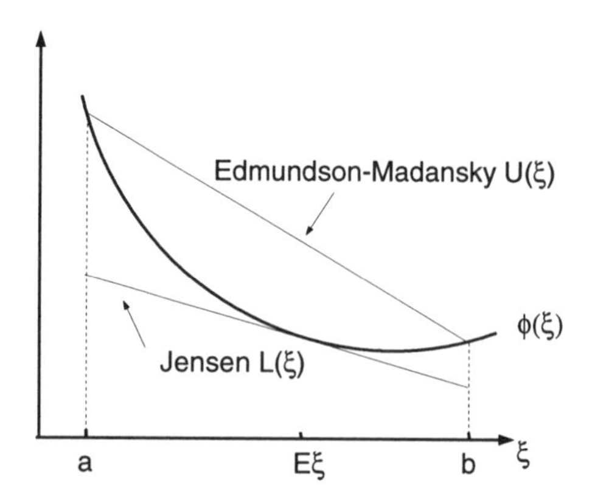

Figure 14 The Jensen lower bound and the Edmundson–Madansky upper bound in a minimization problem. Note that x is fixed.

variables, we get a linear function. This polynomial is therefore generated by straight lines. From the viewpoint of *computations*, we do not have to relate to this general polynomial. Instead, we take one (independent) random variable at a time, and calculate (4.2). This way we end up with 2 possible values for each random variable, and hence, 2^k possible values of ξ for which we have to evaluate the recourse function.

Assume that the function $\phi(\xi)$ in Figure 14 is linear. Then it appears from the figure that both the Jensen lower bound and the Edmundson–Madansky upper bound are exact. This is indeed a correct observation: both bounds are exact whenever the function is linear. And, in particular, this means that if the function is linear, the error is zero. In the example (4.1) used to illustrate the Jensen and Edmundson–Madansky bounds we observed that the bounds where equal. This shows that the function $\phi(\xi)$ is linear over the support we used.

One special use of the Jensen lower bound and Edmundson–Madansky upper bound is worth mentioning. Assume we have a random vector, containing a number of independent random variables, and a function that is convex with respect to that random vector, but the random vector either has a continuous distribution, or a discrete distribution with a very large number of outcomes. In both cases we might have to simplify the distribution before making any attempts to attack the problem.

The principle we are going to use is as follows. Take one random variable at a time. First partition the support of the variable into a finite number of intervals. Then apply the principle of the Edmundson–Madansky bound on *one interval at a time*. Since we are inside an interval, we use *conditional distributions*, rather than the original one. This will in effect replace the

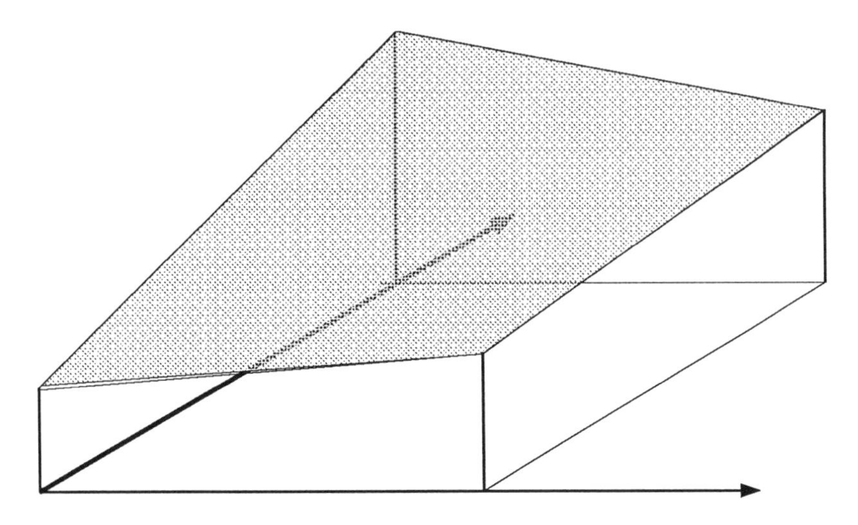

Figure 15 Illustration of the Edmundson–Madansky upper bound in two dimensions. The function itself is not drawn. The Jensen lower bound, which is simply a plane, is also not drawn.

distribution over the interval by a distribution that has probability mass only at the end points. This is illustrated in Figure 16, where we have shown the case for one random variable. The support of $\tilde{\xi}$ has been partitioned into two parts, called *cells*. For each of these cells, we have drawn the straight lines corresponding to the Jensen lower bound and the Edmundson–Madansky upper bound. Corresponding to each cell, there is a one-point distribution that gives a lower bound, and a two-point distribution that gives an upper bound, just as we have outlined earlier.

If the random variables have continuous (but bounded) distributions, we use these conditional bounds to replace the original distribution with discrete distributions. If the distribution is already discrete, we can remove some of the outcomes by using the Edmundson–Madansky inequality conditionally on parts of the support, again pushing probability mass to the end points of the intervals. Of course, the Jensen inequality can be used in the same way to construct conditional lower bounds. The point with these changes is not to create bounds per se, but to simplify distributions in such a way that we have control over what we have done to the problem when simplifying. The idea is outlined in Figure 17. Whatever the original distribution was, we now have two distributions: one giving an overall lower bound, the other an overall upper bound.

Since the random variables in the vector were assumed to be independent, this operation has produced discrete distributions for the random vector as

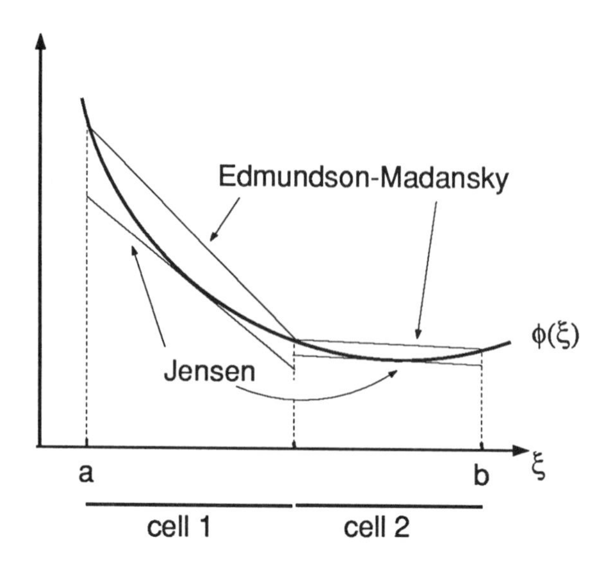

Figure 16 Illustration of the effect on the Jensen lower bound and the Edmundson–Madansky upper bound of partitioning the support into two cells.

Figure 17 Simplifying distributions by using Jensen and Edmundson–Madansky on subintervals of the support. The stars represent conditional expectations, and hence a distribution resulting in a lower bound. The bars are endpoints of intervals, representing a distribution yielding an upper bound.

well.

3.4.3 Combinations

If we have randomness in the objective function, but not in the right-hand side (so $h(\xi)-T(\xi)x \equiv h_0 - T_0x$), then, by simple linear programming duality, we can obtain the dual of $Q(x,\xi)$ with all randomness again in the right-hand side, but now in a setting of maximization. In such a setting the Jensen bound is an upper bound and the Edmundson–Madansky bound a lower bound.

If we have randomness in both the objective and the right-hand side, and the random variables affecting these two positions are different and independent, then we get a lower bound by applying the Jensen rule on the right-hand side random variables and the Edmundson–Madansky rule in the objective. If we do it the other way around, we get an overall upper bound.

3.4.4 A Piecewise Linear Upper Bound

Although the Edmundson–Madansky distribution is very useful, it still requires that $Q(x, \xi)$ be evaluated at an exponential number of points. That is, if there are k random variables, we must work with 2^k points. This means that with more than about 10 random variables we are not in business. In order to facilitate upper bounds, a number of approaches that are not of exponential complexity in terms of the number of random variables have been designed. In what follows we shall briefly demonstrate how to obtain a piecewise linear upper bound that does not exhibit this exponential characterization.

The idea behind the development is as follows. The recourse function $Q(\hat{x}, \xi)$ is convex in ξ (for a fixed $\hat{x}$). We might envisage it as a bowl. The Jensen lower bound represents a supporting hyperplane below the recourse function, like a table on which the bowl sits. Any supporting hyperplane would give a lower bound, but, as we have seen, the one that touches $Q(\hat{x}, \xi)$ in $E\tilde{\xi}$ gives the highest lower bound. The Edmundson–Madansky upper bound, on the other hand, is much like a lid on the bowl. They are both illustrated in Figure 14. The purpose of the piecewise linear upper bound is to find another bowl that fits inside the bowl $Q(\hat{x}, \xi)$, but at the same time has more curvature than the Edmundson–Madansky lid. Also, this new bowl must represent a function that is easy to integrate.

The piecewise linear upper bound has exactly these properties. It should be noted that it is impossible to compare the piecewise linear upper bound with the Edmundson–Madansky bound, in the sense that either one can be best in a given example. In particular, the new bound may be $+\infty$ even if the problem is feasible (meaning that $Q(\hat{x}, \xi) < \infty$ for all possible ξ). This can never happen to the Edmundson–Madansky upper bound. It seems that the new bound is reasonably good on "loose" problems, i.e. problems that are very far from being infeasible, such as problems with complete recourse. The Edmundson–Madansky bound is better on "tight" problems.

Let us illustrate the method in a simplified setting. We shall only consider randomness in the right-hand side of $Wy = b$, and leave the discussion of randomness in the upper bound c to Chapter 6. Define $\phi(\xi)$ by

$$\phi(\xi) = \min_y \{q^{\mathrm{T}} y \mid Wy = b + \xi,\ 0 \leq y \leq c\}$$

where all components in the random vector $\tilde{\xi}^{\mathrm{T}} = (\tilde{\xi}_1, \tilde{\xi}_2, \ldots)$ are mutually independent. Furthermore, let the support be given by $\Xi(\tilde{\xi}) = [A, B]$. For convenience, but without any loss of generality, we shall assume that $E\tilde{\xi} = 0$. The goal is to create a piecewise linear, separable and convex function in ξ:

$$U(\xi) = \phi(0) + \sum_i \begin{cases} d_i^+ \xi_i & \text{if } \xi_i \geq E\tilde{\xi}_i = 0 \\ d_i^- \xi_i & \text{if} \xi_i < E\tilde{\xi}_i = 0. \end{cases} \tag{4.3}$$

There is a very good reason for such a choice. Note how $U(\xi)$ is separable in its components ξ_i. Therefore, for almost all distribution functions, U is simple to integrate.

To appreciate the bound, we must understand its basic motivation. If we take some minimization problem, like the one here, and *add extra constraints*, the resulting problem will bound the original problem from above. What we shall do is to add restrictions with respect to the upper bounds c. We shall do this by viewing $\phi(\xi)$ as a parametric problem in ξ, and reserve portions of the upper bound c for the individual random variables ξ_i. We may, for example, end up by saying that two units of c_j are reserved for variable ξ_i, meaning that these two units can be used in the parametric analysis, *only when we consider* ξ_i. For all other variables ξ_k these two units will be viewed as nonexisting. The clue of the bound is to introduce the best possible set of such constraints, such that the resulting problem is easy to solve (and gives a good bound).

First, let us calculate $\phi(E\tilde{\xi}) = \phi(0)$ by finding

$$\phi(0) = \min_y \{q^{\mathrm{T}} y \mid Wy = b,\ 0 \le y \le c\} = q^{\mathrm{T}} y^0.$$

This can be interpreted as the *basic* setting, and all other values of ξ will be seen as *deviations* from $E\tilde{\xi} = 0$. (Of course, any other starting point will also do—for example solving $Q(A)$, where, as stated before, A is the lowest possible value of ξ.) Note that since y^0 is "always" there, we can in the following operate with bounds $-y^0 \le y \le c - y^0$. For this purpose, we define $\alpha^1 = -y^0$ and $\beta^1 = c - y^0$. Let e_i be a unit vector of appropriate size with a $+1$ in position i.

Next, define a counter r and let $r := 1$. Now check out the case when $\xi_r > 0$ by solving (remembering that B_r is the maximal value of ξ_r)

$$\min_y \{q^{\mathrm{T}} y \mid Wy = e_r B_r,\ \alpha^r \le y \le \beta^r\} = q^{\mathrm{T}} y^{r+} = d_r^+ B_r. \tag{4.4}$$

Note that d_r^+ represents the *per unit cost* of increasing the right-hand side from 0 to $e_r B_r$. Similarly, check out the case with $\xi_r < 0$ by solving

$$\min_y \{q^{\mathrm{T}} y \mid Wy = e_r A_r,\ \alpha^r \le y \le \beta^r\} = q^{\mathrm{T}} y^{r-} = d_r^- A_r. \tag{4.5}$$

Now, based on $y^{r\pm}$, we shall assign portions of the bounds to the random variable $\tilde{\xi}_r$. These portions of the bounds will be given to $\tilde{\xi}_r$ and left unused by other random variables, even when $\tilde{\xi}_r$ does not need them. That is done by means of the following problem, where we calculate what is left for the next random variable:

$$\alpha_i^{r+1} = \alpha_i^r - \min\{y_i^{r+}, y_i^{r-}, 0\}. \tag{4.6}$$

What we are doing here is to find, for each variable, how much $\tilde{\xi}_r$, in the worst case, uses of the bound on variable i in the negative direction. That is then subtracted off what we had before. There are three possibilities. Both (4.4) and (4.5) may yield non-negative values for the variable y_i. In that case nothing is used of the available "negative bound" α_i^r. Then $\alpha_i^{r+1} = \alpha_i^r$. Alternatively, if (4.4) has $y_i^{r+} < 0$, then it will in the worst case use y_i^{r+} of the available "negative bound". Finally, if (4.5) has $y_i^{r-} < 0$ then in the worst case we use y_i^{r-} of the bound. Therefore α_i^{r+1} is what is left for the next random variable. Similarly, we find

$$\beta_i^{r+1} = \beta_i^r - \max\{y_i^{r+}, y_i^{r-}, 0\}, \tag{4.7}$$

where β_i^{r+1} shows how much is still available of bound i in the forward (positive) direction.

We next increase the counter r by one and repeat (4.4)–(4.7). This takes care of the piecewise linear functions in ξ.

Note that it is possible to solve (4.4) and (4.5) by *parametric* linear programming, thereby getting not just one linear piece above $E\tilde{\xi}$ and one below, but rather piecewise linearity on both sides. Then (4.6) and (4.7) must be updated to "worst case" analysis of bound usage. That is simple to do.

Let us turn to our example (4.1). Since we have developed the piecewise linear upper bound for equality constraints, we shall repeat the problem with slack variables added explicitly.

$$\begin{array}{lllllll}
\phi(\xi_1,\xi_2) = \min\{2x_{raw1} + 3x_{raw2}\} & & & & & & \\
\text{s.t.} \quad x_{raw1} + & x_{raw2} & + s_1 & & & = & 100, \\
\quad 2x_{raw1} + & 6x_{raw2} & & - s_2 & & = & 180 + \xi_1, \\
\quad 3x_{raw1} + & 3x_{raw2} & & & - s_3 & = & 162 + \xi_2, \\
\quad x_{raw1} & & & & & \geq & 0, \\
 & x_{raw2} & & & & \geq & 0, \\
 & & s_1 & & & \geq & 0, \\
 & & & s_2 & & \geq & 0, \\
 & & & & s_3 & \geq & 0.
\end{array}$$

In this setting, what we need to develop is the following:

$$U(\xi_1,\xi_2) = \phi(0,0) + \begin{cases} d_1^+\xi_1 & \text{if } \xi_1 \geq 0, \\ d_1^-\xi_1 & \text{if } \xi_1 < 0, \end{cases} + \begin{cases} d_2^+\xi_2 & \text{if } \xi_2 \geq 0, \\ d_2^-\xi_2 & \text{if } \xi_2 < 0. \end{cases}$$

First, we have already calculated $\phi(0,0) = 126$ with $x_{raw1} = 36$, $x_{raw2} = 18$ and $s_1 = 46$. Next, let us try to find $d_1^{\pm}$. To do that, we need α^1, which equals $(-36, -18, -46, 0, 0)$. We must then formulate (4.4), using $\xi_1 \in [-30.91, 30.91]$:

$$
\begin{array}{llllll}
\min\{2x_{raw1} + 3x_{raw2}\} & & & & & \\
\text{s.t.} \quad x_{raw1} + & x_{raw2} + s_1 & & & = 0, \\
\quad 2x_{raw1} + 6x_{raw2} & & - s_2 & & = 30.91, \\
\quad 3x_{raw1} + 3x_{raw2} & & & - s_3 & = 0, \\
\quad x_{raw1} & & & & \geq -36, \\
& x_{raw2} & & & \geq -18, \\
& & s_1 & & \geq -46, \\
& & & s_2 & \geq 0, \\
& & & & s_3 \geq 0.
\end{array}
$$

The solution to this is $y^{1+} = (-7.7275, 7.7275, 0, 0, 0)^{\mathrm{T}}$, with a total cost of 7.7275. This gives us

$$
d_1^+ = \frac{(2,3,0,0,0)y^{1+}}{30.91} = 0.25.
$$

Next, we solve the same problem, just with 30.91 replaced by −30.91. This amounts to problem (4.5), and gives us the solution is $y^{1-} = (7.7275, -7.7275, 0, 0, 0)^{\mathrm{T}}$, with a total cost of −7.7275. Hence, we get

$$
d_1^- = \frac{(2,3,0,0,0)y^{1-}}{-30.91} = 0.25.
$$

The next step is to update α according to (4.6) to find out how much is left of the negative bounds on the variables. For x_{raw1} we get

$$
\alpha_{raw1}^2 = -36 - \min\{-7.7275, 7.7275, 0\} = -28.2725.
$$

For x_{raw2} we get in a similar manner

$$
\alpha_{raw2}^2 = -18 - \min\{7.7275, -7.7275, 0\} = -10.2725.
$$

For the three other variables, α_i^2 equals α_i^1. We can now turn to (4.4) for random variable 2. The problem to solve is as follows, when we remember the $\xi_2 \in [-23.18, 23.18]$.

$$
\begin{array}{llllll}
\min\{2x_{raw1} + 3x_{raw2}\} & & & & & \\
\text{s.t.} \quad x_{raw1} + & x_{raw2} + s_1 & & & = 0, \\
\quad 2x_{raw1} + 6x_{raw2} & & - s_2 & & = 0, \\
\quad 3x_{raw1} + 3x_{raw2} & & & - s_3 & = 23.18, \\
\quad x_{raw1} & & & & \geq -28.2725, \\
& x_{raw2} & & & \geq -10.2725, \\
& & s_1 & & \geq -46, \\
& & & s_2 & \geq 0, \\
& & & & s_3 \geq 0.
\end{array}
$$

The solution to this is $y^{2+} = (11.59, -3.863, -7.727, 0, 0)^{\mathrm{T}}$, with a total cost of 11.59. This gives us

$$d_2^+ = \frac{(2,3,0,0,0)y^{2+}}{23.18} = 0.5.$$

Next, we solve the same problem, just with 23.18 replaced by −23.18. This amounts to problem (4.5), and gives us the solution $y^{2-} = (-11.59, 3.863, 7.727, 0, 0)^{\mathrm{T}}$, with a total cost of −11.59. Hence we get

$$d_2^- = \frac{(2,3,0,0,0)y^{2-}}{-23.18} = 0.5.$$

This finishes the calculation of the (piecewise) linear functions in the upper bound. What we have now found is that

$$U(\xi_1,\xi_2) = 126 + \begin{cases} \frac{1}{4}\xi_1 & \text{if } \xi_1 \geq 0, \\ \frac{1}{4}\xi_1 & \text{if } \xi_1 < 0, \end{cases} + \begin{cases} \frac{1}{2}\xi_2 & \text{if } \xi_2 \geq 0, \\ \frac{1}{2}\xi_2 & \text{if } \xi_2 < 0, \end{cases}$$

which we easily see can be written as

$$U(\xi_1,\xi_2) = 126 + \tfrac{1}{4}\xi_1 + \tfrac{1}{2}\xi_2.$$

In other words, as we already knew from calculating the Edmundson–Madansky upper bound and Jensen lower bound, the recourse function is linear in this example. Let us, for illustration, integrate with respect to ξ_1.

$$\int_{-30.91}^{30.91} \tfrac{1}{4}\xi_1 f(\xi_1)\, d\xi_1 = \tfrac{1}{4}E\tilde{\xi}_1 = 0.$$

This is how it should be for linearity, the contribution from a random variable over which U (and therefore ϕ) is linear is zero. We should of course get the same result with respect to ξ_2, and therefore the upper bound is 126, which equals the Jensen lower bound.

Now that we have seen how things go in the linear case, let us try to see how the results will be when linearity is not present. Hence assume that we have now developed the necessary parameters $d_i^{\pm}$ for (4.3). Let us integrate with respect to the random variable $\tilde{\xi}_i$, assuming that $\Xi_i = [A_i, B_i]$:

$$\begin{aligned} &\int_{A_i}^{0} d_1^- \xi_i f(\xi_i) d\xi_i + \int_0^{B_i} d_1^+ \xi_i f(\xi_i) d\xi_i \\ &= \quad d_i^- E\{\tilde{\xi}_i \mid \tilde{\xi}_i \leq 0\} P\{\tilde{\xi}_i \leq 0\} + d_i^+ E\{\tilde{\xi}_i \mid \tilde{\xi}_i > 0\} P\{\tilde{\xi}_i > 0\}. \end{aligned}$$

This result should not come as much of a surprise. When one integrates a linear function, one gets the function evaluated at the expected value of the random variable. We recognize this integration from the Jensen calculations. From this, we also see, as we have already claimed a few times, that if $d_i^+ = d_i^-$ for all i, then the contribution to the upper bound from $\tilde{\xi}$ equals $\phi(E\tilde{\xi})$, which equals the contribution to the Jensen lower bound.

Let us repeat why this is an upper bound. What we have done is to distribute the bounds c on the variables among the different random variables. They have been given separate pieces, which they will not share with others, even if they, for a given realization of $\tilde{\xi}$, do not need the capacities themselves. This partitioning of the bounds among the random variables represents a set of extra constraints on the problem, and hence, since we have a minimization problem, the extra constraints yield an upper bound. If we run out of capacities before all random variables have received their parts, we must conclude that the upper bound is $+\infty$. This cannot happen with the Edmundson–Madansky upper bound. If $\phi(\xi)$ is feasible for all ξ then the Edmundson–Madansky bound is always finite. However, as for the Jensen and Edmundson–Madansky bounds, the piecewise linear upper bound is also exact when the recourse function turns out to be linear.

As mentioned before, we shall consider random upper bounds in Chapter 6, in the setting of networks.

3.5 Approximations

3.5.1 Refinements of the "Wait-and-See" Solution

Let us, also in this section, assume that $x = \hat{x}$, and as before define $\phi(\xi) = Q(\hat{x}, \xi)$. Using any of the above (or other) methods we can find bounds on the recourse function. Assume we have calculated $\mathcal{L}$ and $\mathcal{U}$ such that

$$\mathcal{L} \leq E\phi(\tilde{\xi}) \leq \mathcal{U}.$$

We can now look at $\mathcal{U} - \mathcal{L}$ to see if we are happy with the result or not. If we are not, there are basically two approaches that can be used. Either we might resort to a better bounding procedure (probably more expensive in terms of CPU time) or we might start using the old bounding methods on a partition of the support, thereby making the bounds tighter. Since we know only finitely many different methods, we shall eventually be left with only the second option.

The set-up for such an approach to bounding will be as follows. First, partition the support of the random variables into an arbitrary selection of *cells*—possibly only one cell initially. We shall only consider cells that are rectangles, so that they can be described by intervals on the individual random variables. Figure 18 shows an example in two dimensions with five cells. Now, apply the bounding procedures on each of the cells, and add up the results.

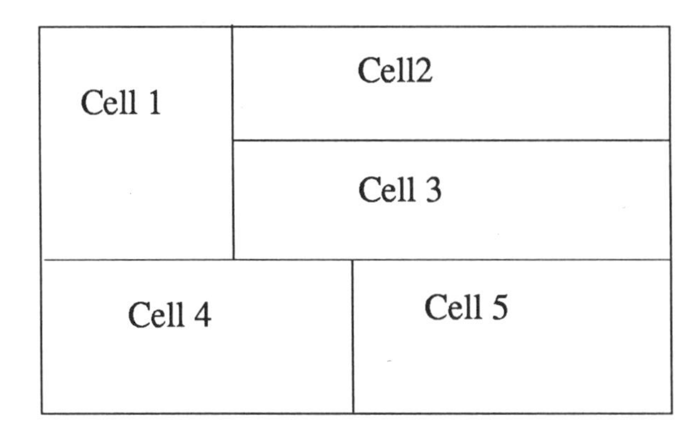

Figure 18 Partitioning of cells.

For example, in Figure 18, we need to find five conditional expectations, so that we can calculate the Jensen lower bound on each cell. Adding these up, using as weights the probability of being in the cell, we get an overall lower bound. In the same way, an upper bound can be calculated on each cell and added up to produce an overall upper bound. If the error $\mathcal{U} - \mathcal{L}$ is too large, one or more of the cells must be partitioned. It is natural to chose the cell(s) with the largest error(s), but along which coordinate(s) should it/they be partitioned, and through which point(s) in the cell(s)? Note that this use of conditional expectations is very similar to the way we created discrete distributions towards the end of Section 3.4.2. In particular, check out the discussion of Figure 16 on page 172.

Not much is known about what is a good *point* for partitioning. Obvious possibilities are the middle of the support and the (conditional) mean or median. Our experience is that the middle of the support is good, so we shall use that. However, this subject is clearly open for discussion. Let us therefore turn to the problem of picking the correct *coordinate* (random variable). For example, if we have picked Cell 1 in Figure 18 to be partitioned, should we draw a vertical or horizontal line? This might seem like a minor question at first sight. However, this is not at all the case. To see why, assume there is a random variable that is never of any importance, such as a random upper bound on a variable that, because of its high cost, is never used. Hence the realized value of this random variable is totally uninteresting. Assume that, for some reason, we pick this random variable for partitioning. The effect will be that when we calculate the bounds again on the two new cells and add them up, we have exactly the same error as before. But—and that is crucial—we now have two cells instead of one. From a *practical* point of view, these cells are exactly equal. They are only different with respect to a random variable that could as well have been dropped. Hence, in effect, we now have *increased our work load.* It is now harder to achieve a given error bound than it was before the partition. And note, we shall never recover from the error, in the sense that intelligent choices later on will not counteract this one bad choice. Each time we make a bad partition, the workload from there onwards basically doubles for the cell from which we started. Since we do not want to unnecessarily increase the workload too often, we must be careful with how we partition.

Now that we know that bad choices can increase the workload, what should we do? The first observation is that chosing at random is not a good idea, because, every now and then, we shall make bad choices. On the other hand, it is clear that the partitioning procedure will have to be a heuristic. Hence, we must make sure that we have a heuristic rule that we hope never makes really bad choices.

By knowing our problem well, we may be able to order the random variables according to their importance in the problem. Such an ordering could be used as is, or in combination with other ideas. For some network problems, such as the PERT problem (see Section 6.6), the network structure may present us with such a list. If we can compile the list, it seems reasonable to ask, from a modelling point of view, if the random variables last on the list should really have been there in the first place. They do not appear to be important.

Over the years, some attempts to understand the problem of partitioning have been made. Most of them are based on the assumption that the Edmundson–Madansky bound was used to calculate the upper bound. The reason is that the dual variables associated with the solution of the recourse function tell us something about its curvature. With the Edmundson–Madansky bound, we solve the recourse problem at all extreme points of the support, and thus get a reasonably good idea of what the function looks like.

To introduce some formality, assume we have only one random variable $\tilde{\xi}$, with support $\Xi = [A, B]$. When finding the Edmundson–Madansky upper bound, we calculated $\phi(A) = Q(\hat{x}, A)$ and $\phi(B) = Q(\hat{x}, B)$, obtaining dual solutions π^A and π^B. We know from duality that

$$\phi(A) = (\pi^A)^{\mathrm{T}}[h(A) - T(A)\hat{x}],$$
$$\phi(B) = (\pi^B)^{\mathrm{T}}[h(B) - T(B)\hat{x}].$$

We also know that, as long as $q(\xi) \equiv q_0$ (which we are assuming in this section), a π that is dual feasible for one ξ is dual feasible for all ξ, since ξ does not enter the dual constraints. Hence, we know that

$$\alpha = \phi(A) - (\pi^B)^{\mathrm{T}}[h(A) - T(A)\hat{x}] \geq 0$$

and

$$\beta = \phi(B) - (\pi^A)^{\mathrm{T}}[h(B) - T(B)\hat{x}] \geq 0.$$

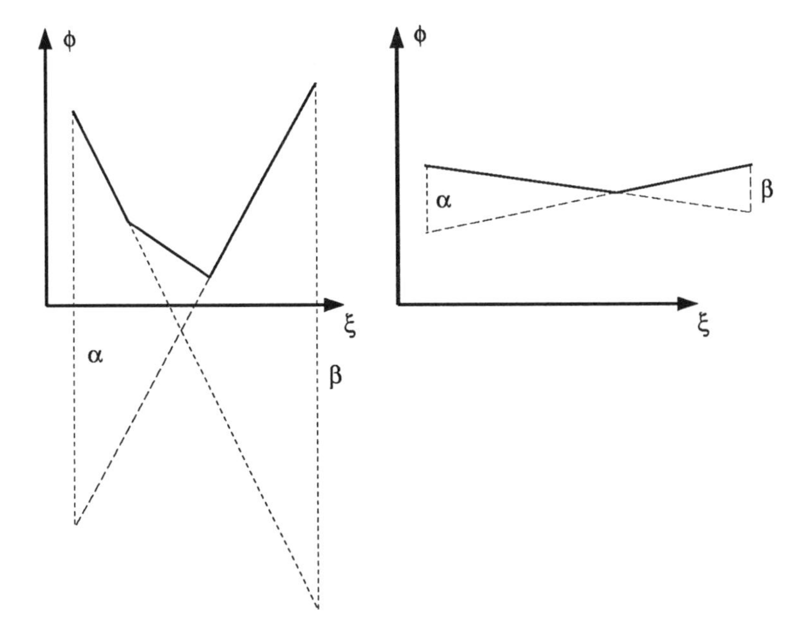

Figure 19 An illustration of a situation where both α and β give good information about curvature.

The parameters α and β contain information about the curvature of $\phi(\xi)$. In particular, note that if, for example, $\alpha = 0$ then π^2 is an optimal dual solution corresponding to $\xi = A$. If $\pi^A \neq \pi^B$ in such a case, we are simply facing dual degeneracy. In line with this argument, a small α (or β) should mean little curvature. But we may, for example, have α large and β small. So what is going on?

Figure 19 shows two different cases (both in one dimension), where both α and β are good indicators of how important a random variable is. Intuitively, it seems reasonable to say that the left part of the figure indicates an important random variable, and the right part an unimportant one. And, indeed, in the left part both α and β will be large, whereas in the right part both will be small.

But then consider Figure 20. Intuitively, the random variable is unimportant, but, in fact, the slopes at the end points are the same as in the left part of Figure 19, and the slopes describe how the objective changes as a function of the random variable. However, in this case α is very small, whereas β is large. What is happening is that α and β pick up two properties of the recourse function. If the function is very flat (as in the right part of Figure 19) then both parameters will be small. If the function is very non-

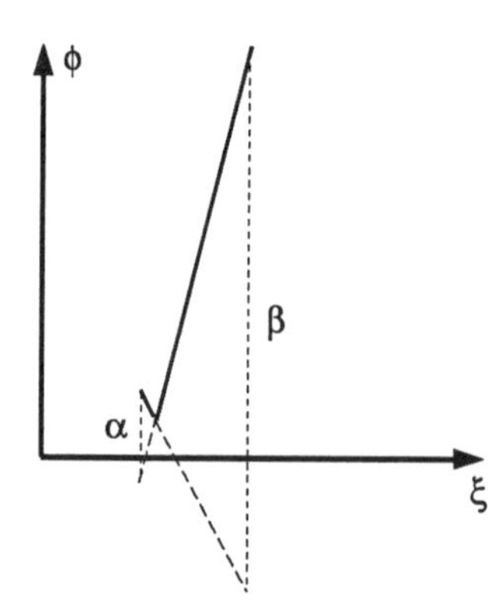

Figure 20 An illustration of a case where α is small and β is large.

linear (as in the left part of Figure 19), both parameters will be large. But if we have much curvature in the sense of the slopes of ϕ at the end point, but still almost linearity (as in Figure 20), then the smaller of the two parameters will be small. Hence the conclusion seems to be to calculate both α and β, pick the smaller of the two, and use that as a measure of nonlinearity.

Using α and β, we have a good measure of nonlinearity in one dimension. However, with more than one dimension, we must again be careful. We can certainly perform tests corresponding to those illustrated in Figures 19 and 20, for one random variable at a time. But the question is what value should we give the other random variables during the test. If we have k random variables, and have the Edmundson–Madansky calculations available, there are 2^{k-1} different ways we can fix all but one variable and then compare dual solutions. There are at least two possible approaches.

A first possibility is to calculate α and β for all neighbouring pairs of extreme points in the support, and pick the one for which the minimum of α and β is the largest. We then have a random variable for which ϕ is very nonlinear, at least in parts of the support. We may, of course, have picked a variable for which ϕ is linear most of the time, and this will certainly happen once in a while, but the idea is tested and found sound.

An alternative, which tries to check average nonlinearity rather than maximal nonlinearity, is to use all 2^{k-1} pairs of neighbouring extreme points involving variation in only one random variable, find the minimum of α and β for each such pair, and then calculate the average of these minima. Then we pick the random variable for which this average is maximal.

The number of pairs of neighbouring extreme points is fairly large. With k random variables, we have $k2^{k-1}$ pairs to compare. Each comparison requires the calculation of two inner products. We have earlier indicated that the

Edmundson–Madansky upper bound cannot be used for much more than 10 random variables. In such a case we must perform 5120 pairwise comparisons.

Looking back at Figures 19 and 20, we note that an important feature of the recourse function is its slope, as a function of the random variable. We alluded to the slope when discussing the parameters α and β, but we did not really show how to find the slopes.

We know from linear programming duality that the optimal value of a dual variable shows how the objective function will change (locally) as the corresponding right-hand side element increases. Given that we use the Edmundson–Madansky upper bound, these optimal dual solutions are available to us at all extreme points of the support. If ξ^j is the value of $\tilde{\xi}$ at such an extreme point, we have

$$\phi(\xi^j) = Q(\hat{x}, \xi^j) = q^{\mathrm{T}} y(\xi^j) = \pi(\xi^j)^{\mathrm{T}}[h(\xi^j) - T(\xi^j)\hat{x}].$$

What we need to know to utilize this information is how the right-hand side changes as a given random variable $\tilde{\xi}_i$ changes. This is easy to calculate, since all we have to do is to find the derivative of

$$h(\xi) - T(\xi)\hat{x} = h_0 + \sum_j h_j \xi_j - \Big(T_0 + \sum_j T_j \xi_j\Big)\hat{x}$$

with respect to ξ_i. This is easily found to be

$$h_i - T_i \hat{x} \equiv \delta_i.$$

Note that this expression is independent of the value of $\tilde{\xi}$, and hence it is the same at all extreme points of the support. Now, if $\pi(\xi^j)$ is the optimal dual solution at an extreme point of the support, represented by ξ^j, then the slope of $\phi(\xi) = Q(\hat{x}, \xi)$ with respect to ξ_i is given by

$$\pi(\xi^j)^{\mathrm{T}} \delta_i.$$

And, more generally, if we let $\delta^{\mathrm{T}} = (\delta_1, \delta_2, \ldots)$, the vector

$$\pi(\xi^j)^{\mathrm{T}} \delta \equiv (\pi^j)^{\mathrm{T}} \tag{5.1}$$

characterizes how $\phi(\xi) = Q(\hat{x}, \xi)$ changes with respect to all random variables. Since these calculations are performed at each extreme point of the support, and each extreme point has a probability according to the Edmundson–Madansky calculations, we can interpret the vectors π^j as outcomes of a random vector $\tilde{\pi}$ that has 2^k possible values and the corresponding Edmundson–Madansky probabilities. If, for example, the random variable $\tilde{\pi}_i$ has only one possible value, we know that $\phi(\xi)$ is linear in ξ_i. If $\tilde{\pi}_i$ has several possible values, its variance will tell us quite a bit about how the slope varies over the support. Since the random variables $\tilde{\xi}_i$ may have very different units, and the dual variables measure changes in the objective function *per unit* change in a right-hand side element, it seems reasonable to try to account for differences in units. A possible (heuristic) approach is to multiply the outcomes of $\tilde{\pi}_i$ by the length of the support of $\tilde{\xi}_i$, before calculating means and variances. (Assume, for example, that we are selling apples and bananas, and the demands are uncertain. For some reason, however, we are measuring bananas in tons and apples in kilograms. Now, if $\tilde{\pi}_1$ refers to bananas and $\tilde{\pi}_2$ to apples, would you see these products as equally important if $\tilde{\pi}_1$ and $\tilde{\pi}_2$ had the same variance?)

Computationally, this is easier than the approach based on α and β, because it requires that only 2^k inner products are made. The distribution of $\tilde{\pi}$ can be calculated as we visit the extreme points of the support (for finding the Edmundson–Madansky bound), and we never have to store the inner products.

All the above ideas are based on information from the Edmundson–Madansky upper bound, and therefore the solution of 2^k linear programs. As we have pointed out several times, for much more than 10 random variables, we are not able to find the Edmundson–Madansky upper bound. And if so, we shall not be able to use the partitioning ideas above either. Therefore we should have ideas of how to partition that do not depend on which upper bound we use. This does not imply, though, that the ideas that are to follow cannot be used with success for the Edmundson–Madansky upper bound as well.

One idea, which at first sight looks rather stupid, is the following: perform *all possible bi-partitions* (i.e. with k random variables, perform all k, one at a time) and pick the one that is best. By "best", we here mean with the smallest error in the next step. More formally, let

$$\mathcal{U}_i^\ell - \mathcal{L}_i^\ell$$

be the error on the "left" cell if we partition random variable i, and let

$$\mathcal{U}_i^r - \mathcal{L}_i^r$$

be the error on the "right" cell. If p_i^ℓ is the probability of being in the left cell, given that we are in the original cell, when we partition coordinate i, we chose to partition the random variable i that minimizes

$$(\mathcal{U}_i^\ell - \mathcal{L}_i^\ell)p_i^\ell + (\mathcal{U}_i^r - \mathcal{L}_i^r)(1 - p_i^\ell). \tag{5.2}$$

In other words, we perform all possible partitions, keep the best, and discard the remaining information. If the upper bound we are using is expensive in terms of CPU time, such an idea of "look-ahead" has two effects, which pull in different directions. On one hand, the information we are throwing away has cost a lot, and that seems like a waste. On the other hand, the very fact that the upper bound is costly makes it crucial to have few cells in the end. With a cheap (in terms of CPU time) upper bound, the approach seems more reasonable, since checking all possibilities is not particularly costly, but, even so, bad partitions will still double the work load locally. Numerical tests indicate that this approach is very good even with the Edmundson–Madansky upper bound, and the reason seems to be that it produces so few cells. Of course, without Edmundson–Madansky, we cannot calculate α, β and $\tilde{\pi}$, so if we do not like the look-ahead, we are in need of a new heuristic.

We have pointed out before that the piecewise linear upper bound can obtain the value $+\infty$. That happens if one of the problems (4.4) or (4.5) becomes infeasible. If that takes place, the random variable being treated when it happens is clearly a candidate for partitioning.

So far we have not really defined what constitutes a good partition. We shall return to that after the next subsection. But first let us look at an example illustrating the partitioning ideas.

Example 3.2 Consider the following function:

$$\begin{array}{rrrl}
\phi(\xi_1,\xi_2) = & \multicolumn{3}{l}{\max\{x+2y\}} \\
\text{s.t.} & -x + & y & \leq 6, \\
& 2x - & 3y & \leq 21, \\
& -3x + & 7y & \leq 49, \\
& x + & 12y & \leq 120, \\
& 2x + & 3y & \leq 45, \\
& x & & \leq \xi_1, \\
& & y & \leq \xi_2.
\end{array}$$

Let us assume that $\Xi_1 = [0, 20]$ and $\Xi_2 = [0, 10]$. For simplicity, we shall assume uniform and independent distributions. We do that because the form of the distribution is rather unimportant for the heuristics we are to explain.

The feasible set for the problem, except the upper bounds, is given in Figure 21. The circled numbers refer to the numbering of the inequalities. For all problems we have to solve (for varying values of ξ), it is reasonably easy to read the solution directly from the figure.

Since we are maximizing, the Jensen bound is an upper bound, and the Edmundson–Madansky bound a lower bound. We easily find the Jensen upper bound from

$$\phi(10, 5) = 20.$$

To find a lower bound, and also to calculate some of the information needed to use the heuristics, we first calculate ϕ at all extreme points of the support. Note that in what follows we view the upper bounds on the variables

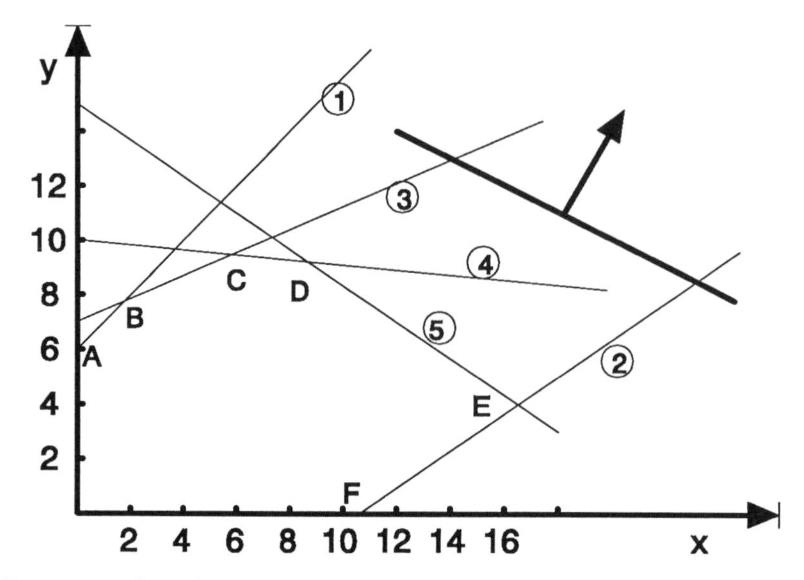

Figure 21 Set of feasible solutions for the example used to illustrate the piecewise linear upper bound.

as ordinary constraints. The results for the extreme-point calculations are summed up in Table 1.

The first idea we wish to test is based on comparing pairs of extreme points, to see how well the optimal dual solution (which is dual feasible for all right-hand sides) at one extreme-point works at a neighbouring extreme point. We use the indexing L and U to indicate Low and Up of the support.

LL:UL We first must test the optimal dual solution π^{LL} together with the right-hand side b^{UL}. We get

$$\begin{aligned}\alpha &= (\pi^{LL})^{\mathrm{T}} b^{UL} \\ &= (0,0,0,0,0,1,2)(6,21,49,120,45,20,0)^{\mathrm{T}} - \phi(U,L) \\ &= 20 - 10.5 = 9.5.\end{aligned}$$

We then do the opposite, to find

$$\begin{aligned}\beta &= (\pi^{UL})^{\mathrm{T}} b^{LL} \\ &= (0,\tfrac{1}{2},0,0,0,0,\tfrac{7}{2})(6,21,49,120,45,0,0)^{\mathrm{T}} - \phi(L,L) \\ &= 10.5 - 0 = 10.5.\end{aligned}$$

The minimum is therefore 9.5 for the pair LL:UL.

Table 1 Important characteristics of the solution of $\phi(\xi_1, \xi_2)$ at the four extreme points of the support.

(ξ_1, ξ_2)	x	y	ϕ	Optimal dual solution (π)
$(0, 0) = (L, L)$	0.000	0.000	0.000	$(0, 0, 0, 0, 0, 1, 2)$
$(20, 0) = (U, L)$	10.500	0.000	10.500	$(0, \frac{1}{2}, 0, 0, 0, 0, \frac{7}{2})$
$(0, 10) = (L, U)$	0.000	6.000	12.000	$(2, 0, 0, 0, 0, 3, 0)$
$(20, 10) = (U, U)$	8.571	9.286	27.143	$(0, 0, 0, 0.0476, 0.476, 0, 0)$

LL:LU Following a similar logic, we get the following:

$$\begin{aligned}\alpha &= (\pi^{LL})^{\mathrm{T}} b^{LU} \\ &= (0, 0, 0, 0, 0, 1, 2)(6, 21, 49, 120, 45, 0, 10)^{\mathrm{T}} - \phi(L, U) \\ &= 20 - 12 = 8, \\ \beta &= (\pi^{LU})^{\mathrm{T}} b^{LL} \\ &= (2, 0, 0, 0, 0, 3, 0)(6, 21, 49, 120, 45, 0, 0)^{\mathrm{T}} - \phi(L, L) \\ &= 12 - 0 = 12.\end{aligned}$$

The minimal value for the pair LL:LU is therefore 8.

LU:UU For this pair we get the following:

$$\begin{aligned}\alpha &= (\pi^{UU})^{\mathrm{T}} b^{LU} \\ &= (0, 0, 0, 0.0476, 0.476, 0, 0)(6, 21, 49, 120, 45, 0, 10)^{\mathrm{T}} - \phi(L, U) \\ &= 27.143 - 12 = 15.143 \\ \beta &= (\pi^{LU})^{\mathrm{T}} b^{UU} \\ &= (2, 0, 0, 0, 0, 3, 0)(6, 21, 49, 120, 45, 20, 10)^{\mathrm{T}} - \phi(U, U) \\ &= 72 - 27.143 = 44.857.\end{aligned}$$

The minimal value for the pair LU:UU is therefore 15.143.

UL:UU For the final pair the results are given by

$$\begin{aligned}\alpha &= (\pi^{UU})^{\mathrm{T}} b^{UL} \\ &= (0, 0, 0, 0.0476, 0.476, 0, 0)(6, 21, 49, 120, 45, 20, 0)^{\mathrm{T}} - \phi(U, L) \\ &= 27.143 - 10.5 = 16.643, \\ \beta &= (\pi^{UL})^{\mathrm{T}} b^{UU} \\ &= (0, \tfrac{1}{2}, 0, 0, 0, 0, \tfrac{7}{2})(6, 21, 49, 120, 45, 20, 10)^{\mathrm{T}} - \phi(U, U) \\ &= 46.5 - 27.143 = 18.357.\end{aligned}$$

The minimal value for the pair UL:UU is therefore 16.643.

If we were to pick the pair with the largest minimum of α and β, we should pick the pair UL:UU, over which it is ξ_2 that varies. In such a case we have tried to find that part of the function that is the most nonlinear. When we look at Figure 21, we see that as ξ_2 increases (with $\xi_1 = 20$), the optimal solution moves from F to E and then to D, where it stays when ξ_2 comes above the y coordinate in D. It is perhaps not so surprising that this is the most serious nonlinearity in ϕ.

If we try to find the random variable with the highest average nonlinearity, by summing the errors over those pairs for which the given random variable varies, we find that for $\tilde{\xi}_1$ the sum is $9.5 + 15.143 = 24.643$, and for $\tilde{\xi}_2$ it is $8 + 16.643$, which also equals 24.643. In other words, we have no conclusion.

The next approach we suggested was to look at the dual variables as in (5.1). The right-hand side structure is very simple in our example, so it is easy to find the connections. We define two random variables: $\tilde{\pi}_1$ for the row constraining x, and $\tilde{\pi}_2$ for the row constraining y. With the simple kind of uniform distributions we have assumed, each of the four values for $\tilde{\pi}_1$ and $\tilde{\pi}_2$ will have probability 0.25. Using Table 1, we see that the possible values for $\tilde{\pi}_1$ are 0, 1 and 3 (with 0 appearing twice), while for $\tilde{\pi}_2$ they are 0, 2 and 3.5 (also with 0 appearing twice). There are different ideas we can follow.

1. We can find out how the dual variables vary between the extreme points. The largest individual change is that $\tilde{\pi}_2$ falls from 3.5 to 0 as we go from UL to UU. This should again confirm that $\tilde{\xi}_2$ is a candidate for partitioning.
2. We can calculate $E\tilde{\pi} = (1, \frac{11}{8})$, and the individual variances to 1.5 and 2.17. If we choose based on variance, we pick $\tilde{\xi}_2$.
3. We also argued earlier that the size of the support was of some importance. A way of accommodating that is to multiply all outcomes with the length of the support. (That way, all dual variables are, in a sense, a measure of change *per total support.*) That should make the dual variables comparable. The calculations are left to the reader. We now end up with $\tilde{\pi}_1$ having the largest variance. (And if we now look at the biggest change in dual variable over pairs of neighboring extreme points, $\tilde{\xi}_1$ will be the one to partition.)

No conclusions should be made based on these numbers in terms of what is a good heuristic. We have presented these numbers to illustrate the computations and to indicate how it is possible to make arguments about partitioning. Before we conclude, let us consider the "look-ahead" strategy (5.2). In this case there are two possibilities: either we split at $\xi_1 = 10$ or we split at $\xi_2 = 5$. If we check what we need to compute in this case, we will find that some calculations are required in addition to those in Table 1, and $\phi(E\tilde{\xi}) = \phi(10, 5) = 20$, which we have already found. The additional numbers are presented in Table 2.

Table 2 Function values needed for the "look-ahead"strategy.

(5,5)	15
(15,5)	25
(10,10)	27.143
(20,5)	25

(10,7.5)	25
(10,2.5)	15
(0,5)	10
(10,0)	10

Based on this, we can find the total error after splitting to be about 4.5 for $\tilde{\xi}_1$ and 5 for $\tilde{\xi}_2$. Therefore, based on "look-ahead", we should chose $\tilde{\xi}_1$, since that reduces the error the most.

□

3.5.2 Using the L-shaped Method within Approximation Schemes

We have now investigated how to bound $\mathcal{Q}(x)$ for a fixed x. We have done that by combining upper and lower bounding procedures with partitioning of the support of $\tilde{\xi}$. On the other hand, we have earlier discussed (exact) solution procedures, such as the L-shaped decomposition method (Section 3.2) and the scenario aggregation (Section 2.6). These methods take a full event/scenario tree as input and solve this (at least in principle) to optimality. We shall now see how these methods can be combined.

The starting point is a set-up like Figure 18. We set up an initial partition of the support, possibly containing only one cell. We then find all conditional expectations (in the example there are five), and give each of them a probability equal to that of being in their cell, and we view this as our "true" distribution. The L-shaped method is then applied. Let ξ^i denote the conditional expectation of $\tilde{\xi}$, given that $\tilde{\xi}$ is contained in the ith cell. Then the partition gives us the support $\{\xi^1, \ldots \xi^\ell\}$. We then solve

$$\left.\begin{array}{rl} \min c^{\mathrm{T}}x + \mathcal{L}(x) & \\ \text{s.t. } Ax = b, & \\ x \geq 0, & \end{array}\right\} \quad (5.3)$$

where

$$\mathcal{L}(x) = \sum_{j=1}^{\ell} p^j Q(x, \xi^j),$$

with p^j being the probability of being in cell j. Let $\hat{x}$ be the optimal solution to (5.3). Clearly if $\overline{x}$ is the optimal solution to the original problem then

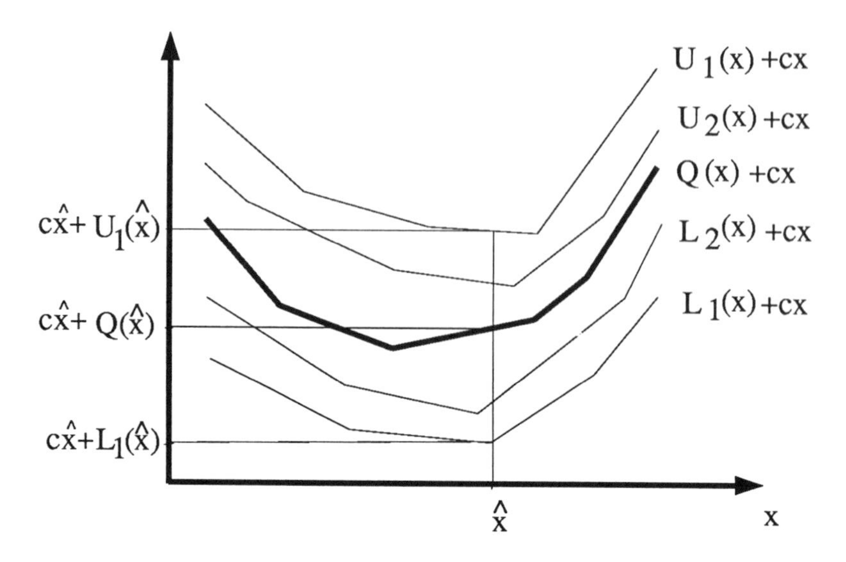

Figure 22 Example illustrating the use of bounds in the L-shaped decomposition method. An initial partition corresponds to the lower bounding function $\mathcal{L}_1(x)$ and the upper bounding function $\mathcal{U}_1(x)$. For all x we have $\mathcal{L}_1(x) \leq \mathcal{Q}(x) \leq \mathcal{U}_1(x)$. We minimize $cx + \mathcal{L}_1(x)$ to obtain $\hat{x}$. We find the error $\mathcal{U}_1(\hat{x}) - \mathcal{L}_1(\hat{x})$, and we decide to refine the partition. This will cause $\mathcal{L}_1$ to be replaced by $\mathcal{L}_2$ and $\mathcal{U}_1$ by $\mathcal{U}_2$. Then the process can be repeated.

$$c^{\mathrm{T}}\hat{x} + \mathcal{L}(\hat{x}) \leq c^{\mathrm{T}}\overline{x} + \mathcal{L}(\overline{x}) \leq c^{\mathrm{T}}\overline{x} + \mathcal{Q}(\overline{x}),$$

so that the optimal value we found by solving (5.3) is really a lower bound on $\min c^{\mathrm{T}}x + \mathcal{Q}(x)$. The first inequality follows from the observation that $\hat{x}$ minimizes $c^{\mathrm{T}}x + \mathcal{L}(x)$. The second inequality holds because $\mathcal{L}(x) \leq \mathcal{Q}(x)$ for all x (Jensen's inequality). Next, we use some method to calculate $\mathcal{U}(\hat{x})$, for example the Edmundson–Madansky or piecewise linear upper bound. Note that

$$c^{\mathrm{T}}\overline{x} + \mathcal{Q}(\overline{x}) \leq c^{\mathrm{T}}\hat{x} + \mathcal{Q}(\hat{x}) \leq c^{\mathrm{T}}\hat{x} + \mathcal{U}(\hat{x}),$$

so $c^{\mathrm{T}}\hat{x}+\mathcal{U}(\hat{x})$ is indeed an upper bound on $c^{\mathrm{T}}\overline{x}+\mathcal{Q}(\overline{x})$. Here the first inequality holds because $\overline{x}$ minimizes $c^{\mathrm{T}}x + \mathcal{Q}(x)$, and the second because, for all x, $\mathcal{Q}(x) \leq \mathcal{U}(x)$.

We then have a solution $\hat{x}$ and an error $\mathcal{U}(\hat{x}) - \mathcal{L}(\hat{x})$. If we are not satisfied with the precision, we refine the partition of the support, and repeat the use of the L-shaped method. It is worth noting that the old optimality cuts generated in the L-shaped method are still valid, but generally not tight. The reason is that, with more cells, and hence a larger ℓ, the function $\mathcal{L}(x)$ is now closer to $\mathcal{Q}(x)$. Feasibility cuts are still valid and tight. Figure 22 illustrates how the approximating functions $\mathcal{L}(x)$ and $\mathcal{U}(x)$ change as the partition is refined.

In total, this gives us the procedure in Figure 23. The **procedure** refine(Ξ) will not be detailed, since there are so many options. We refer to our earlier discussion of the subject in Section 3.5.1. Note that, for simplicity, we have assumed that, after a partitioning, the procedure starts all over again in the **repeat** loop. That is of course not needed, since we already have checked the present $\hat{x}$ for feasibility. If we replace the set $\mathcal{A}$ by Ξ in the call to **procedure** feascut, the **procedure** Bounding L-shaped must stay as it is. In many cases this may be a useful change, since $\mathcal{A}$ might be very large. (In this case old feasibility cuts might no longer be tight.)

3.5.3 What is a Good Partition?

We have now seen partitioning used in two different settings. In the first we just wanted to bound a one-stage stochastic program, while in the second we used it in combination with the L-shaped decomposition method. The major difference is that in the latter case we solve a two-stage stochastic program between each time we partition. Therefore, in contrast to the one-stage setting, the same partition (more and more refined) is used over and over again.

In the two-stage setting a new question arises. How many partitions should we make between each new call to the L-shaped decomposition method? If we make only one, the overall CPU time will probably be very large because a new LP (only slightly changed from last time) must be solved each time we make a new cell. On the other hand, if we make many partitions per call to L-shaped, we might partition extensively in an area where it later turns out that partitioning is not needed (remember that x enters the right-hand side of the second-stage constraints, moving the set of possible right-hand sides around). We must therefore strike a balance between getting enough cells and not getting them in the wrong places.

This brings us to the question of what is a good partitioning strategy. It should clearly be one that minimizes CPU time for solving the problem at hand. Tests indicate that for the one-stage setting, using the idea of the variance of the (random) dual variables on page 183, is a good idea. It creates quite a number of cells, but because it is cheap (given that we already use the Edmundson–Madansky upper bound) it is quite good overall. But, in the setting of the L-shaped decomposition method, this large number of cells become something of a problem. We have to carry them along from iteration to iteration, repeatedly finding upper and lower bounds on each of them. Here it is much more important to have few cells for a given error level. And that is best achieved by looking ahead using (5.2). Our general advice is therefore

```
procedure Bounding L-shaped(ε₁, ε₂:real);
begin
    K := 0, L := 0; Ξ̂ := {Eξ̃};
    θ̂ := −∞, LP(A, b, c, x̂, feasible);
    stop := not (feasible);
    while not (stop) do begin
        feascut(𝒜, x̂,newcut);
        if not (newcut) then begin
            Find ℒ(x̂);
            newcut := (ℒ(x̂) − θ̂ > ε₁);
            if newcut then begin
                (* Create an optimality cut—see page 155 *)
                L := L + 1;
                Construct the cut −β_L^T x + θ ≥ α_L;
            end;
        end;
        if newcut then begin
            master(K, L, x̂, θ̂,feasible);
            stop := not (feasible);
        end
        else begin
            Find 𝒰(x̂);
            stop := (𝒰(x̂) − ℒ(x̂) ≤ ε₂);
            if not (stop) then refine(Ξ̂);
        end;
    end;
end;
```

Figure 23 The L-shaped decomposition algorithm in a setting of approximations and bounds. The procedures that we refer to start on page 156, and the set $\mathcal{A}$ was defined on page 150.

that in the setting of two (or more) stages one should seek a strategy that minimizes the final number of cells, and that it is worthwhile to pay quite a lot per iteration to achieve this goal.

3.6 Simple Recourse

Let us consider the particular simple recourse problem

$$\min\{c^{\mathrm{T}}x + E_{\tilde{\xi}}Q(x,\tilde{\xi}) \mid Ax = b,\ x \geq 0\}, \tag{6.1}$$

where

$$Q(x,\xi) = \min\{q^{+\mathrm{T}}y^{+} + q^{-\mathrm{T}}y^{-} \mid y^{+} - y^{-} = \xi - Tx,\ y^{+} \geq 0, y^{-} \geq 0\}.$$

Hence we assume

$$\begin{aligned} W &= (I,-I),\\ T(\xi) &\equiv T \text{ (constant)},\\ h(\xi) &\equiv \xi, \end{aligned}$$

and in addition

$$\overline{q} = q^{+} + q^{-} \geq 0.$$

In other words, we consider the case where only the right-hand side is random, and we shall see that in this case, using our former presentation $h(\xi) = h^0 + \sum_i h^i\xi_i$, we only need to know the marginal distributions of the components $h_j(\xi)$ of $h(\xi)$. However, stochastic dependence or independence of these components does not matter at all. This justifies the above setting $h(\xi) \equiv \xi$.

By linear programming duality, we have for the recourse function

$$\begin{aligned} &Q(x,\xi)\\ &\quad= \min\{q^{+\mathrm{T}}y^{+} + q^{-\mathrm{T}}y^{-} \mid y^{+} - y^{-} = \xi - Tx,\ y^{+} \geq 0, y^{-} \geq 0\}\\ &\quad= \max\{(\xi - Tx)^{\mathrm{T}}\pi \mid -q^{-} \leq \pi \leq q^{+}\}. \end{aligned} \tag{6.2}$$

Observe that our assumption $\overline{q} \geq 0$ is equivalent to solvability of the second-stage problem. Defining

$$\chi := Tx,$$

the dual solution $\pi^\star$ of (6.2) is obvious:

$$\pi_i^\star = \begin{cases} q_i^{+} & \text{if } \xi_i - \chi_i > 0,\\ -q_i^{-} & \text{if } \xi_i - \chi_i \leq 0. \end{cases}$$

Hence, with

$$\hat{Q}_i(\chi_i,\xi_i) = \begin{cases} (\xi_i - \chi_i)q_i^{+} & \text{if } \chi_i < \xi_i,\\ -(\xi_i - \chi_i)q_i^{-} & \text{if } \chi_i \geq \xi_i, \end{cases}$$

we have

$$Q(x,\xi) = \sum_i \hat{Q}_i(\chi_i,\xi_i) \text{ with } \chi = Tx.$$

The expected recourse follows immediately:

$$\begin{aligned} E_{\tilde{\xi}}Q(x,\tilde{\xi}) &= \int_{\Xi} Q(x,\xi)P_{\tilde{\xi}}(d\xi) \\ &= \sum_i \int_{\Xi} \hat{Q}_i(\chi_i,\xi_i)P_{\tilde{\xi}}(d\xi) \\ &= \sum_i \left\{ q_i^+ \int_{\xi_i>\chi_i} (\xi_i-\chi_i)P_{\tilde{\xi}}(d\xi) - q_i^- \int_{\xi_i\le\chi_i} (\xi_i-\chi_i)P_{\tilde{\xi}}(d\xi) \right\}. \end{aligned}$$

The last expression shows that knowledge of the marginal distributions of the $\tilde{\xi}_i$ is sufficient to evaluate the expected recourse. Moreover, $E_{\tilde{\xi}}Q(x,\tilde{\xi})$ is a so-called *separable* function in $(\chi_1,\cdots,\chi_{m_1})$, i.e. $E_{\tilde{\xi}}Q(x,\tilde{\xi}) = \sum_{i=1}^{m_1} \mathcal{Q}_i(\chi_i)$, where, owing to $q^+ + q^- = \overline{q}$,

$$\left.\begin{aligned} \mathcal{Q}_i(\chi_i) &= q_i^+ \textstyle\int_{\xi_i>\chi_i}(\xi_i-\chi_i)P_{\tilde{\xi}}(d\xi) - q_i^- \int_{\xi_i\le\chi_i}(\xi_i-\chi_i)P_{\tilde{\xi}}(d\xi) \\ &= q_i^+ \textstyle\int_{\Xi}(\xi_i-\chi_i)P_{\tilde{\xi}}(d\xi) - (q_i^+ + q_i^-)\int_{\xi_i\le\chi_i}(\xi_i-\chi_i)P_{\tilde{\xi}}(d\xi) \\ &= q_i^+\overline{\xi}_i - q_i^+\chi_i - \overline{q}_i \textstyle\int_{\xi_i\le\chi_i}(\xi_i-\chi_i)P_{\tilde{\xi}}(d\xi) \end{aligned}\right\} \quad (6.3)$$

with $\overline{\xi}_i = E_{\tilde{\xi}}\tilde{\xi}_i$.

The reformulation (6.3) reveals the shape of the functions $\mathcal{Q}_i(\chi_i)$. Assume that Ξ is bounded such that $\alpha_i < \xi_i \le \beta_i \ \forall i, \xi \in \Xi$. Then we have

$$\mathcal{Q}_i(\chi_i) = \begin{cases} q_i^+\overline{\xi}_i - q_i^+\chi_i & \text{if } \chi_i \le \alpha_i, \\ q_i^+\overline{\xi}_i - q_i^+\chi_i - \overline{q}_i \displaystyle\int_{\xi_i\le\chi_i}(\xi_i-\chi_i)P_{\tilde{\xi}}(d\xi) & \text{if } \alpha_i < \chi_i < \beta_i, \\ -q_i^-\overline{\xi}_i + q_i^-\chi_i & \text{if } \chi_i \ge \beta_i, \end{cases} \quad (6.4)$$

showing that for $\chi_i < \alpha_i$ and $\chi_i > \beta_i$ the functions $\mathcal{Q}_i(\chi_i)$ are linear (see Figure 24). In particular, we have

$$\mathcal{Q}_i(\chi_i) = \hat{Q}_i(\chi_i,\overline{\xi}_i) \text{ if } \chi_i \le \alpha_i \text{ or } \chi_i \ge \beta_i. \quad (6.5)$$

Following the approximation scheme described in Section 3.5.1, the relation (6.5) allows us to determine an error bound without computing the E–M bound.[3] To see this, consider any fixed $\hat{\chi}_i$. If $\hat{\chi}_i \le \alpha_i$ or $\hat{\chi}_i \ge \beta_i$ then, by (6.5),

$$\mathcal{Q}_i(\hat{\chi}_i) = \hat{Q}_i(\hat{\chi}_i,\overline{\xi}_i).$$

[3] By "E–M" we mean the Edmundson–Madansky bound described in Section 3.4.2.

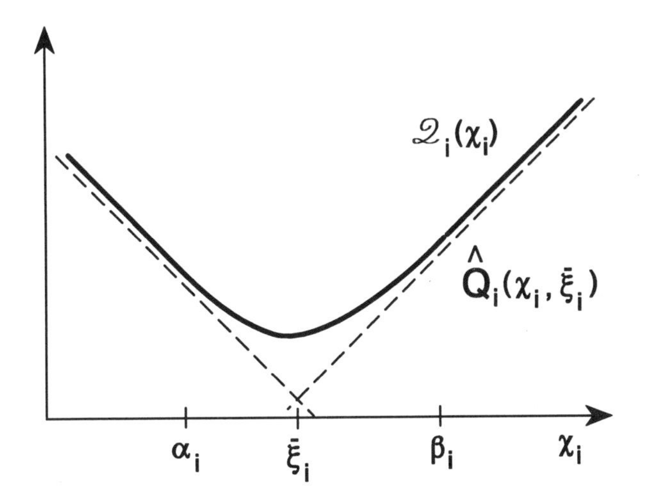

Figure 24 Simple recourse: supporting $\mathcal{Q}_i(\chi_i)$ by $\hat{Q}_i(\chi_i, \overline{\xi}_i)$.

If, on the other hand, $\alpha_i < \hat{\chi}_i < \beta_i$, we partition the interval $(\alpha_i, \beta_i]$ into the two subintervals $(\alpha_i, \hat{\chi}_i]$ and $(\hat{\chi}_i, \beta_i]$ with the conditional expectations

$$\overline{\xi}_i^1 = E_{\tilde{\xi}}(\tilde{\xi}_i \mid \xi_i \in (\alpha_i, \hat{\chi}_i]), \quad \overline{\xi}_i^2 = E_{\tilde{\xi}}(\tilde{\xi}_i \mid \xi_i \in (\hat{\chi}_i, \beta_i]).$$

Obviously relation (6.5) also applies analogously to the conditional expectations

$$\mathcal{Q}_i^1(\hat{\chi}_i) = E_{\tilde{\xi}}(\hat{Q}_i(\hat{\chi}_i, \tilde{\xi}_i) \mid \xi_i \in (\alpha_i, \hat{\chi}_i])$$

and

$$\mathcal{Q}_i^2(\hat{\chi}_i) = E_{\tilde{\xi}}(\hat{Q}_i(\hat{\chi}_i, \tilde{\xi}_i) \mid \xi_i \in (\hat{\chi}_i, \beta_i]).$$

Therefore

$$\mathcal{Q}_i^1(\hat{\chi}_i) = \hat{Q}_i(\hat{\chi}_i, \overline{\xi}_i^1), \quad \mathcal{Q}_i^2(\hat{\chi}_i) = \hat{Q}_i(\hat{\chi}_i, \overline{\xi}_i^2),$$

and, with $p_i^1 = P(\xi_i \in (\alpha_i, \hat{\chi}_i])$ and $p_i^2 = P(\xi_i \in (\hat{\chi}_i, \beta_i])$,

$$\begin{aligned}\mathcal{Q}_i(\hat{\chi}_i) &= p_i^1 \mathcal{Q}_i^1(\hat{\chi}_i) + p_i^2 \mathcal{Q}_i^2(\hat{\chi}_i) \\ &= p_i^1 \hat{Q}_i(\hat{\chi}_i, \overline{\xi}_i^1) + p_i^2 \hat{Q}_i(\hat{\chi}_i, \overline{\xi}_i^2).\end{aligned}$$

Hence, instead of using the E–M upper bound, we can easily determine the exact value $\mathcal{Q}_i(\hat{\chi}_i)$. With $\hat{Q}_i(\chi_i, \overline{\xi}_i^1, \overline{\xi}_i^2) := p_i^1 \hat{Q}_i(\chi_i, \overline{\xi}_i^1) + p_i^2 \hat{Q}_i(\chi_i, \overline{\xi}_i^2)$, the resulting situation is demonstrated in Figure 25.

Assume now that for a partition of the intervals $(\alpha_i, \beta_i]$ into subintervals $I_{i\nu} := (\delta_{i\nu}, \delta_{i\nu+1}]$, $\nu = 0, \cdots, N_i - 1$, with $\alpha_i = \delta_{i0} < \delta_{i1} < \cdots < \delta_{iN_i} = \beta_i$,

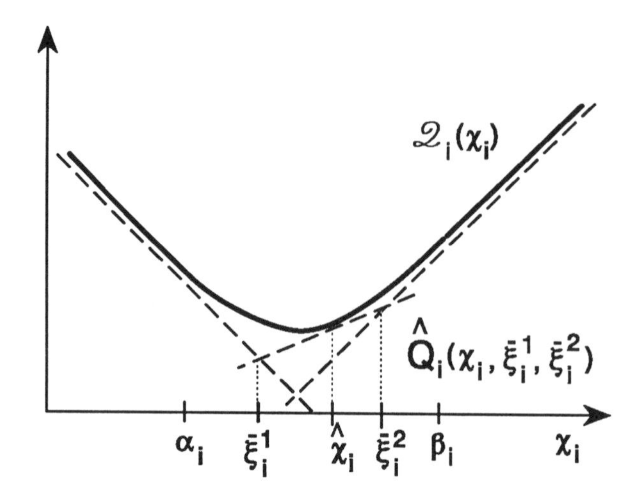

Figure 25 Simple recourse: supporting $\mathcal{Q}_i(\chi_i)$ by $\hat{Q}_i(\chi_i, \overline{\xi}_i^1, \overline{\xi}_i^2)$.

we have minimized the Jensen lower bound (see Section 3.4.1), letting $p_{i\nu} = P(\xi_i \in I_{i\nu})$, $\overline{\xi}_{i\nu} = E_{\tilde{\xi}}(\tilde{\xi}_i \mid \xi_i \in I_{i\nu})$:

$$\begin{aligned}
\min\nolimits_{x,\chi} & \left[c^{\mathrm{T}}x + \sum_{i=1}^{k} \sum_{\nu=0}^{N_i-1} p_{i\nu} \hat{Q}_i(\chi_i, \overline{\xi}_{i\nu})\right] \\
\text{s.t. } Ax & = b, \\
Tx - \chi & = 0, \\
x & \geq 0,
\end{aligned}$$

yielding the solution $\hat{x}$ and $\hat{\chi} = T\hat{x}$. Obviously relation (6.5) holds for conditional expectations $\mathcal{Q}_{i\nu}(\hat{\chi}_i)$ (with respect to $I_{i\nu}$) as well. Then for each component of $\hat{\chi}$ there are three possibilities.

(a) If $\hat{\chi}_i \leq \alpha_i$, then

$$\hat{Q}_i(\hat{\chi}_i, \overline{\xi}_{i\nu}) = \mathcal{Q}_{i\nu}(\hat{\chi}_i) = E_{\tilde{\xi}}(\hat{Q}_i(\hat{\chi}_i, \tilde{\xi}_i) \mid \xi_i \in I_{i\nu}), \ \nu = 0, \cdots, N_i - 1,$$

and hence

$$\mathcal{Q}_i(\hat{\chi}_i) = \sum_{\nu=0}^{N_i-1} p_{i\nu} \hat{Q}_i(\hat{\chi}_i, \overline{\xi}_{i\nu}),$$

i.e. there is no error with respect to this component.

(b) If $\hat{\chi}_i \geq \beta_i$, then it again follows from (6.5) that

$$\mathcal{Q}_i(\hat{\chi}_i) = \sum_{\nu=0}^{N_i-1} p_{i\nu} \hat{Q}_i(\hat{\chi}_i, \overline{\xi}_{i\nu}).$$

(c) If $\hat{\chi}_i \in I_{i\mu}$ for exactly one μ, with $0 \le \mu < N_i$, then there are two cases. First, if $\delta_{i\mu} < \hat{\chi}_i < \delta_{i\mu+1}$, partition $I_{i\mu} = (\delta_{i\mu}, \delta_{i\mu+1}]$ into

$$J^1_{i\mu} = (\delta_{i\mu}, \hat{\chi}_i] \text{ and } J^2_{i\mu} = (\hat{\chi}_i, \delta_{i\mu+1}].$$

Now, again owing to (6.5), it follows that

$$\mathcal{Q}_i(\hat{\chi}_i) = \sum_{\nu \neq \mu} p_{i\nu} \hat{Q}_i(\hat{\chi}_i, \overline{\xi}_{i\nu}) + \sum_{\rho=1}^{2} p^\rho_{i\mu} \hat{Q}_i(\hat{\chi}_i, \overline{\xi}^\rho_{i\mu}),$$

where

$$p^\rho_{i\mu} = P(\xi_i \in J^\rho_{i\mu}),\ \ \overline{\xi}^\rho_{i\mu} = E_{\tilde{\xi}}(\tilde{\xi}_i \mid \xi_i \in J^\rho_{i\mu}),\ \rho = 1, 2.$$

If, on the other hand, $\hat{\chi}_i = \delta_{i\mu+1}$, we again have

$$\mathcal{Q}_i(\hat{\chi}_i) = \sum_{\nu=0}^{N_i-1} p_{i\nu} \hat{Q}_i(\hat{\chi}_i, \overline{\xi}_{i\nu}).$$

In conclusion, having determined the minimal point $\hat{\chi}$ for the Jensen lower bound, we immediately get the exact expected recourse at this point and decide whether for all components the relative error fits into a prescribed tolerance, or in which component the refinement (partitioning the subinterval containing $\hat{\chi}_i$ by deviding it exactly at $\hat{\chi}_i$) seems appropriate for a further improvement of the approximate solution of (6.1). Many empirical tests have shown this approach to be very efficient. In particular, for this special problem type higher dimensions of $\tilde{\xi}$ do not cause severe computational difficulties, as they did for general stochastic programs with recourse, as discussed in Section 3.5 .

3.7 Integer First Stage

This book deals almost exclusively with convex problems. The only exception is this section, where we discuss, very briefly, some aspects of integer programming. The main reason for doing so is that some solution procedures for integer programming fit very well with some decomposition procedures for (continuous) stochastic programming. Because of that we can achieve two goals: we can explain some connections between stochastic and integer programming, and we can combine the two subject areas. This allows us to arrive at a method for stochastic integer programming. Note that talking about stochastic and integer programming as two distinct areas is really meaningless, since stochastic programs can contain integrality constraints, and integer programs can be stochastic. But we still do it, with some hesitation, since the splitting is fairly common within the mathematical programming community.

To get started, let us first formulate a deterministic integer programming problem in a very simple format, and then outline a common solution procedure, namely *branch-and-bound*. An integer program can be formulated as

$$\left.\begin{aligned} &\min c^{\mathrm{T}}x \\ &\text{s.t. } Ax = b \\ &\qquad x_i \in \{a_i, a_i+1, \ldots, b_i-1, b_i\} \text{ for all } i, \end{aligned}\right\} \tag{7.1}$$

where $\{a_i, a_i+1, \ldots, b_i-1, b_i\}$ is the set of all integers from a_i to b_i.

The branch-and-bound procedure is based on replacing $x_i \in \{a_i, a_i + 1, \ldots, b_i - 1, b_i\}$ by $a_i \le x_i \le b_i$ for all i, and solving the corresponding *relaxed* linear program to obtain $\hat{x}$. If $\hat{x}$ happens to be integral, we are done, since integrality is satisfied without being enforced. If $\hat{x}$ is not integral, we have obtained a lower bound $\underline{z} = c^{\mathrm{T}}\hat{x}$ on the true optimal objective, since dropping constraints in a minimization problem yields a lower bound.

To continue from here, we pick one variable x_j, called the *branching variable*, and one integer d_j. Normally d_j is chosen as the largest integer less than or equal to the value of x_j in the LP solution, and x_j is normally a variable that was nonintegral in the LP solution. We then replace our original problem (7.1) by two similar problems:

$$\left.\begin{aligned} &\left.\begin{aligned} &\min c^{\mathrm{T}}x \\ &\text{s.t. } Ax = b, \\ &\qquad x_i \in \{a_i, \ldots, b_i\} \text{ for all } i \ne j, \\ &\qquad x_j \in \{a_j, \ldots, d_j\}, \end{aligned}\right\} \\ &\text{and} \\ &\left.\begin{aligned} &\min c^{\mathrm{T}}x \\ &\text{s.t. } Ax = b, \\ &\qquad x_i \in \{a_i, \ldots, b_i\} \text{ for all } i \ne j, \\ &\qquad x_j \in \{d_j + 1, \ldots, b_j\}. \end{aligned}\right\} \end{aligned}\right\} \tag{7.2}$$

What we have done is to *branch*. We have replaced the original problem by two similar problems that each investigate their part of the solution space. The two problems are now put into a collection of *waiting nodes*. The term "waiting node" is used because the branching can be seen as building up a tree, where the original problem sits in the root and the new problems are stored in child nodes. Waiting nodes are then leaves in the tree, waiting to be analysed. Leaves can also be *fathomed* or *bounded*, as we shall see shortly.

We next continue to work with the problem in one of these waiting nodes.

We shall call this problem the *present* problem. When doing so, a number of different situations can occur.

1. The present problem may be infeasible, in which case it is simply dropped, or *fathomed.*
2. The present problem might turn out to have an *integral* optimal solution $\hat{x}$, in other words a solution that is truly feasible. If so, we compare $c^{\mathrm{T}}\hat{x}$ with the *best-so-far* objective value $\overline{z}$ (we initiate $\overline{z}$ at $+\infty$). If the new objective value is better, we keep $\hat{x}$ and update $\overline{z}$ so that $\overline{z} = c^{\mathrm{T}}\hat{x}$. We then fathom the present problem.
3. The present problem might have a nonintegral solution $\hat{x}$ with $c^{\mathrm{T}}\hat{x} \geq \overline{z}$. In this case the present problem cannot possibly contain an optimal integral solution, and it is therefore dropped, or *bounded.* (This is the process that gives half of the name of the method.)
4. The present problem has a solution $\hat{x}$ that does not satisfy any of the above criteria. If so, we branch as we did in (7.2), creating two child nodes. We then add them to the tree, making them waiting nodes.

An example of an intermediate stage for a branch-and-bound tree can be found in Figure 26. Three branchings have taken place, and we are left with two fathomed, one bounded and one waiting node. The next step will now be to branch on the waiting node.

Note that as branching proceeds, the interval over which we solve the continuous version must eventually contain only one point. Therefore, sooner or later, we come to a situation where the problem is either infeasible, or we are faced with an integral solution. We cannot go on branching forever. Hence the algorithm will eventually stop, either telling us that no integral solution exists, or giving us such a solution.

Much research in integer programming concerns how to pick the correct variable x_j for branching, how to pick the branching value d_j, how to *formulate* the problem so that branching becomes simpler, and how to obtain a good initial (integer) solution so as to have a $\overline{z} < \infty$ to start out with. To have a good integer programming algorithm, these subjects are crucial. We shall not, however, discuss those subjects here. Instead, we should like to draw attention to some analogies between the branch-and-bound algorithm for integer programs and the problem of bounding a stochastic (continuous) program.

- In the integer case we partition the solution space, and in the stochastic case the input data (support of random variables).
- In the integer case we must find a branching variable, and in the stochastic case a random variable for partitioning.

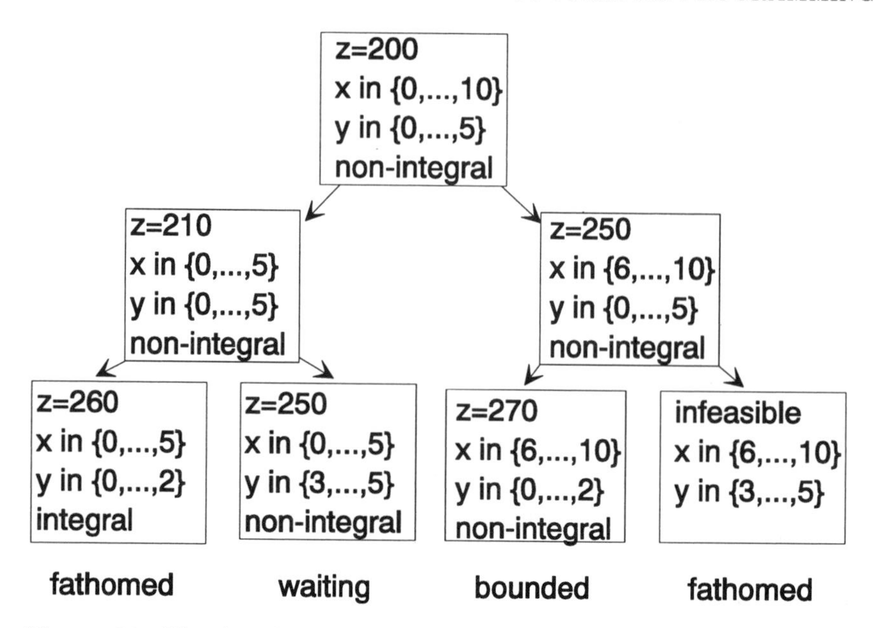

Figure 26 The situation after three branchings in a branch-and-bound tree. One waiting node is left.

- In the integer case we must find a value d_j of x_j (see (7.2)) for the branching, and in the stochastic case we must determine a point d_j in the support through which we want to partition.
- Both methods therefore operate with a situation as depicted in Figure 18, but in one case the rectangle is the solution space, while in the other it is the support of the random variables.
- Both problems can be seen as building up a tree. For integer programming we build a branch-and-bound tree. For stochastic programming we build a splitting tree. The branch-and-bound-tree in Figure 26 could have been a splitting tree as well. In that case we should store the error rather than the objective value.
- In the integer case we fathom a problem (corresponding to a cell in Figure 18, or a leaf in the tree) when it has nothing more to tell us, in the stochastic case we do this when the bounds (in the cell or leaf) are close enough.

From this, it should be obvious that anyone who understands the ins and outs of integer programming, will also have a lot to say about bounding stochastic programs. Of course there are differences, but they are smaller than one might think.

So far, what we have compared is really the problem of bounding the recourse function with the problem of solving an integer program by branch-and-bound. Next, let us consider the *cutting-plane* methods for integer programs, and compare them with methods like the L-shaped decomposition method for (continuous) stochastic programs. It must be noted that cutting-plane methods are hardly ever used in their pure form for solving integer programs. They are usually combined with other methods. For the sake of exposition, however, we shall biefly sketch some of the ideas.

When we solve the relaxed linear programming version of (7.1), we have difficulties because we have increased the solution space. However, all points that we have added are non-integral. In principle, it is possible to add extra constraints to the linear programming relaxation to cut off some of these noninteger solutions, namely those that are not convex combinations of feasible integral points. These cuts will normally be added in an iterative manner, very similarly to the way we added cuts in the L-shaped decomposition method. In fact, the L-shaped decomposition method is known as Benders' decomposition in other areas of mathematical programming, and its original goal was to solve (mixed) integer programming problems. However, it was not cast in the way we are presenting cuts below.

So, in all its simplicity, a cutting-plane method will run through two major steps. The first is to solve a relaxed linear program; the second is to evaluate the solution, and if it is not integral, add cuts that cut away nonintegral points (including the present solution). These cuts are then added to the relaxed linear program, and the cycle is repeated. Cuts can be of different types. Some come from straightforward arithmetic operations based on the LP solution and the LP constraints. These are not necessarily very tight. Others are based on structure. For a growing number of problems, knowledge about some or all *facets* of the (integer) solution space is becoming available. By a facet in this case, we understand the following. The solution space of the relaxed linear program contains all integral feasible points, and none extra. If we add a *minimal* number of new inequalities, such that no integral points are cut off, and such that all extreme points of the new feasible set are integers, then the intersection between a hyperplane representing such an inequality and the new set of feasible solutions is called a facet. Facets are sometimes added as they are found to be violated, and sometimes before the procedure is started.

How does this relate to the L-shaped decomposition procedure? Let us be a bit formal. If all costs in a recourse problem are zero, and we choose to use the L-shaped decomposition method, there will be no optimality cuts, only feasibility cuts. Such a stochastic linear program could be written as

$$\left.\begin{array}{ll}\min c^{\mathrm{T}}x & \\ \text{s.t.} & Ax = b, \\ & x \geq 0, \\ & Wy(\xi) = h(\xi) - T(\xi)x,\ y(\xi) \geq 0.\end{array}\right\} \tag{7.3}$$

To use the L-shaped method to solve (7.3), we should begin solving the problem

$$\begin{array}{ll}\min c^{\mathrm{T}}x & \\ \text{s.t.} & Ax = b, \\ & x \geq 0,\end{array}$$

i.e. (7.3) without the last set of constraints added. Then, if the resulting $\hat{x}$ makes the last set of constraints in (7.3) feasible for all ξ, we are done. If not, an implied feasibility cut is added.

An integer program, on the other hand, could be written as

$$\left.\begin{array}{ll}\min c^{\mathrm{T}}x & \\ \text{s.t.} & Ax = b, \\ & x_i \in \{a_i, \ldots, b_i\} \text{ for all } x_i.\end{array}\right\} \tag{7.4}$$

A cutting-plane procedure for (7.4) will solve the problem with the constraints $a \leq x \leq b$ so that the integrality requirement is relaxed. Then, if the resulting $\hat{x}$ is integral in all its elements, we are done. If not, an integrality cut is added. This cut will, if possible, be a facet of the solution space with all extreme points integer.

By now, realizing that integrality cuts are also feasibility cuts, the connection should be clear. Integrality cuts in integer programming are just a special type of feasibility cuts.

For the bounding version of the L-shaped decomposition method we combined bounding (with partitioning of the support) with cuts. In the same way, we can combine branching and cuts in the branch-and-cut algorithm for integer programs (still deterministic). The idea is fairly simple (but requires a lot of details to be efficient). For all waiting nodes, before or after we have solved the relaxed LP, we add an appropriate number of cuts, before we (re)solve the LP. How many cuts we add will often depend on how well we know the facets of the (integer) solution space. This new LP will have a smaller (continuous) solution space, and is therefore likely to give a better result—either in terms of a nonintegral optimal solution with a higher objective value (increasing the probability of bounding), or in terms of an integer solution.

So, finally, we have reached the ultimate question. How can all of this be used to solve integer stochastic programs? Given the simplification that we have integrality only in the first-stage problem, the procedure is given in Figure 27. In the procedure we operate with a set of waiting nodes $\mathcal{P}$. These are nodes in the cut-and-branch tree that are not yet fathomed or bounded.

The **procedure** feascut was presented earlier in Figure 9, whereas the new **procedure** intcut is outlined in Figure 28. Let us try to compare the L-shaped integer programming method with the continuous one presented in Figure 10.

3.7.1 Initialization

In the continuous case we started by assuming the existence of an $\hat{x}$, feasible in the first stage. It can be found, for example, by solving the expected value problem. This is not how we start in the integer case. The reason is partly that finding a feasible solutions is more complicated in that setting. On the other hand, it might be argued that if we hope to solve the integer stochastic problem, we should be able to solve the expected value problem (or at least find a feasible solution to the master problem), thereby being able to start out with a feasible solution (and a $\overline{z}$ better than ∞). But, even in this case, we shall not normally be calling **procedure** master with a feasible solution at hand. If we have just created a feasibility cut, the present $\hat{x}$ is not feasible. Therefore the difference in initialization is natural. This also affects the generation of feasibility cuts.

3.7.2 Feasibility Cuts

Both approaches operate with feasibility cuts. In the continuous case these are all implied constraints, needed to make the second-stage problem feasible for all possible realizations of the random variables. For the integer case, we still use these, and we add any cuts that are commonly used in branch-and-cut procedures in integer programming, preferably facets of the solution space with integral extreme points. To reflect all possible kinds of such cuts (some concerning second-stage feasibility, some integrality), we use a call to **procedure** feascut *plus* the new **procedure** intcut. Typically, implied constraints are based on an $\hat{x}$ that is nonintegral, and therefore infeasible. In the end, though, integrality will be there, based on the branching part of the algorithm, and then the cuts will indeed be based on a feasible (integral) solution.

3.7.3 Optimality Cuts

Creation of optimality cuts are the same in both cases, since in the integer case we create such cuts only for feasible (integer) solutions.

3.7.4 Stopping Criteria

The stopping criteria are basically the same, except that what halts the whole procedure in the continuous case just fathoms a node in the integer case.

```
procedure L-shaped Integer;
begin
    Let z̄ := ∞, the best solution so far;
    Let 𝒫 := {initial problem with K := L := 0};
    while 𝒫 ≠ ∅ do begin
        Pickproblem(𝒫, P); 𝒫 := 𝒫 \ {P};
        repeat (* for problem P *)
            master(K, L, x̂, θ̂,feasible);
            fathom := not (feasible) or (cᵀx̂ + θ̂ > z̄);
            if not (fathom) then begin
                feascut(𝒜, x̂,newcut);
                if not (newcut) then intcut(x̂, newcut);
                if not (newcut) then begin
                    if x̂ integral then begin
                        Find 𝒬(x̂);
                        z̄ := min{z̄, cᵀx̂ + 𝒬(x̂)};
                        fathom := (θ̂ ≥ 𝒬(x̂));
                        if not (fathom) then begin
                            L := L + 1;
                            Create the cut −β_Lᵀx + θ ≥ α_L;
                        end;
                    end
                    else begin
                        Use branching to create 2 new problems P1 and P2;
                        Let 𝒫 := 𝒫 ∪ {P1, P2};
                    end;
                end;
            end;
        until fathom;
    end; (* while *)
end;
```

Figure 27 The L-shaped decomposition method when the first-stage problem contains integers.

```
procedure intcut(x̂:real; newcut:boolean);
begin
   if violated integrality constraints found then begin
      K := K + 1;
      Create a cut −γ_K^T x + θ ≥ δ_K;
      newcut := true;
   end
   else newcut := false;
end;
```

Figure 28 Procedure for generating cuts based on integrality.

3.8 Stochastic Decomposition

Throughout this book we are trying to reserve superscripts on variables and parameters for outcomes/realizations, and subscripts for time and components of vectors. This creates difficulties in this section. Since whatever we do will be wrong compared with our general rules, we have chosen to use the indexing of the original authors of papers on stochastic decomposition.

The L-shaped decomposition method, outlined in Section 3.2, is a deterministic method. By that, we mean that if the algorithm is repeated with the same input data, it will give the same results each time. In contrast to this, we have what are called stochastic methods. These are methods that ideally will not give the same results in two runs, even with the same input data. We say "ideally" because it is impossible in the real world to create truly random numbers, and hence, in practice, it is possible to repeat a run. Furthermore, these methods have stopping criteria that are statistical in nature. Normally, they converge with probability 1.

The reason for calling these methods random is that they are guided by some random effects, for example samples. In this section we are presenting the method called stochastic decomposition (SD). The approach, as we present it, requires relatively complete recourse.

We have until now described the part of the right-hand side in the recourse problem that does not depend on x by $h_0 + H\xi$. This was done to combine two different effects, namely to allow certain right-hand side elements to be dependent, but at the same time to be allowed to work on independent random variables. SD does not require independence, and hence we shall replace $h_0 + H\xi$ by just ξ, since we no longer make any assumptions about independence between components of ξ. We do assume, however, that

$q(\xi) \equiv q_0$, so all randomness is in the right-hand side. The problem to solve is therefore the following:

$$\begin{array}{ll} \min\{\phi(x) \equiv c^{\mathrm{T}}x + \mathcal{Q}(x)\} \\ \text{s.t.} \;\; Ax = b, \\ \qquad\;\; x \geq 0, \end{array}$$

where

$$\mathcal{Q}(x) = \int Q(x, \xi) f(\xi)\, d\xi$$

with f being the density function for $\tilde{\xi}$ and

$$Q(x, \xi) = \min\{q_0^{\mathrm{T}} y \mid Wy = \xi - T(\xi)x, \;\; y \geq 0\}.$$

Using duality, we get the following alternative formulation of $Q(x, \xi)$:

$$Q(x, \xi) = \max\{\pi^{\mathrm{T}}[\xi - T(\xi)x] \mid \pi^{\mathrm{T}} W \leq q_0^{\mathrm{T}}\}.$$

Again we note that ξ and x do not enter the constraints of the dual formulation, so that if a given ξ and x produce a solvable problem, the problem is dual feasible for *all* ξ and x. Furthermore, if π^0 is a dual feasible solution then

$$Q(x, \xi) \geq (\pi^0)^{\mathrm{T}}[\xi - T(\xi)x]$$

for any ξ and x, since π^0 is feasible but not necessarily optimal in a maximization problem. This observation is a central part of SD. Refer back to our discussion of how to interpret the Jensen lower bound in Section 3.4.1, where we gave three different interpretations, one of which was approximate optimization using a finite number of dual feasible bases, rather than all possible dual feasible bases. In SD we shall build up a collection of dual feasible bases, and in some of the optimizations use this subset rather than all possible bases. In itself, this will produce a lower-bounding solution.

But SD is also a sampling technique. By ξ_k, we shall understand the sample made in iteration k. At the same time, x_k will refer to the iterate (i.e. the presently best guess of the optimal solution) in iteration k. The first thing to do after a new sample has been made available is to evaluate $Q(x_k, \xi_j)$ for the new iterate and all samples ξ_j found so far. First we solve for the newest sample ξ_k,

$$Q(x_k, \xi_k) = \max\{\pi^{\mathrm{T}}[\xi_k - T(\xi_k)x_k] \mid \pi^{\mathrm{T}} W \leq q_0^{\mathrm{T}}\},$$

to obtain an optimal dual solution π_k. Note that this optimization, being the first involving ξ_k, is exact. If we let V be the collection of all dual feasible solutions obtained so far, we now add π_k to V. Next, instead of evaluating $Q(x_k, \xi_j)$ for $j = 1, \ldots, k-1$ (i.e. for the old samples) exactly, we simply solve

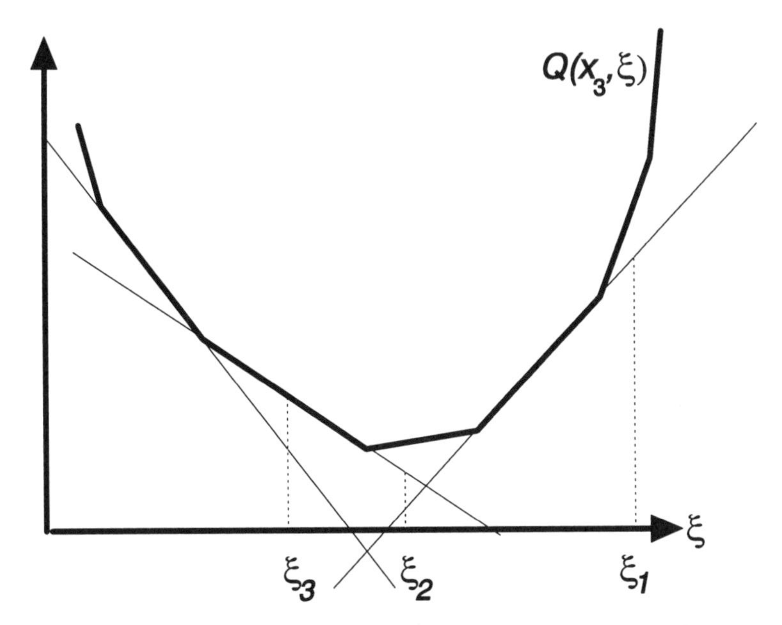

Figure 29 Illustration of how stochastic decomposition performs exact optimization for the latest (third) sample point, but inexact optimization for the two old points.

$$\max_{\pi}\{\pi^{\mathrm{T}}(\xi_j - T(\xi_j)x_k) \mid \pi \in V\}$$

to obtain π_j^k. Since V contains a finite number of vectors, this operation is very simple. Note that for all samples but the new one we perform approximate optimization using a limited set of dual feasible bases. The situation is illustrated in Figure 29. There we see the situation for the third sample point. We first make an exact optimization for the new sample point, ξ_3, obtaining a true optimal dual solution π_3. This is represented in Figure 29 by the supporting hyperplane through $\xi_3, Q(x_3, \xi_3)$. Afterwards, we solve inexactly for the two old sample points. There are three bases available for the inexact optimization. These bases are represented by the three thin lines. As we see, neither of the two old sample points find their true optimal basis.

If $\Xi(\tilde{\xi}) = \{\xi_1, \xi_2, \xi_3\}$, with each outcome having the same probability $\frac{1}{3}$, we could now calculate a lower bound on $\mathcal{Q}(x_3)$ by computing

$$\mathcal{L}(x_3) = \frac{1}{3}\sum_{j=1}^{3}(\pi_j^3)^{\mathrm{T}}(\xi_j - T(\xi_j)x_3).$$

This would be a lower bound because of the inexact optimization performed for the old sample points. However, the three sample points probably do not represent the true distribution well, and hence what we have is only something that in expectation is a lower bound. Since, eventually, this term will converge towards $\mathcal{Q}(x)$, we shall in what follows write

$$\mathcal{Q}(x_k) = \frac{1}{k} \sum_{j=1}^{k} (\pi_j^k)^{\mathrm{T}} (\xi_j - T(\xi_j) x_k).$$

Remember, however, that this is not the true value of $\mathcal{Q}(x_k)$—just an estimate.

In other words, we have now observed two major differences from the exact L-shaped method (page 159). First, we operate on a sample rather than on all outcomes, and, secondly, what we calculate is an estimate of a lower bound on $\mathcal{Q}(x_k)$ rather than $\mathcal{Q}(x_k)$ itself. Hence, since we have a lower bound, what we are doing is more similar to what we did when we used the L-shaped decomposition method within approximation schemes, (see page 192). However, the reason for the lower bound is somewhat different. In the bounding version of L-shaped, the lower bound was based on conditional expectations, whereas here it is based on inexact optimization. On the other hand, we have earlier pointed out that the Jensen lower bound has three different interpretations, one of which is to use conditional expectations (as in **procedure** Bounding L-shaped) and another that is inexact optimization (as in SD). So what is actually the principal difference?

For the three interpretations of the Jensen bound to be equivalent, the limited set of bases must come from solving the recourse problem in the points of conditional expectations. That is not the case in SD. Here the points are random (according to the sample ξ_j). Using a limited number of bases still produces a lower bound, but not the Jensen lower bound.

Therefore SD and the bounding version of L-shaped are really quite different. The reason for the lower bound is different, and the objective value in SD is only a lower bound in terms of expectations (due to sampling). One method picks the limited number of points in a very careful way, the other at random. One method has an exact stopping criteria (error bound), the other has a statistically based stopping rule. So, more than anything else, they are alternative approaches. If one cannot solve the exact problem, one either resorts to bounds or to sample-based methods.

In the L-shaped method we demonstrated how to find optimality cuts. We can now find a cut corresponding to x_k (which is not binding and might even not be a lower bound, although it represents an estimate of a lower bound). As for the L-shaped method, we shall replace $\mathcal{Q}(x)$ in the objective by θ, and then add constraints. The cut generated in iteration k is given by

$$\theta \geq \frac{1}{k}\sum_{j=1}^{k}(\pi_j^k)^{\mathrm{T}}[\xi_j - T(\xi_j)x] = \alpha_k^k + (\beta_k^k)^{\mathrm{T}}x.$$

The double set of indices on α and β indicate that the cut was generated in iteration k (the subscript) and that it has been updated in iteration k (the superscript).

In contrast to the L-shaped decomposition method, we must now also look at the old cuts. The reason is that, although we expect these cuts to be loose (since we use inexact optimization), they may in fact be far too tight (since they are based on a sample). Also, being old, they are based on a sample that is smaller than the present one, and hence, probably not too good. We shall therefore want to phase them out, but not by throwing them away. Assume that there exists a lower bound on $Q(x,\xi)$ such that $Q(x,\xi) \geq \underline{Q}$ for all x and ξ. Then the old cuts

$$\theta \geq \alpha_j^{k-1} + (\beta_j^{k-1})^{\mathrm{T}}x \quad \text{for } j = 1,\ldots,k-1$$

will be replaced by

$$\left.\begin{aligned} \theta &\geq \frac{k-1}{k}[\alpha_j^{k-1} + (\beta_j^{k-1})^{\mathrm{T}}x] + \frac{1}{k}\underline{Q} \\ &= \alpha_j^k + (\beta_j^k)^{\mathrm{T}}x \quad \text{for } j = 1,\ldots,k-1. \end{aligned}\right\} \tag{8.1}$$

For technical reasons, $\underline{Q} = 0$ is to be preferred. This inequality is looser than the previous one, since $\underline{Q} \leq Q(x,\xi)$. The master problem now becomes

$$\left.\begin{aligned} &\min c^{\mathrm{T}}x + \theta \\ &\text{s.t.} \qquad Ax = b, \\ &\qquad -(\beta_j^k)^{\mathrm{T}}x + \theta \geq \alpha_j^k \quad \text{for} \quad j = 1,\ldots,k, \\ &\qquad\qquad x \geq 0, \end{aligned}\right\} \tag{8.2}$$

yielding the next iterate x_{k+1}. Note that, since we assume relatively complete recourse, there are no feasibility cuts. The above format is the one to be used for computations. To understand the method better, however, let us show an alternative version of (8.2) that is less useful computationally but is more illustrative (see Figure 30 for an illustration):

$$\begin{aligned} &\min \{\phi_k(x) \equiv c^{\mathrm{T}}x + \max\nolimits_{j\in\{1,\cdots,k\}}[\alpha_j^k + (\beta_j^k)^{\mathrm{T}}x]\} \\ &\text{s.t. } Ax = b, \; x \geq 0. \end{aligned}$$

This defines the function $\phi_k(x)$ and shows more clearly than (8.2) that we do indeed have a function in x that we are minimizing. Also $\phi_k(x)$ is the present estimate of $\phi(x) = c^{\mathrm{T}}x + \mathcal{Q}(x)$.

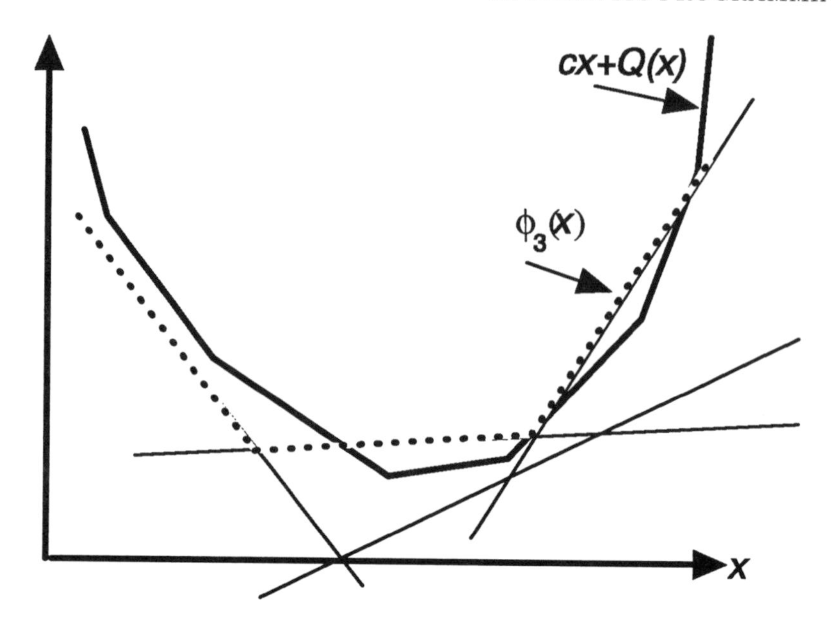

Figure 30 Representation of $c^{\mathrm{T}}x + \mathcal{Q}(x)$ by a piecewise linear function.

The above set-up has one major shortcoming: it might be difficult to extract a converging subsequence from the sequence x_k. A number of changes therefore have to be made. These make the algorithm look more messy, but the principles are not lost. To make it simpler (empirically) to extract a converging subsequence, we shall introduce a sequence of *incumbent* solutions $\overline{x}_k$. Following the incumbent, there will be an index i_k that shows in which iteration the current $\overline{x}_k$ was found.

We initiate the method by setting the counter $k := 0$, choose an $r \in (0,1)$ (to be explained later), and let $\xi_0 := E\tilde{\xi}$. Thus we solve

$$\begin{aligned} \min\, & c^{\mathrm{T}}x + q_0^{\mathrm{T}}y \\ \text{s.t.}\quad & Ax = b, \\ & Wy = \xi_0 - T(\xi_0)x, \ x, y \geq 0, \end{aligned}$$

to obtain an initial x_1. We initiate the incumbent $\overline{x}_0 = x_1$ and show that it was found in iteration 1 by letting $i_0 = 1$.

Next, let us see what is done in a general iteration of the algorithm. First the counter is increased by letting $k := k + 1$, and a sample ξ_k is found. We now need to find a new cut k as outlined before, and we need to update the cut that corresponds to the current incumbent. First, we solve

$$\max\{\pi^{\mathrm{T}}[\xi_k - T(\xi_k)x_k] \mid \pi^{\mathrm{T}}W \leq q_0^{\mathrm{T}}\}$$

to obtain π_k. Next, we solve

$$\max\{\pi^{\mathrm{T}}(\xi_k - T(\xi_k)\overline{x}_{k-1}) \mid \pi^{\mathrm{T}} W \leq q_0^{\mathrm{T}}\}$$

to obtain $\overline{\pi}_k$. As before, we then update the set of dual feasible bases by letting $V := V \cup \{\pi_k, \overline{\pi}_k\}$.

We then need to make one new cut and update the old cuts. First, the new cut is made exactly as before. We solve

$$\max\{\pi^{\mathrm{T}}[\xi_j - T(\xi_j)x_k] \mid \pi \in V\}$$

to obtain π_j for $j = 1, \ldots, k-1$, and then create the kth cut as

$$\theta \geq \frac{1}{k}\sum_{j=1}^{k}(\pi_j)^{\mathrm{T}}[\xi_j - T(\xi_j)x] = \alpha_k^k + (\beta_k^k)^{\mathrm{T}}x.$$

In addition, we need to update the incumbent cut i_k. This is done just the way we found cut k. We solve

$$\max\{\pi^{\mathrm{T}}[\xi_j - T(\xi_j)\overline{x}_{k-1}] \mid \pi \in V\}$$

to obtain $\overline{\pi}_j$, and replace the old cut i_k by

$$\theta \geq \frac{1}{k}\sum_{j=1}^{k}(\overline{\pi}_j)^{\mathrm{T}}[\xi_j - T(\xi_j)x] = \alpha_{i_{k-1}}^k + (\beta_{i_{k-1}}^k)^{\mathrm{T}}x.$$

The remaining cuts are updated as before by letting

$$\theta \geq \frac{k-1}{k}[\alpha_j^{k-1} + (\beta_j^{k-1})^{\mathrm{T}}x] + \frac{1}{k}\underline{Q} = \alpha_j^k + (\beta_j^k)^{\mathrm{T}}x \text{ for } j = 1, \ldots, k-1,\ j \neq i_{k-1}.$$

Now, it is time to check if the incumbent should be changed. We shall use Figure 31 for illustration, and we shall use the function $\phi_k(x)$ defined earlier. In the figure we have $k = 3$. When we entered iteration k, our approximation of $\phi(x)$ was given by $\phi_{k-1}(x)$. Our incumbent solution was $\overline{x}_{k-1}$ and our iterate was x_k. We show this in the top part of Figure 31 as $\overline{x}_2$ and x_3. The position of $\overline{x}_2$ is somewhat arbitrary, since we cannot know how things looked in the previous iteration. Therefore $\phi_{k-1}(x_k) - \phi_{k-1}(\overline{x}_{k-1}) \leq 0$ was our approximation of how much we should gain by making x_k our new incumbent. However, x_k might be in an area where $\phi_{k-1}(x)$ is a bad approximation of $\phi(x)$. The function $\phi_k(x)$, on the other hand, was developed around x_k, and should therefore be good in that area (in addition to being approximately as good as $\phi_{k-1}(x)$ around $\overline{x}_{k-1}$). This can be seen in the bottom part of Figure 31, where $\phi_3(x)$ is given. The function $\phi_3(x)$ is based on three cuts. One is new, the other two are updates of the two cuts in the top part of the

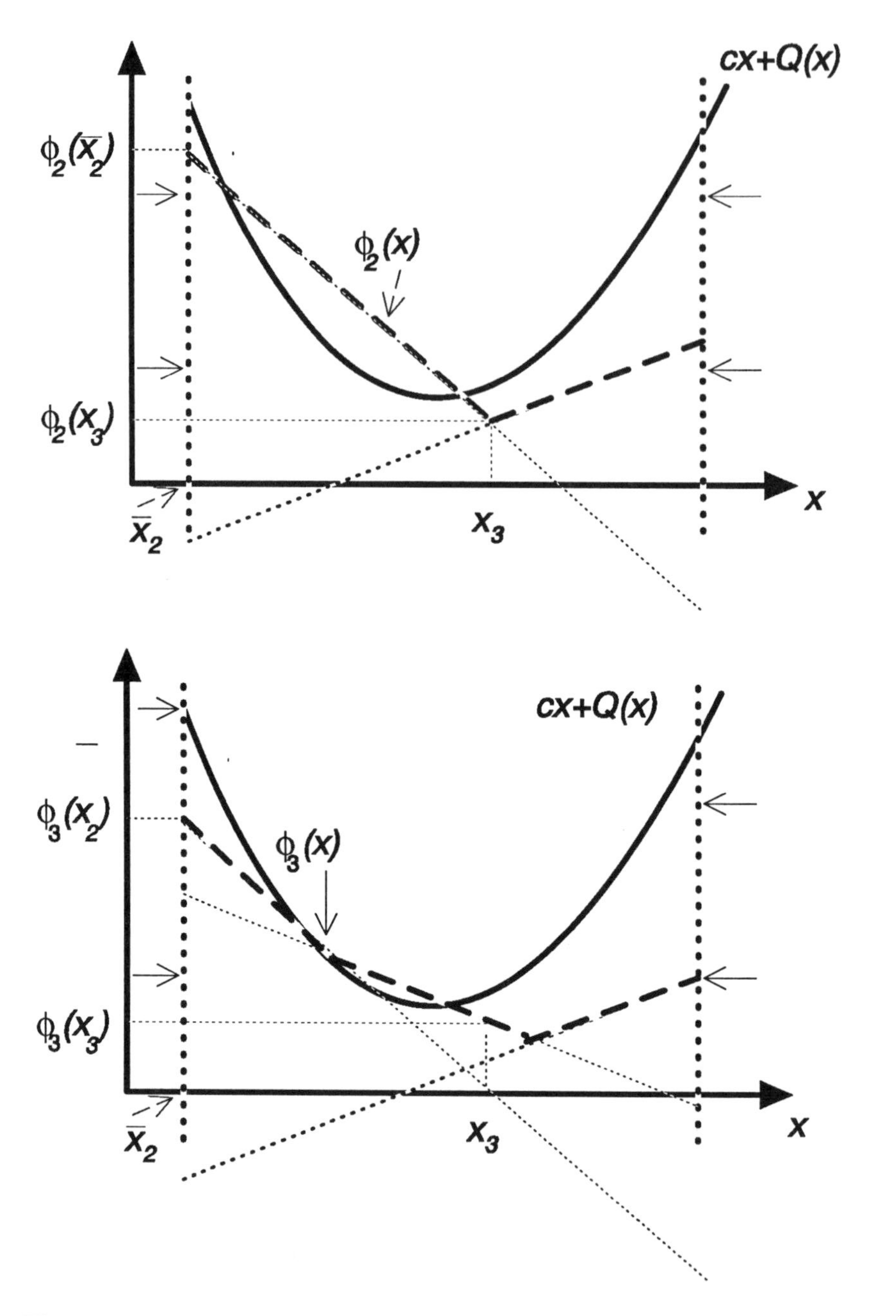

Figure 31 Calculations to find out if the incumbent should be changed.

figure, according to (8.1). Hence $\phi_k(x_k) - \phi_k(\overline{x}_{k-1})$ is a measure of how much we actually gained. If

$$\phi_k(x_k) - \phi_k(\overline{x}_{k-1}) < r[\phi_{k-1}(x_k) - \phi_{k-1}(\overline{x}_{k-1})],$$

we gained at least a portion r of what we hoped for, and we let $\overline{x}_k := x_k$ and $i_k := k$. If not, we were not happy with the change, and we let $\overline{x}_k := \overline{x}_{k-1}$ and $i_k := i_{k-1}$. When we have updated the incumbent, we solve a new master problem to obtain x_{k+1} and repeat the process.

The stopping criterion for SD is of a statistical nature, and its complexity is beyond the scope of this book. For a reference, see the end of the chapter.

3.9 Stochastic Quasi-Gradient Methods

We are still dealing with recourse problems stated in the somewhat more general form

$$\min_{x \in X} \left[f(x) + \int_{\Xi} Q(x, \xi) \; P_{\tilde{\xi}}(d\xi) \right]. \tag{9.1}$$

This formulation also includes the stochastic linear program with recourse, letting

$$\begin{aligned} X &= \{x \mid Ax = b, \; x \geq 0\}, \\ f(x) &= c^{\mathrm{T}} x, \\ Q(x, \xi) &= \min\{(q(\xi))^{\mathrm{T}} y \mid Wy = h(\xi) - T(\xi)x, \; y \geq 0\}. \end{aligned}$$

To describe the so-called *stochastic quasi-gradient method* (SQG), we simplify the notation by defining

$$F(x, \xi) := f(x) + Q(x, \xi)$$

and hence considering the problem

$$\min_{x \in X} E_{\tilde{\xi}} F(x, \tilde{\xi}), \tag{9.2}$$

for which we assume that

$$\begin{aligned} &E_{\tilde{\xi}} F(x, \tilde{\xi}) \quad \text{is finite and convex in } x, && \text{(9.3 i)} \\ &X \qquad\qquad\quad \text{is convex and compact.} && \text{(9.3 ii)} \end{aligned}$$

Observe that for stochastic linear programs with recourse the assumptions (9.3) are satisfied if, for instance,

- we have relatively complete recourse, the recourse function $Q(x,\xi)$ is a.s. finite $\forall x$, and the components of $\tilde{\xi}$ are square-integrable (i.e. their second moments exist);
- $X = \{x \mid Ax = b,\ x \geq 0\}$ is bounded.

Then, starting from some feasible point $x^0 \in X$, we may define an iterative process by

$$x^{\nu+1} = \Pi_X(x^\nu - \rho_\nu v^\nu), \tag{9.4}$$

where v^ν is a random vector, $\rho_\nu \geq 0$ is some step size and Π_X is the projection onto X, i.e. for $y \in \mathbb{R}^n$, with $||\cdots||$ the Euclidean norm,

$$\Pi_X(y) = \arg\min_{x\in X} ||y - x||. \tag{9.5}$$

By assumption (9.3 i), $\varphi(x) := E_{\tilde{\xi}} F(x,\tilde{\xi})$ is convex in x. If this function is also differentiable with respect to x at any arbitrary point z with the gradient $g := \nabla\varphi(z) = \nabla_x E_{\tilde{\xi}} F(z,\tilde{\xi})$, then $-g$ is the direction of steepest descent of $\varphi(x) = E_{\tilde{\xi}} F(x,\tilde{\xi})$ in z, and we should probably like to choose $-g$ as the search direction to decrease our objective. However, this does not seem to be a practical approach, since, as we know already, evaluating $\varphi(x) = E_{\tilde{\xi}} F(x,\tilde{\xi})$, as well as $\nabla\varphi(z) = \nabla_x E_{\tilde{\xi}} F(z,\tilde{\xi})$, is a rather cumbersome task.

In the differentiable case we know from Proposition 1.21 that, for a convex function φ,

$$(x - z)^{\mathrm{T}} \nabla\varphi(z) \leq \varphi(x) - \varphi(z) \tag{9.6}$$

has to hold $\forall x, z$ (see Figure 26 in Chapter 1). But, even if the convex function φ is not differentiable at some point z, e.g. if it has a kink there, it is shown in convex analysis that there exists at least one vector g such that

$$(x - z)^{\mathrm{T}} g \leq \varphi(x) - \varphi(z)\ \forall x. \tag{9.7}$$

Any vector g satisfying (9.7) is called a *subgradient* of φ at z, and the set of all vectors satisfying (9.7) is called the *subdifferential* of φ at z and is denoted by $\partial\varphi(z)$. If φ is differentiable at z then $\partial\varphi(z) = \{\nabla\varphi(z)\}$; otherwise, i.e. in the nondifferentiable case, $\partial\varphi(z)$ may contain more than one element as shown for instance in Figure 32. Furthermore, in view of (9.7), it is easily seen that $\partial\varphi(z)$ is a convex set.

If φ is convex and $g \neq 0$ is a subgradient of φ at z then, by (9.7) for $\lambda > 0$, it follows that

$$\begin{aligned}
\varphi(z + \lambda g) &\geq \varphi(z) + g^{\mathrm{T}}(x - z) \\
&= \varphi(z) + g^{\mathrm{T}}(\lambda g) \\
&= \varphi(z) + \lambda ||g||^2 \\
&> \varphi(z).
\end{aligned}$$

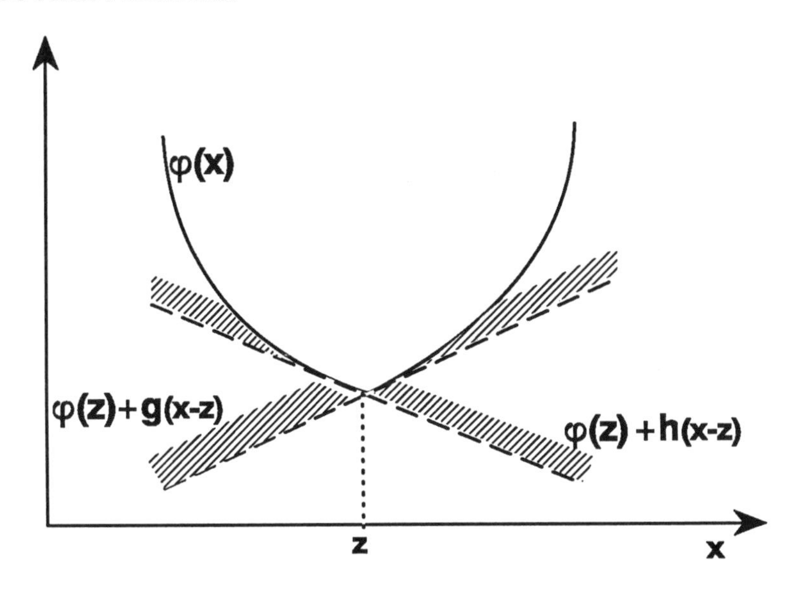

Figure 32 Nondifferentiable convex function: subgradients.

Hence any subgradient, $g \in \partial\varphi$, such that $g \neq 0$ is a direction of ascent, although not necessarily the direction of steepest ascent as the gradient would be if φ were differentiable in z. However, in contrast to the differentiable case, $-g$ need not be a direction of strict descent for φ in z. Consider for example the convex function in two variables

$$\psi(u, v) := |u| + |v|.$$

Then for $\hat{z} = (0, 3)^{\mathrm{T}}$ we have $g = (1, 1)^{\mathrm{T}} \in \partial\psi(\hat{z})$, since for all $\varepsilon > 0$ the gradient $\nabla\psi(\varepsilon, 3)$ exists and is equal to g. Hence, by (9.6), we have, for all (u, v),

$$\begin{aligned}\left[\begin{pmatrix} u \\ v \end{pmatrix} - \begin{pmatrix} \varepsilon \\ 3 \end{pmatrix}\right]^{\mathrm{T}} g &= \begin{pmatrix} u - \varepsilon \\ v - 3 \end{pmatrix}^{\mathrm{T}} \begin{pmatrix} 1 \\ 1 \end{pmatrix} \\ &= u - \varepsilon + v - 3 \\ &\leq |u| + |v| - |\varepsilon| - |3|,\end{aligned}$$

which is obviously true $\forall\varepsilon \geq 0$, such that g is a subgradient in $(0, 3)^{\mathrm{T}}$. Then for $0 < \lambda < 3$ and $\hat{z} - \lambda g = (-\lambda, 3 - \lambda)^{\mathrm{T}}$ it follows that

$$\psi(\hat{z} - \lambda g) = 3 = \psi(\hat{z}),$$

and therefore, in this particular case, $-g$ is not a strict descent direction for ψ in $\hat{z}$. Nevertheless, as we see in Figure 33, moving from $\hat{z}$ along the ray

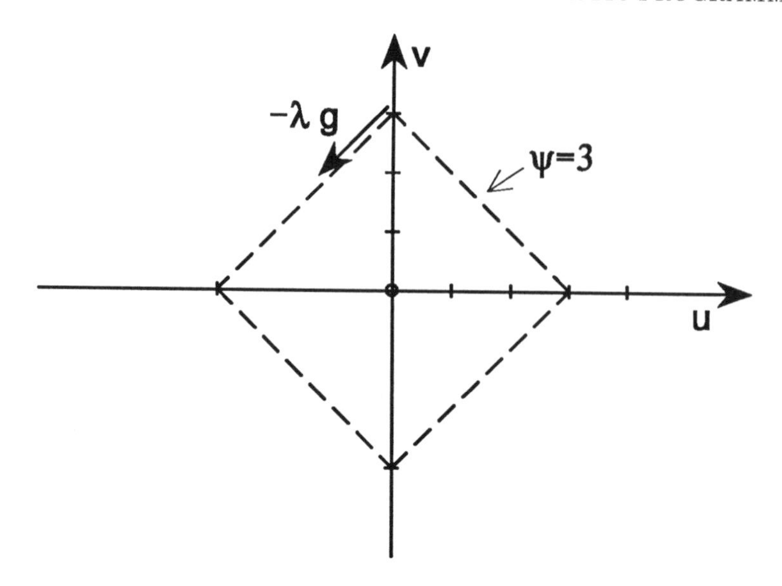

Figure 33 Decreasing the distance to $\arg\min \psi$ using a subgradient.

$\hat{z} - \lambda g$, $\lambda > 0$, for any $\lambda < 3$ we would come closer—with respect to the Euclidean norm—to $\arg\min \psi = \{(0,0)^{\mathrm{T}}\}$ than we are at $\hat{z}$.

It is worth noting that this property of a subgradient of a convex function holds in general, and not only for our particular example. Let φ be a convex function and assume that $g \in \partial\varphi(z)$, $g \neq 0$. Assume further that $z \notin \arg\min \varphi$ and $x^\star \in \arg\min \varphi$. Then we have for $\rho > 0$ with the Euclidean norm, using (9.7),

$$\begin{aligned} \|(z - \rho g) - x^\star\|^2 &= \|(z - x^\star) - \rho g\|^2 \\ &= \|z - x^\star\|^2 + \rho^2\|g\|^2 - 2\rho g^{\mathrm{T}}(z - x^\star) \\ &\leq \|z - x^\star\|^2 + \rho^2\|g\|^2 - 2\rho[\varphi(z) - \varphi(x^\star)]. \end{aligned}$$

Since, by our assumption, $\varphi(z) - \varphi(x^\star) > 0$, we may choose a step size $\rho = \bar{\rho} > 0$ such that

$$\bar{\rho}^2\|g\|^2 - 2\bar{\rho}[\varphi(z) - \varphi(x^\star] < 0,$$

implying that $z - \bar{\rho}g$ is closer to $x^\star \in \arg\min \varphi$ than z. This property provides the motivation for the iterative procedures known as *subgradient methods*, which minimize convex functions even in the nondifferentiable case.

Obviously for the above procedure (9.4) we may not expect any reasonable convergence statement without further assumptions on the search direction v^ν and on the step size ρ_ν. Therefore let v^ν be a so-called *stochastic quasi-*

gradient, i.e. assume that

$$E(v^\nu \mid x^0, \cdots, x^\nu) \in \partial_x E_{\tilde{\xi}} F(x^\nu, \tilde{\xi}) + b^\nu, \tag{9.8}$$

where ∂_x denotes the subdifferential with respect to x, as mentioned above, coinciding with the gradient in the differentiable case.

Let us recall what we are doing here. Starting with some x^ν, we choose for (9.4) a random vector v^ν. It seems plausible to assume that v^ν depends in some way on $\tilde{\xi}$ (e.g. on an observation ξ^ν or on a sample $\{\xi^{\nu 1}, \cdots, \xi^{\nu N_\nu}\}$ of $\tilde{\xi}$) and on x^ν. Then, after the choice of the step size ρ_ν, by (9.4) the next iterate $x^{\nu+1}$ depends on x^ν. It follows that v^ν is itself random. This implies that the tuples $(x^0, x^1, \cdots, x^\nu)$ are random $\forall \nu \geq 1$. Hence (9.8) is not yet much of a requirement. It just says that the expected value of v^ν, under the condition of the path of iterates generated so far, $(x^0, \cdots, x^\nu)$, is to be written as the sum of a subgradient $g^\nu \in \partial_x E_{\tilde{\xi}} F(x^\nu, \tilde{\xi})$ and some vector b^ν.

Since, by the convexity according to (9.3 i) and applying (9.7),

$$E_{\tilde{\xi}} F(x^*, \tilde{\xi}) - E_{\tilde{\xi}} F(x^\nu, \tilde{\xi}) \geq g^{\nu\mathrm{T}}(x^* - x^\nu) \tag{9.9}$$

for any solution x^* of (9.2) and any $g^\nu \in \partial_x E_{\tilde{\xi}} F(x^\nu, \tilde{\xi})$, we have from (9.8) that

$$0 \geq E_{\tilde{\xi}} F(x^*, \tilde{\xi}) - E_{\tilde{\xi}} F(x^\nu, \tilde{\xi}) \geq E(v^\nu \mid x^0, \cdots, x^\nu)^{\mathrm{T}}(x^* - x^\nu) + \gamma_\nu, \tag{9.10}$$

where

$$\gamma_\nu = -b^{\nu\mathrm{T}}(x^* - x^\nu). \tag{9.11}$$

Intuitively, if we assume that $\{x^\nu\}$ converges to $x^\star$ and all v^ν are uniformly bounded, i.e. $|v^\nu| \leq \alpha$ for some constant α, we should require that $\|b^\nu\| \overset{\nu\to\infty}{\longrightarrow} 0$, implying $\gamma_\nu \overset{\nu\to\infty}{\longrightarrow} 0$ as well. Observe that the particular choice of a *stochastic subgradient*

$$v^\nu \in \partial_x F(x^\nu, \xi^\nu), \tag{9.12}$$

or more generally

$$v^\nu = \frac{1}{N_\nu} \sum_{\mu=1}^{N_\nu} w^\mu, \quad w^\mu \in \partial_x F(x^\nu, \xi^{\nu\mu}), \tag{9.13}$$

where the ξ^ν or $\xi^{\nu\mu}$ are independent samples of $\tilde{\xi}$, would yield $b^\nu = 0$, $\gamma_\nu = 0 \ \forall \nu$, provided that the operations of integration and differentiation may be exchanged, as asserted for example by Proposition 1.2 for the differentiable case.

Finally, assume that for the step size ρ_ν together with v^ν and γ_ν we have

$$\rho_\nu \geq 0, \ \sum_{\nu=0}^{\infty} \rho_\nu = \infty, \ \sum_{\nu=0}^{\infty} E_{\tilde{\xi}}(\rho_\nu |\gamma_\nu| + \rho_\nu^2 \|v^\nu\|^2) < \infty. \tag{9.14}$$

With the choices (9.12) or (9.13), for uniformly bounded v^ν this assumption could obviously be replaced by the step size assumption

$$\rho_\nu \geq 0, \sum_{\nu=0}^{\infty} \rho_\nu = \infty, \sum_{\nu=0}^{\infty} \rho_\nu^2 < \infty. \tag{9.15}$$

With these prerequisites, it can be shown that, under the assumptions (9.3), (9.8) and (9.14) (or (9.3), (9.12) or (9.13), and (9.15)) the iterative method (9.4) converges almost surely (a.s.) to a solution of (9.2).

3.10 Solving Many Similar Linear Programs

In both the L-shaped (continuous and integer) and stochastic decomposition methods we are faced with the problem of solving many similar LPs. This is most obvious in the L-shaped method: cut formation requires the solution of many LPs that differ only in the right-hand side and objective. This amount of work, which is typically enormous, must be performed in each major iteration. For stochastic decomposition, it is perhaps less obvious that we are facing such a large workload, but, added over all iterations, we still end up with a large number of similar LPs.

The problem of solving a large number of similar LPs has attracted attention for quite a while, in particular when there is only right-hand side randomness. Therefore let us proceed under the assumption that $q(\xi) \equiv q_0$.

The major idea is that of *bunching*. This is a simple idea. If we refer back to the discussion of the L-shaped decomposition method, we observed that the dual formulation of the recourse problem was given by

$$\max_\pi \{\pi^{\mathrm{T}}(h(\xi) - T(\xi)x) \mid \pi^{\mathrm{T}} W \leq q_0^{\mathrm{T}}\}. \tag{10.1}$$

What we observe here is that the part that varies, $h(\xi) - T(\xi)x$, appears only in the objective. As a consequence, if (10.1) is feasible for one value of x and ξ, it is feasible for all values of x and ξ. Of course, the problem might be unbounded (meaning that the primal is infeasible) for some x and ξ. For the moment we shall assume that that does not occur. (But if it does, it simply shows that we need a feasibility cut, not an optimality cut).

In a given iteration of the L-shaped decomposition method, x will be fixed, and all we are interested in is the selection of right-hand sides resulting from all possible values of ξ. Let us therefore simplify notation, and assume that we have a selection of right-hand sides $\mathcal{B}$, so that, instead of (10.1), we solve

$$\max_\pi \{\pi^{\mathrm{T}} h \mid \pi^{\mathrm{T}} W \leq q_0^{\mathrm{T}}\} \tag{10.2}$$

for all $h \in \mathcal{B}$. Assume (10.2) is solved for one value of $h \in \mathcal{B}$ with optimal basis B. Then B is a dual feasible basis for all $h \in \mathcal{B}$. Therefore, for all $h \in \mathcal{B}$ for which $B^{-1}h \geq 0$, the basis B is also primal feasible, and hence optimal. The idea behind bunching is simply to start out with some $h \in \mathcal{B}$, find the optimal basis B, and then check $B^{-1}h$ for all other $h \in \mathcal{B}$. Whenever $B^{-1}h \geq 0$, we have found the optimal solution for that h, and these right-hand sides are bunched together. We then remove these right-hand sides from $\mathcal{B}$, and repeat the process, of course with a warm start from B, using the dual simplex method, for one of the remaining right-hand sides in $\mathcal{B}$. We continue until all right-hand sides are bunched. That gives us all information needed to find $\mathcal{Q}$ and the necessary optimality cut.

This procedure has been followed up in several directions. An important one is called *trickling down*. Again, we start out with $\mathcal{B}$, and we solve (10.2) for some right-hand side to obtain a dual feasible basis B. This basis is stored in the root of a search tree that we are about to make. Now, for one $h \in \mathcal{B}$ at a time do the following. Start in the root of the tree, and calculate $B^{-1}h$. If $B^{-1}h \geq 0$, register that this right-hand side belongs to the bunch associated with B, and go to the next $h \in \mathcal{B}$. If $B^{-1}h \not\geq 0$, pick a row for which primal feasibility is not satisfied. Perform a dual pivot step to obtain a new basis B' (still dual feasible). Create a new node in the search tree associated with this new B'. If the pivot was made in row i, we let the new node be the ith child of the node containing the previous basis. Continue until optimality is found. This situation is illustrated in Figure 34, where a total of eight bases are stored. The numbers on the arc refer to the row where pivoting took place, the B in the nodes illustrate that there is a basis stored in each node.

This might not seem efficient. However, the real purpose comes after some iterations. If a right-hand side h is such that $B^{-1}h \not\geq 0$, and one of the negative primal variables corresponds to a row index i such that the ith child of the given node in the search tree already exists, we simply move to that child *without having to price*. This is why we use the term *trickling down*. We try to trickle a given h as far down in the tree as possible, and only when there is no negative primal variable that corresponds to a child node of the present node do we price and pivot explicitly, thereby creating a new branch in the tree.

Attempts have been made to first create the tree, and then trickle down the right-hand sides in the finished tree. This was not successful for two reasons. If we try to enumerate *all* dual feasible bases, then the tree grows out of hand (this corresponds to extreme point enumeration), and if we try to find the correct selection of such bases, then that in itself becomes an overwhelming problem. Therefore a pre-defined tree does not seem to be a good idea.

It is worth noting that the idea of storing a selection of dual feasible bases, as was done in the stochastic decomposition method, is also related to the above approach. In that case the result is a lower bound on $\mathcal{Q}(x)$.

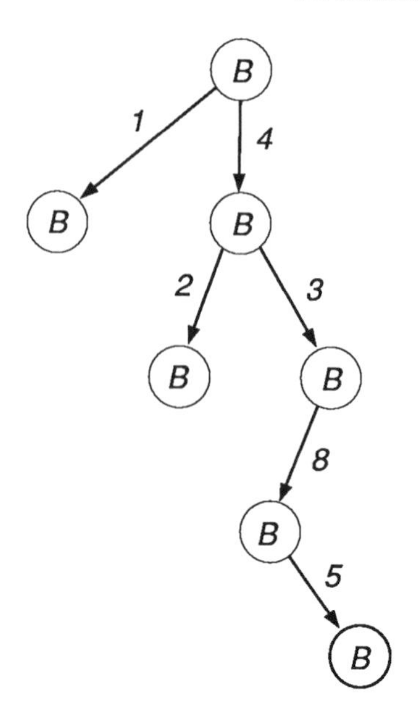

Figure 34 Example of a bunching tree.

A variant of these methods is as follows. Start out with one dual feasible basis B as in the trickling down procedure. Pick a *leading* right-hand side. Now solve the problem corresponding to this leading right-hand side using the dual simplex method. On pivoting along, create a branch of the search tree just as for trickling down. The difference is as follows. For each basis B encountered, check $B^{-1}h$ for all $h \in \mathcal{B}$. Then split the right-hand sides remaining in $\mathcal{B}$ into three sets. Those that have $B^{-1}h \geq 0$ are bunched with that B, and removed from $\mathcal{B}$. Those that have a primal infeasibility in the same row as the one chosen to be the pivot row for the leading problem are kept in $\mathcal{B}$ and hence carried along at least one more dual pivot step. The remaining right-hand sides are left behind in the given node, to be picked up later on.

When the leading problem has been solved to optimality, and bunching has been performed with respect to its optimal basis, check if there are any right-hand sides left in $\mathcal{B}$. If there are, let one of them be the leading right-hand side, and continue the process. Eventually, when a leading problem has been solved to optimality, $\mathcal{B} = \emptyset$. At that time, start backtracking the search tree. Whenever a selection of right-hand sides left behind is encountered, pick one of them as the leading problem, and repeat the process. On returning to the root, and finding there are no right-hand sides left behind there, the process is finished. All right-hand sides are bunched. Technically, what has now been done is to traverse the search tree in pre-order.

What remains to be discussed is what to store in the search tree. We have already seen that the minimal amount to store at any arc in the tree is the index of the leaving basic column (represented by the numbering of the children), and the entering column. If that is all we store, we have to pivot, but not price out, in each step of the trickling down. If we have enough storage, it is more efficient to store for example the eta-vector (from the revised simplex method) or the Schur complement (it is not important here if you do not know what the eta-vector or the Schur complement is). Of course, we could in principle store B^{-1}, but for all practical problems that is too much to store.

3.10.1 Randomness in the Objective

The discussion of trickling down etc. was carried out in a setting of right-hand side randomness only. However, as with many other problems we have faced in this book, pure objective function randomness can be changed into pure right-hand side randomness by using linear programming duality. Therefore the discussions of right-hand side randomness apply to objective function randomness as well.

Then, one may ask what happens if there is randomness in both the objective and the right-hand side. Trickling down cannot be performed the way we have outlined it in that case. This is because a basis that was optimal for one ξ will, in general, be neither primal nor dual feasible for some other ξ. On the other hand, the basis may be good, not far from the optimal one. Hence warm starts based on an old basis, performing a combination of primal and dual simplex steps, will almost surely be better than solving the individual LPs from scratch.

3.11 Bibliographical Notes

Benders' [1] decomposition is the basis for all decomposition methods in this chapter. In stochastic programming, as we have seen, it is more common to refer to Benders' decomposition as the L-shaped decomposition method. That approach is outlined in detail in Van Slyke and Wets [63]. An implementation of the L-shaped decomposition method, called MSLiP, is presented in Gassmann [31]. It solves multistage problems based on nested decomposition. Alternative computational methods are also discussed in Kall [44].

The regularized decomposition method has been implemented under the name QDECOM. For further details on the method and QDECOM, in particular for a special technique to solve the master (3.6), we refer to the original publication of Ruszczyński [61]; the presentation in this chapter is close to the description in his recent paper [62].

Some attempts have also been made to use interior point methods. As examples consider Birge and Qi [7], Birge and Holmes [6], Mulvey and Ruszczyński [60] and Lustig, Mulvey and Carpenter [55]. The latter two combine interior point methods with parallel processing.

Parallel techniques have been tried by others as well; see e.g. Berland [2] and Jessup, Yang and Zenios [42]. We shall mention some others in Chapter 6.

The idea of combining branch-and-cut from integer programming with primal decomposition in stochastic programming was developed by Laporte and Louveaux [53]. Although the method is set in a strict setting of integrality only in the first stage, it can be expanded to cover (via a reformulation) multistage problems that possess the so-called block-separable recourse property, see Louveaux [54] for details.

Stochastic quasi-gradient methods were developed by Ermoliev [20, 21], and implemented by, among others, Gaivoronski [27, 28]. Besides stochastic quasi-gradients several other possibilities for constructing stochastic descent directions have been investigated, e.g. in Marti [57] and in Marti and Fuchs [58, 59].

The Jensen lower bound was developed in 1906 [41]. The Edmundson–Madansky upper bound is based on work by Edmundson [19] and Madansky [56]. It has been extended to the multidimensional case by Gassmann and Ziemba [33]; see also Hausch and Ziemba [36] and Edirisinghe and Ziemba [17, 18]. Other references in this area include Huang, Vertinsky and Ziemba [39] and Huang, Ziemba and Ben-Tal [40]. The Edmundson–Madansky bound was generalized to the case of stochastically dependent components by Frauendorfer [23].

The piecewise linear upper bound is based on two independent approaches, namely those of Birge and Wets [11] and Wallace [66]. These were later combined and strengthened in Birge and Wallace [8].

There is a large collection of bounds based on extreme measures (see e.g. Dulá [12, 13], Hausch and Ziemba [36], Huang, Ziemba and Ben-Tal [40] and Kall [48]). Both the Jensen and Edmundson–Madansky bounds can be put into this category. For a fuller description of these methods, consult Birge and Wets [10], Dupačová [14, 15, 16] and Kall [47]; more on extreme measures may be found in Karr [51] and Kemperman [52].

Bounds can also be found when limited information is available. Consult e.g. Birge and Dulá [5]. An upper bound based on structure can be found in Wallace and Yan [68].

Stochastic decomposition was developed by Higle and Sen [37, 38].

The ideas presented about trickling down and similar methods come from different authors, in particular Wets [70, 72], Haugland and Wallace [35], Wallace [65, 64] and Gassmann and Wallace [32]. A related approach is that of Gartska and Rutenberg [29], which is based on parametric optimization.

Partitioning has been discussed several times during the years. Some general ideas are presented in Birge and Wets [9]. More detailed discussions (with numerical results), on which the discussions in this book are based, can be found in Frauendorfer and Kall [26] and Berland and Wallace [3, 4]. Other texts about approximation by discretization include for example those of Kall [43, 45, 46], and Kall, Ruszczyński and Frauendorfer [49].

When partitioning the support to tighten bounds, it is possible to use more complicated cells than we have done. For example, Frauendorfer [24, 25] uses simplices. It is also possible to use more general polyhedra.

For simple recourse, the separability of the objective, which facilitates computations substantially, was discovered by Wets [69]. The ability to replace the Edmundson–Madansky upper bound by the true objective's value was discussed in Kall and Stoyan [50]. Wets [71] has derived a special pivoting scheme that avoids the tremendous increase of the problem size known from general recourse problems according to the number of blocks (i.e. realizations). See also discussions by Everitt and Ziemba [22] and Hansotia [34].

The fisheries example in the beginning of the chapter comes from Wallace [67]. Another application concerning natural resources is presented by Gassmann [30].

Exercises

1. The second-stage constraints of a two-stage problem look as follows:

$$\begin{pmatrix} 1 & 3 & -1 & 0 \\ 2 & -1 & 2 & 1 \end{pmatrix} y = \begin{pmatrix} -6 \\ -4 \end{pmatrix} \xi + \begin{pmatrix} 5 & -1 & 0 \\ 0 & 2 & 4 \end{pmatrix} x$$
$$y \geq 0$$

   where $\tilde{\xi}$ is a random variable with support $\Xi = [0, 1]$. Write down the LP (both primal and dual formulation) needed to check if a given x produces a feasible second-stage problem. Do it in such a way that if the problem is not feasible, you obtain an inequality in x that cuts off the given x. If you have access to an LP code, perform the computations, and find the inequality explicitly for $\hat{x} = (1, 1, 1)^{\mathrm{T}}$.

2. Look back at problem (4.1) we used to illustrate the bounds. Add one extra constraint, namely

$$x_{raw1} \leq 40.$$

   (a) Find the Jensen lower bound after this constraint has been added.
   (b) Find the Edmundson–Madansky upper bound.
   (c) Find the piecewise linear upper bound.

(d) Try to find a good variable for partitioning.

3. Assume that you are facing a decision problem where randomness is involved. You have no idea about the distribution of the random variables involved. However, you can obtain samples from the distribution by running an expensive experiment. You have decided to use stochastic decomposition to solve the problem, but are concerned that you may not be able to perform enough experiments for convergence to take place. The cost of a single experiment is much higher than the costs involved in the arithmetic operations of the algorithm.

   (a) Argue why (or why not) it is reasonable to use stochastic decomposition under the assumptions given. (You can assume that all necessary convexity is there.)

   (b) What changes could you suggest in stochastic decomposition in order to (at least partially) overcome the fact that samples are so expensive?

4. Let φ be a convex function. Show that

$$x^\star \in \arg\min \varphi \text{ iff } 0 \in \partial\varphi(x^\star).$$

   (See the definition following (9.7).

5. Show that for a convex function φ and any arbitrary z the subdifferential $\partial\varphi(z)$ is a convex set. [Hint: For any subgradient (9.7) has to hold.]

6. Assume that you are faced with a large number of linear programs that you need to solve. They represent all recourse problems in a two-stage stochastic program. There is randomness in both the objective function and the right-hand side, but the random variables affecting the objective are different from, and independent of, the random variables affecting the right-hand side.

   (a) Argue why (or why not) it is a good idea to use some version of bunching or trickling down to solve the linear programs.

   (b) Given that you *must* use bunching or trickling down in some version, how would you organize the computations?

7. First consider the following integer programming problem:

$$\min_x\{cx \mid Ax \le h,\ x_i \in \{0, \ldots, b_i\}\ \forall i\}.$$

   Next, consider the problem of finding $E\phi(\tilde{x})$, with

$$\phi(x) = \min_y\{cy \mid Ay \le h,\ 0 \le y \le x\}.$$

(a) Assume that you solve the integer program with branch-and-bound. Your first step is then to solve the integer program above, but with $x_i \in \{0, \ldots, b_i\}$ $\forall i$ replaced by $0 \leq x \leq b$. Assume that you get $\hat{x}$. Explain why $\hat{x}$ can be a good partitioning point if you wanted to find $E\phi(\tilde{x})$ by repeatedly partitioning the support, and finding bounds on each cell. [Hint: It may help to draw a little picture.]

(b) We have earlier referred to Figure 18, stating that it can be seen as both the partitioning of the support for the stochastic program, and partitioning the solution space for the integer program. Will the number of cells be largest for the integer or the stochastic program above? Note that there is not necessarily a clear answer here, but you should be able make arguments on the subject. Question (a) may be of some help.

8. Look back at Figure 17. There we replaced one distribution by two others: one yielding an upper bound, and one a lower bound. The possible values for these two new distributions were not the same. How would you use the ideas of Jensen and Edmundson–Madansky to achieve, as far as possible, the same points? You can assume that the distribution is bounded. [Hint: The Edmundson–Madansky distribution will have two more points than the Jensen distribution.]

References

[1] Benders J. F. (1962) Partitioning procedures for solving mixed-variables programming problems. *Numer. Math.* 4: 238–252.

[2] Berland N. J. (1993) *Stochastic optimization and parallel processing.* PhD thesis, Department of Informatics, University of Bergen.

[3] Berland N. J. and Wallace S. W. (1993) Partitioning of the support to tighten bounds on stochastic PERT problems. Working paper, Department of Managerial Economics and Operations Research, Norwegian Institute of Technology, Trondheim.

[4] Berland N. J. and Wallace S. W. (1993) Partitioning the support to tighten bounds on stochastic linear programs. Working paper, Department of Managerial Economics and Operations Research, Norwegian Institute of Technology, Trondheim.

[5] Birge J. R. and Dulá J. H. (1991) Bounding separable recourse functions with limited distribution information. *Ann. Oper. Res.* 30: 277–298.

[6] Birge J. R. and Holmes D. (1992) Efficient solution of two stage stochastic linear programs using interior point methods. *Comp. Opt. Appl.* 1: 245–276.

[7] Birge J. R. and Qi L. (1988) Computing block-angular Karmarkar

projections with applications to stochastic programming. *Management Sci.* pages 1472–1479.

[8] Birge J. R. and Wallace S. W. (1988) A separable piecewise linear upper bound for stochastic linear programs. *SIAM J. Control and Optimization* 26: 725–739.

[9] Birge J. R. and Wets R. J.-B. (1986) Designing approximation schemes for stochastic optimization problems, in particular for stochastic programs with recourse. *Math. Prog. Study* 27: 54–102.

[10] Birge J. R. and Wets R. J.-B. (1987) Computing bounds for stochastic programming problems by means of a generalized moment problem. *Math. Oper. Res.* 12: 149–162.

[11] Birge J. R. and Wets R. J.-B. (1989) Sublinear upper bounds for stochastic programs with recourse. *Math. Prog.* 43: 131–149.

[12] Dulá J. H. (1987) An upper bound on the expectation of sublinear functions of multivariate random variables. Preprint, CORE.

[13] Dulá J. H. (1992) An upper bound on the expectation of simplicial functions of multivariate random variables. *Math. Prog.* 55: 69–80.

[14] Dupačová J. (1976) Minimax stochastic programs with nonconvex nonseparable penalty functions. In Prékopa A. (ed) *Progress in Operations Research*, pages 303–316. North-Holland, Amsterdam.

[15] Dupačová J. (1980) Minimax stochastic programs with nonseparable penalties. In Iracki K., Malanowski K., and Walukiewicz S. (eds) *Optimization Techniques, Part I*, volume 22 of *Lecture Notes in Contr. Inf. Sci.*, pages 157–163. Springer-Verlag, Berlin.

[16] Dupačová J. (1987) The minimax approach to stochastic programming and an illustrative application. *Stochastics* 20: 73–88.

[17] Edirisinghe N. C. P. and Ziemba W. T. (1994) Bounding the expectation of a saddle function, with application to stochastic programming. *Math. Oper. Res.* 19: 314–340.

[18] Edirisinghe N. C. P. and Ziemba W. T. (1994) Bounds for two-stage stochastic programs with fixed recourse. *Math. Oper. Res.* 19: 292–313.

[19] Edmundson H. P. (1956) Bounds on the expectation of a convex function of a random variable. Technical Report Paper 982, The RAND Corporation.

[20] Ermoliev Y. (1983) Stochastic quasigradient methods and their application to systems optimization. *Stochastics* 9: 1–36.

[21] Ermoliev Y. (1988) Stochastic quasigradient methods. In Ermoliev Y. and Wets R. J.-B. (eds) *Numerical Techniques for Stochastic Optimization*, pages 143–185. Springer-Verlag.

[22] Everitt R. and Ziemba W. T. (1979) Two-period stochastic programs with simple recourse. *Oper. Res.* 27: 485–502.

[23] Frauendorfer K. (1988) Solving SLP recourse problems with arbitrary multivariate distributions—the dependent case. *Math. Oper. Res.* 13: 377–394.

[24] Frauendorfer K. (1989) A simplicial approximation scheme for convex two-stage stochastic programming problems. Manuscript, Inst. Oper. Res., University of Zurich.

[25] Frauendorfer K. (1992) *Stochastic Two-Stage Programming*, volume 392 of *Lecture Notes in Econ. Math. Syst.* Springer-Verlag, Berlin.

[26] Frauendorfer K. and Kall P. (1988) A solution method for SLP recourse problems with arbitrary multivariate distributions – the independent case. *Probl. Contr. Inf. Theory* 17: 177–205.

[27] Gaivoronski A. (1988) Interactive program SQG-PC for solving stochastic programming problems on IBM PC/XT/AT compatibles—user guide. Working Paper WP-88-11, IIASA, Laxenburg.

[28] Gaivoronski A. (1988) Stochastic quasigradient methods and their implementation. In Ermoliev Y. and Wets R. J.-B. (eds) *Numerical Techniques for Stochastic Optimization*, pages 313–351. Springer-Verlag.

[29] Gartska S. J. and Rutenberg D. P. (1973) Computation in discrete stochastic programs with recourse. *Oper. Res.* 21: 112–122.

[30] Gassmann H. I. (1989) Optimal harvest of a forest in the presence of uncertainty. *Can. J. Forest Res.* 19: 1267–1274.

[31] Gassmann H. I. (1990) MSLiP: A computer code for the multistage stochastic linear programming problem. *Math. Prog.* 47: 407–423.

[32] Gassmann H. I. and Wallace S. W. (1993) Solving linear programs with multiple right hand sides: Pivoting and ordering schemes. Working paper, Department of Economics, Norwegian Institute of Technology, Trondheim.

[33] Gassmann H. and Ziemba W. T. (1986) A tight upper bound for the expectation of a convex function of a multivariate random variable. *Math. Prog. Study* 27: 39–53.

[34] Hansotia B. J. (1980) Stochastic linear programs with simple recourse: The equivalent deterministic convex program for the normal, exponential and Erlang cases. *Naval. Res. Logist. Quart.* 27: 257–272.

[35] Haugland D. and Wallace S. W. (1988) Solving many linear programs that differ only in the righthand side. *Eur. J. Oper. Res.* 37: 318–324.

[36] Hausch D. B. and Ziemba W. T. (1983) Bounds on the value of information in uncertain decision problems, II. *Stochastics* 10: 181–217.

[37] Higle J. L. and Sen S. (1991) Stochastic decomposition: An algorithm for two stage stochastic linear programs with recourse. *Math. Oper. Res.* 16: 650–669.

[38] Higle J. L. and Sen S. (1991) Statistical verification of optimality conditions for stochastic programs with recourse. *Ann. Oper. Res.* 30: 215–240.

[39] Huang C. C., Vertinsky I., and Ziemba W. T. (1977) Sharp bounds on the value of perfect information. *Oper. Res.* 25: 128–139.

[40] Huang C. C., Ziemba W. T., and Ben-Tal A. (1977) Bounds on the expectation of a convex function of a random variable: With applications to stochastic programming. *Oper. Res.* 25: 315–325.

[41] Jensen J. L. (1906) Sur les fonctions convexes et les inégalités entre les valeurs moyennes. *Acta Math.* 30: 173–177.

[42] Jessup E. R., Yang D., and Zenios S. A. (1993) Parallel factorization of structured matrices arising in stochastic programming. Report 93-02, Department of Public and Business Administartion, University of Cyprus, Nicosia, Cyprus.

[43] Kall P. (1974) Approximations to stochastic programs with complete fixed recourse. *Numer. Math.* 22: 333–339.

[44] Kall P. (1979) Computational methods for solving two-stage stochastic linear programming problems. *Z. Angew. Math. Phys.* 30: 261–271.

[45] Kall P. (1986) Approximation to optimization problems: An elementary review. *Math. Oper. Res.* 11: 9–18.

[46] Kall P. (1987) On approximations and stability in stochastic programming. In Guddat J., Jongen H. T., Kummer B., and Nožička F. (eds) *Parametric Optimization and Related Topics*, pages 387–407. Akademie-Verlag, Berlin.

[47] Kall P. (1988) Stochastic programming with recourse: Upper bounds and moment problems—a review. In Guddat J., Bank B., Hollatz H., Kall P., Klatte D., Kummer B., Lommatzsch K., Tammer K., Vlach M., and Zimmermann K. (eds) *Advances in Mathematical Optimization (Dedicated to Prof. Dr. Dr. hc. F. Nožička)*, pages 86–103. Akademie-Verlag, Berlin.

[48] Kall P. (1991) An upper bound for SLP using first and total second moments. *Ann. Oper. Res.* 30: 267–276.

[49] Kall P., Ruszczyński A., and Frauendorfer K. (1988) Approximation techniques in stochastic programming. In Ermoliev Y. M. and Wets R. J.-B. (eds) *Numerical Techniques for Stochastic Optimization*, pages 33–64. Springer-Verlag, Berlin.

[50] Kall P. and Stoyan D. (1982) Solving stochastic programming problems with recourse including error bounds. *Math. Operationsforsch. Statist., Ser. Opt.* 13: 431–447.

[51] Karr A. F. (1983) Extreme points of certain sets of probability measures, with applications. *Math. Oper. Res.* 8: 74–85.

[52] Kemperman J. M. B. (1968) The general moment problem, a geometric approach. *Ann. Math. Statist.* 39: 93–122.

[53] Laporte G. and Louveaux F. V. (1993) The integer l-shaped method for stochastic integer programs. *Oper. Res. Lett.* 13: 133–142.

[54] Louveaux F. V. (1986) Multistage stochastic linear programs with block separable recourse. *Math. Prog. Study* 28: 48–62.

[55] Lustig I. J., Mulvey J. M., and Carpenter T. J. (1991) Formulating two-stage stochastic programs for interior point methods. *Oper. Res.* 39: 757–770.

[56] Madansky A. (1959) Bounds on the expectation of a convex function of a multivariate random variable. *Ann. Math. Statist.* 30: 743–746.

[57] Marti K. (1988) *Descent Directions and Efficient Solutions in Discretely*

Distributed Stochastic Programs, volume 299 of *Lecture Notes in Econ. Math. Syst.* Springer-Verlag, Berlin.

[58] Marti K. and Fuchs E. (1986) Computation of descent directions and efficient points in stochastic optimization problems without using derivatives. *Math. Prog. Study* 28: 132–156.

[59] Marti K. and Fuchs E. (1986) Rates of convergence of semi-stochastic approximation procedures for solving stochastic optimization problems. *Optimization* 17: 243–265.

[60] Mulvey J. M. and Ruszczyński A. (1992) A new scenario decomposition method for large-scale stochastic optimization. Technical Report SOR-91-19, Princeton University, Princeton, New Jersey.

[61] Ruszczyński A. (1986) A regularized decomposition method for minimizing a sum of polyhedral functions. *Math. Prog.* 35: 309–333.

[62] Ruszczyński A. (1993) Regularized decomposition of stochastic programs: Algorithmic techniques and numerical results. Working Paper WP-93-21, IIASA, Laxenburg.

[63] Van Slyke R. and Wets R. J.-B. (1969) L-shaped linear programs with applications to optimal control and stochastic linear programs. *SIAM J. Appl. Math.* 17: 638-663.

[64] Wallace S. W. (1986) Decomposing the requirement space of a transportation problem into polyhedral cones. *Math. Prog. Study* 28: 29–47.

[65] Wallace S. W. (1986) Solving stochastic programs with network recourse. *Networks* 16: 295–317.

[66] Wallace S. W. (1987) A piecewise linear upper bound on the network recourse function. *Math. Prog.* 38: 133–146.

[67] Wallace S. W. (1988) A two-stage stochastic facility location problem with time-dependent supply. In Ermoliev Y. and Wets R. J.-B. (eds) *Numerical Techniques in Stochastic Optimization*, pages 489–514. Springer-Verlag, Berlin.

[68] Wallace S. W. and Yan T. (1993) Bounding multistage stochastic linear programs from above. *Math. Prog.* 61: 111–130.

[69] Wets R. (1966) Programming under uncertainty: The complete problem. *Z. Wahrsch. theorie u. verw. Geb.* 4: 316–339.

[70] Wets R. (1983) Stochastic programming: Solution techniques and approximation schemes. In Bachem A., Grötschel M., and Korte B. (eds) *Mathematical Programming: The State-of-the-Art, Bonn 1982*, pages 566–603. Springer-Verlag, Berlin.

[71] Wets R. J.-B. (1983) Solving stochastic programs with simple recourse. *Stochastics* 10: 219–242.

[72] Wets R. J.-B. (1988) Large scale linear programming techniques. In Ermoliev Y. and Wets R. J.-B. (eds) *Numerical Techniques for Stochastic Optimization*, pages 65–93. Springer-Verlag.

4

Probabilistic Constraints

As we have seen in Sections 1.4 and 1.5, at least under appropriate assumptions, chance-constrained problems such as (3.21), or particularly (3.23), as well as recourse problems such as (3.11), or particularly (3.16), (all from Chapter 1), appear as ordinary convex smooth mathematical programming problems. This might suggest that these problems may be solved using known nonlinear programming methods. However, this viewpoint disregards the fact that in the direct application of those methods to problems like

$$\begin{aligned} &\min\nolimits_{x \in X} E_{\tilde{\xi}} c^T(\tilde{\xi}) x \\ &\text{s.t.} \;\; P(\{\xi \mid T(\xi)x \geq h(\xi)\}) \geq \alpha \end{aligned}$$

or

$$\min_{x \in X} E_{\tilde{\xi}}\{c^T x + Q(x, \tilde{\xi})\}$$

where

$$Q(x, \xi) = \min\{q^T y \mid Wy \geq h(\xi) - T(\xi)x, \; y \in Y\},$$

we had repeatedly to obtain gradients and evaluations for functions like

$$P(\{\xi \mid T(\xi)x \geq h(\xi)\})$$

or

$$E_{\tilde{\xi}}\{c^T x + Q(x, \tilde{\xi})\}.$$

Each of these evaluations requires multivariate numerical integration, so that up to now this seems to be outside of the set of efficiently solvable problems. Hence we may try to follow the basic ideas of some of the known nonlinear programming methods, but at the same time we have to find ways to evade the exact evaluation of the integral functions contained in these problems.

On the other hand we also know from the example illustrated in Figure 17 of Chapter 1 that chance constraints may easily define nonconvex feasible sets. This leads to severe computational problems if we intend to find a global optimum. There is one exception to this general problem worth mentioning.

Proposition 4.1 *The feasible set*

$$\mathcal{B}(1) := \{x \mid P(\{\xi \mid T(\xi)x \geq h(\xi)\}) \geq 1\}$$

is convex.

Proof Assume that $x, y \in \mathcal{B}(1)$ and that $\lambda \in (0,1)$. Then for $\Xi_x := \{\xi \mid T(\xi)x \geq h(\xi)\}$ and $\Xi_y := \{\xi \mid T(\xi)y \geq h(\xi)\}$ we have $P(\Xi_x) = P(\Xi_y) = 1$. As is easily shown, this implies for $\Xi_\cap := \Xi_x \cap \Xi_y$ that $P(\Xi_\cap) = 1$. Obviously, for $z := \lambda x + (1-\lambda)y$ we have $T(\xi)z \geq h(\xi)$ $\forall \xi \in \Xi_\cap$ such that $\{\xi \mid T(\xi)z \geq h(\xi)\} \supset \Xi_\cap$. Hence we have $z \in \mathcal{B}(1)$. □

Considering once again the example illustrated in Figure 17 in Section 1.5, we observe that if we had required a reliability $\alpha > 93\%$, the feasible set would have been convex. This is a consequence of Proposition 4.1 for discrete distributions, and may be stated as follows.

Proposition 4.2 *Let $\tilde{\xi}$ have a finite discrete distribution described by $P(\xi = \xi^j) = p_j$, $j = 1, \cdots, r$ ($p_j > 0$ $\forall j$). Then for $\alpha > 1 - \min_{j\in\{1,\cdots,r\}} p_j$ the feasible set*

$$\mathcal{B}(\alpha) := \{x \mid P(\{\xi \mid T(\xi)x \geq h(\xi)\}) \geq \alpha\}$$

is convex.

Proof: The assumption on α implies that $\mathcal{B}(\alpha) = \mathcal{B}(1)$ (see Exercises at the end of this chapter). □

In conclusion, for discrete distributions and reliability levels chosen "high enough" we have a convex problem. Replacing $E_{\tilde{\xi}} c(\tilde{\xi})$ by c, we then simply have to solve the linear program (provided that X is convex polyhedral)

$$\begin{array}{l} \min_{x\in X} c^{\mathrm{T}}x \\ \text{s.t. } T(\xi^j)x \geq h(\xi^j),\ j = 1, \cdots, r. \end{array}$$

This observation may be helpful for some particular chance-constrained problems with discrete distributions. However, it also tells us that for chance-constrained problems stated with continuous-type distributions and requiring a reliability level $\alpha < 1$, we cannot expect—as discussed in Section 3.5 for the recourse problem—approximating the continuous distribution by successively refined discrete ones to be a successful approach. The reason should now be obvious: refining the discrete (approximating) distributions would imply at some stage that $\min_j p_j < 1-\alpha$ such that the "approximating" problems were likely to become nonconvex—even if the original problem with its continuous distribution were convex. And approximating convex problems by nonconvex ones should certainly not be our aim!

In the next two sections we shall describe under special assumptions (multivariate normal distributions) how chance-constrained programs can be treated computationally. In particular, we shall verify that, under our assumptions, a program with joint chance constraints becomes a convex program and that programs with separate chance contraints may be reformulated to become a deterministic convex program amenable to standard nonlinear programming algorithms.

4.1 Joint Chance Constrained Problems

Let us concentrate on the particular stochastic linear program

$$\left.\begin{aligned} &\min c^{\mathrm{T}} x \\ &\text{s.t. } P(\{\xi \mid Tx \geq \xi\}) \geq \alpha \\ &\qquad Dx = d, \\ &\qquad\ \ x \geq 0. \end{aligned}\right\} \tag{1.1}$$

For this problem we know from Propositions 1.5–1.7 in Section 1.5 that if the distribution function F is quasi-concave then the feasible set $\mathcal{B}(\alpha)$ is a closed convex set.

Under the assumption that $\tilde{\xi}$ has a (multivariate) normal distribution, we know that F is even log-concave. We therefore have a smooth convex program. For this particular case there have been attempts to adapt penalty and cutting-plane methods to solve (1.1). Further, variants of the reduced gradient method as sketched in Section 1.7.2 have been designed.

These approaches all attempt to avoid the "exact" numerical integration associated with the evaluation of $F(Tx) = P(\{\xi \mid Tx \geq \xi\})$ and its gradient $\nabla_x F(Tx)$ by relaxing the probabilistic constraint

$$P(\{\xi \mid Tx \geq \xi\}) \geq \alpha.$$

To see how this may be realized, let us briefly sketch one iteration of the reduced gradient method's variant implemented in PROCON, a computer program for minimizing a function under PRObabilistic CONstraints.

With the notation

$$G(x) := P(\{\xi \mid Tx \geq \xi\}),$$

let x be feasible in

$$\left.\begin{aligned} &\min c^{\mathrm{T}} x \\ &\text{s.t. } G(x) \geq \alpha, \\ &\qquad Dx = d, \\ &\qquad\ \ x \geq 0, \end{aligned}\right\} \tag{1.2}$$

and—assuming D to have full row rank—let D be partitioned as $D = (B, N)$ into basic and nonbasic parts and accordingly partition $x^{\mathrm{T}} = (y^{\mathrm{T}}, z^{\mathrm{T}})$, $c^{\mathrm{T}} =$

$(f^{\mathrm{T}}, g^{\mathrm{T}})$ and a descent direction $w^{\mathrm{T}} = (u^{\mathrm{T}}, v^{\mathrm{T}})$. Assume further that for some tolerance $\varepsilon > 0$,

$$y_j > \varepsilon \quad \forall j \quad \text{(strict nondegeneracy).} \tag{1.3}$$

Then the search direction $w^{\mathrm{T}} = (u^{\mathrm{T}}, v^{\mathrm{T}})$ is determined by the linear program

$$\left.\begin{array}{lrl} \max \tau & & \\ \text{s.t.} & f^{\mathrm{T}}u + \phantom{\nabla_z G(x)^{\mathrm{T}}} & g^{\mathrm{T}}v \le -\tau, \\ & \nabla_y G(x)^{\mathrm{T}}u + \nabla_z G(x)^{\mathrm{T}} v & \ge \theta\tau \text{ if } G(x) \le \alpha + \varepsilon, \\ & Bu + \phantom{\nabla_z G(x)^{\mathrm{T}}} & Nv = 0, \\ & & v_j \ge 0 \ \text{ if } z_j \le \varepsilon, \\ & & \|v\|_\infty \le 1, \end{array}\right\} \tag{1.4}$$

where $\theta > 0$ is a fixed parameter as a weight for the directional derivatives of G and $\|v\|_\infty = \max_j\{|v_j|\}$. According to the above assumption, we have from (1.4)

$$u = -B^{-1}Nv,$$

which renders (1.4) into the linear program

$$\left.\begin{array}{ll} \max \tau & \\ \text{s.t.} & r^{\mathrm{T}}v \le -\tau, \\ & s^{\mathrm{T}}v \ge \theta\tau \text{ if } G(x) \le \alpha + \varepsilon, \\ & v_j \ge 0 \text{ if } z_j \le \varepsilon, \\ & \|v\|_\infty \le 1, \end{array}\right\} \tag{1.5}$$

where obviously

$$\begin{aligned} r^{\mathrm{T}} &= g^{\mathrm{T}} - f^{\mathrm{T}}B^{-1}N, \\ s^{\mathrm{T}} &= \nabla_z G(x)^{\mathrm{T}} - \nabla_y G(x)^{\mathrm{T}} B^{-1} N \end{aligned}$$

are the reduced gradients of the objective and the probabilistic constraint function. Problem (1.5)—and hence (1.4)—is always solvable owing to its nonempty and bounded feasible set. Depending on the obtained solution $(\tau^*, u^{*\mathrm{T}}, v^{*\mathrm{T}})$ the method proceeds as follows.

Case 1 When $\tau^* = 0$, ε is replaced by 0 and (1.5) is solved again. If $\tau^* = 0$ again, the feasible solution $x^{\mathrm{T}} = (y^{\mathrm{T}}, z^{\mathrm{T}})$ is obviously optimal. Otherwise the steps of *case 2* below are carried out, starting with the original $\varepsilon > 0$.

Case 2 When $0 < \tau^* \le \varepsilon$, the following cycle is entered:

Step 1 Set $\varepsilon := 0.5\varepsilon$.

Step 2 Solve (1.5). If still $\tau^* \leq \varepsilon$, go to step 1; otherwise, case 3 applies.

Case 3 When $\tau^* > \varepsilon$, $w^{*\mathrm{T}} = (u^{*\mathrm{T}}, v^{*\mathrm{T}})$ is accepted as search direction.

If a search direction $w^{*\mathrm{T}} = (u^{*\mathrm{T}}, v^{*\mathrm{T}})$ has been found, a line search follows using bisection. Since the line search in this case amounts to determining the intersection of the ray $x + \mu w^*, \mu \geq 0$ with the boundary $\mathrm{bd}\mathcal{B}(\alpha)$ within the tolerance ε, the evaluation of $G(x)$ becomes important. For this purpose a special Monte Carlo technique is used, which allows efficient computation of upper and lower bounds of $G(x)$ as well as the gradient $\nabla G(x)$.

If the next iterate $\check{x}$, resulting from the line search, still satisfies strict nondegeneracy, the whole step is repeated with the same partition of D into basic and nonbasic parts; otherwise, a basis exchange is attempted to reinstall strict nondegeneracy for a new basis.

4.2 Separate Chance Constraints

Let us now consider stochastic linear programs with separate (or single) chance constraints as introduced at the end of Section 1.3. Using the formulation given there we are dealing with the problem

$$\left.\begin{array}{l} \min_{x\in X} E_{\tilde{\xi}} c^T(\tilde{\xi})x \\ \text{s.t. } P(\{\xi \mid T_i(\xi)x \geq h_i(\xi)\}) \geq \alpha_i,\ i = 1, \cdots, m, \end{array}\right\} \tag{2.1}$$

where $T_i(\xi)$ is the ith row of $T(\xi)$. The main question is whether or under what assumptions the feasibility set defined by any one of the constraints in (2.1),

$$\{x \mid P(\{\xi \mid T_i(\xi)x \geq h_i(\xi)\} \geq \alpha_i\},$$

is convex. As we know from Section 1.5, this question is very simple to answer for the special case where $T_i(\xi) \equiv T_i$, i.e. where only the right-hand side $h_i(\tilde{\xi})$ is random. That is, with F_i the distribution function of $h_i(\tilde{\xi})$,

$$\begin{aligned} \{x \mid P(\{\xi \mid T_i x \geq h_i(\xi)\}) \geq \alpha_i\} &= \{x \mid F_i(T_i x) \geq \alpha_i\} \\ &= \{x \mid T_i x \geq F_i^{-1}(\alpha_i)\}. \end{aligned}$$

It follows that the feasibility set for this particular chance constraint is just the feasibility set of an ordinary linear constraint.

For the general case let us first simplify the notation as follows. Let

$$\mathcal{B}_i(\alpha_i) := \{x \mid P(\{(t^{\mathrm{T}}, h)^{\mathrm{T}} \mid t^{\mathrm{T}} x \geq h\}) \geq \alpha_i\},$$

where $(\tilde{t}^{\mathrm{T}}, \tilde{h})^{\mathrm{T}}$ is a random vector. Assume now that $(\tilde{t}^{\mathrm{T}}, \tilde{h})^{\mathrm{T}}$ has a joint normal distribution with expectation $\mu \in \mathbb{R}^{n+1}$ and $(n+1) \times (n+1)$

covariance matrix S. For any fixed x, let $\tilde{\zeta}(x) := x^{\mathrm{T}}\tilde{t} - \tilde{h}$. It follows that our feasible set may be rewritten in terms of the random variable $\tilde{\zeta}(x)$ as $\mathcal{B}_i(\alpha_i) = \{x \mid P(\zeta(x) \geq 0) \geq \alpha_i\}$. From probability theory, we know that, because $\tilde{\zeta}(x)$ is a linear combination of jointly normally distributed random variables, it has a (one-dimensional) normal distribution function $F_{\tilde{\zeta}}$ with expectation $m_{\tilde{\zeta}}(x) = \sum_{j=1}^{n} \mu_j x_j - \mu_{n+1}$, and, using the $(n+1)$-vector $z(x) := (x_1, \cdots, x_n, -1)^{\mathrm{T}}$, the variance $\sigma^2_{\tilde{\zeta}}(x) = z(x)^{\mathrm{T}} S z(x)$. Since the covariance matrix S of a (nondegenerate) multivariate normal distribution is positive-definite, it follows that the variance $\sigma^2_{\tilde{\zeta}}(x)$ and, as can be easily shown, the standard deviation $\sigma_{\tilde{\zeta}}(x)$ are convex in x (and $\sigma_{\tilde{\zeta}}(x) > 0 \ \forall x$ in view of $z_{n+1}(x) = -1$). Hence we have

$$\begin{aligned}\mathcal{B}_i(\alpha_i) &= \{x \mid P(\zeta(x) \geq 0) \geq \alpha_i\} \\ &= \left\{x \,\middle|\, P\left(\frac{\zeta(x) - m_{\tilde{\zeta}}(x)}{\sigma_{\tilde{\zeta}}(x)} \geq \frac{-m_{\tilde{\zeta}}(x)}{\sigma_{\tilde{\zeta}}(x)}\right) \geq \alpha_i\right\}.\end{aligned}$$

Observing that for the normally distributed random variable $\tilde{\zeta}(x)$ the random variable $[\tilde{\zeta}(x) - m_{\tilde{\zeta}}(x)]/\sigma_{\tilde{\zeta}}(x)$ has the standard normal distribution function Φ, it follows that

$$\mathcal{B}_i(\alpha_i) = \left\{x \,\middle|\, 1 - \Phi\left(\frac{-m_{\tilde{\zeta}}(x)}{\sigma_{\tilde{\zeta}}(x)}\right) \geq \alpha_i\right\}.$$

Hence

$$\begin{aligned}\mathcal{B}_i(\alpha_i) &= \left\{x \,\middle|\, 1 - \Phi\left(\frac{-m_{\tilde{\zeta}}(x)}{\sigma_{\tilde{\zeta}}(x)}\right) \geq \alpha_i\right\} \\ &= \left\{x \,\middle|\, \Phi\left(\frac{-m_{\tilde{\zeta}}(x)}{\sigma_{\tilde{\zeta}}(x)}\right) \leq 1 - \alpha_i\right\} \\ &= \left\{x \,\middle|\, \frac{-m_{\tilde{\zeta}}(x)}{\sigma_{\tilde{\zeta}}(x)} \leq \Phi^{-1}(1 - \alpha_i)\right\} \\ &= \left\{x \,\middle|\, -\Phi^{-1}(1 - \alpha_i)\sigma_{\tilde{\zeta}}(x) - m_{\tilde{\zeta}}(x) \leq 0\right\}.\end{aligned}$$

Here $m_{\tilde{\zeta}}(x)$ is linear affine in x and $\sigma_{\tilde{\zeta}}(x)$ is convex in x. Therefore the left-hand side of the constraint

$$-\Phi^{-1}(1 - \alpha_i)\sigma_{\tilde{\zeta}}(x) - m_{\tilde{\zeta}}(x) \leq 0$$

is convex iff $\Phi^{-1}(1 - \alpha_i) \leq 0$, which is exactly the case iff $\alpha_i \geq 0.5$. Hence we have, under the assumption of normal distributions and $\alpha_i \geq 0.5$, instead of (2.1) a deterministic convex program with constraints of the type

$$-\Phi^{-1}(1 - \alpha_i)\sigma_{\tilde{\zeta}}(x) - m_{\tilde{\zeta}}(x) \leq 0,$$

which can be solved with standard tools of nonlinear programming.

4.3 Bounding Distribution Functions

In Section 4.1 we mentioned that particular methods have been developed to compute lower and upper bounds for the function

$$G(x) := P(\{\xi \mid Tx \geq \xi\}) = F_{\tilde{\xi}}(Tx)$$

contained in the constraints of problem (1.1). Here $F_{\tilde{\xi}}(\cdot)$ denotes the distribution function of the random vector $\tilde{\xi}$. In the following we sketch some ideas underlying these bounding methods. For a more technical presentation, the reader should consult the references provided below.

To simplify the notation,let us assume that $\tilde{\xi}$ is a random vector with a support $\Xi \subset \mathbb{R}^n$. For any $z \in \mathbb{R}^n$, we have

$$F_{\tilde{\xi}}(z) = P(\{\xi \mid \xi_1 \leq z_1, \cdots, \xi_n \leq z_n\}).$$

Defining the events $A_i := \{\xi \mid \xi_i \leq z_i\}$, $i = 1, \cdots, n$, it follows that

$$F_{\tilde{\xi}}(z) = P(A_1 \cap \cdots \cap A_n).$$

Denoting the complements of the events A_i by

$$B_i := A_i^c = \{\xi \mid \xi_i > z_i\},$$

we know from elementary probability theory that

$$A_1 \cap \cdots \cap A_n = (B_1 \cup \cdots \cup B_n)^c,$$

and consequently

$$\begin{aligned} F_{\tilde{\xi}}(z) &= P(A_1 \cap \cdots \cap A_n) \\ &= P\left((B_1 \cup \cdots \cup B_n)^c\right) \\ &= 1 - P(B_1 \cup \cdots \cup B_n). \end{aligned}$$

Therefore asking for the value of $F_{\tilde{\xi}}(z)$ is equivalent to looking for the probability that at least one of the events $B_1, \cdots, B_n$ occurs. Defining the counter $\tilde{\nu} : \Xi \longrightarrow \mathbb{N}$ by

$$\tilde{\nu}(\xi) := \{\text{number of events out of } B_1, \cdots, B_n \text{ that occur at } \xi\},$$

$\tilde{\nu}$ is clearly a random variable having the range of integers $\{0, 1, \cdots, n\}$. Observing that $P(B_1 \cup \cdots \cup B_n) = P(\tilde{\nu} \geq 1)$, we have

$$F_{\tilde{\xi}}(z) = 1 - P(\tilde{\nu} \geq 1).$$

Hence finding a good approximation for $P(\tilde{\nu} \geq 1)$ yields at the same time a satisfactory approximation of $F_{\tilde{\xi}}(z)$.

With the binomial coefficients for $\mu, k \in \mathbb{N}$ defined for $\mu \geq k$ as

$$\binom{\mu}{k} = \frac{\mu!}{k!(\mu - k)!}$$

(where $0! = 1$ and $\binom{\mu}{k} = 0$ for $\mu < k$) the *binomial moments* of $\tilde{\nu}$ are introduced as

$$S_{k,n} := E_{\tilde{\xi}}\left[\binom{\tilde{\nu}}{k}\right] = \sum_{i=0}^{n} \binom{i}{k} P(\{\xi \mid \tilde{\nu}(\xi) = i\}),\ k = 0, 1, \cdots, n. \quad (3.1)$$

Since $\binom{i}{0} = 1,\ i = 0, 1, \cdots, n$, it follows that $S_{0,n} = 1$. Furthermore, choosing $v \in \mathbb{R}^{n+1}$ according to $v_i := P(\{\xi \mid \tilde{\nu}(\xi) = i\}),\ i = 0, 1, \cdots, n$, it is obvious from (3.1) that v solves the system of linear equations

$$\left.\begin{array}{rcl} v_0 + v_1 + \ v_2 + \cdots \ + & v_n = & S_{0,n}, \\ v_1 + 2v_2 + \cdots \ + & nv_n = & S_{1,n}, \\ v_2 + \cdots \ + & \binom{n}{2} v_n = & S_{2,n}, \\ \ddots & & \vdots \\ \ddots & \vdots & \\ & v_n = & S_{n,n}. \end{array}\right\} \quad (3.2)$$

The coefficient matrix of (3.2) is upper-triangular, with all main diagonal elements equal to 1, and hence with a determinant of 1, such that $v_i = P(\{\xi \mid \tilde{\nu}(\xi) = i\}),\ i = 0, 1, \cdots, n$, is the unique solution of this system of linear equations. However, solving the complete system (3.2) to get $P(\tilde{\nu} \geq 1) = \sum_{i=1}^{n} v_i$ would require the computation of *all* binomial moments. This would be a cumbersome task again.

Instead, we could proceed as follows. Observing that our unique solution, representing probabilities, is nonnegative, it is no restriction to add the conditions $v_i \geq 0,\ \forall i$ to (3.2). In turn, we relax the system by dropping some of the equations (also the first one), in that way getting rid of the need to determine the corresponding binomial moments. Obviously, the above (formerly unique) solution is still feasible to the relaxed system, but no longer unique in general. Hence we get a lower or upper bound on $P(\tilde{\nu} \geq 1)$ by minimizing or maximizing, respectively, the objective $\sum_{i=1}^{n} v_i$ under the relaxed constraints.

To be more specific, let us consider the following relaxation as an example.

For the lower bound we choose

$$\left.\begin{array}{l}\min\{v_1+v_2+\cdots+v_n\}\\ \text{s.t. } v_1+2v_2+\cdots+ \qquad nv_n=S_{1,n},\\ \qquad v_2+\cdots+\binom{n}{2}v_n=S_{2,n},\\ \qquad\qquad v_i\geq 0,\ i=1,\cdots,n.\end{array}\right\} \tag{3.3}$$

and correspondingly for the upper bound we formulate

$$\left.\begin{array}{l}\max\{v_1+v_2+\cdots+v_n\}\\ \text{s.t. } v_1+2v_2+\cdots+ \qquad nv_n=S_{1,n},\\ \qquad v_2+\cdots+\binom{n}{2}v_n=S_{2,n},\\ \qquad\qquad v_i\geq 0,\ i=1,\cdots,n.\end{array}\right\} \tag{3.4}$$

These linear programs are feasible and bounded, and therefore solvable. So, there exist optimal feasible 2×2 bases B.

Consider an arbitrary 2×2 matrix of the form

$$B=\begin{pmatrix} i & i+r\\ \binom{i}{2} & \binom{i+r}{2}\end{pmatrix},$$

where $1\leq i<n$ and $1\leq r\leq n-i$. Computing the determinant of B, we get

$$\begin{aligned}\det B &= i\binom{i+r}{2}-(i+r)\binom{i}{2}\\ &= \tfrac{1}{2}[i(i+r)(i+r-1)-(i+r)i(i-1)]\\ &= \tfrac{1}{2}i(i+r)r\\ &> 0\end{aligned}$$

for all i and r such that $1\leq i<n$ and $1\leq r\leq n-i$. Hence any two columns of the coefficient matrix of (3.3) (or equivalently of (3.4)) form a basis. The question is which one is feasible and optimal. Let us consider the second property first. According to Proposition 1.15, Section 1.6 (page 60), a basis B of (3.3) satisfies the optimality condition if

$$1-e^{\mathrm{T}}B^{-1}N_j\geq 0\ \forall j\neq i,i+r,$$

where $e^{\mathrm{T}}=(1,1)$ and N_j is the jth column of the coefficient matrix of (3.3). Obviously, for (3.4) we have the reverse inequality as optimality condition:

$$1-e^{\mathrm{T}}B^{-1}N_j\leq 0\ \forall j\neq i,i+r.$$

It is straithforward to check[1] that

$$B^{-1} = \begin{pmatrix} \dfrac{i+r-1}{ir} & -\dfrac{2}{ir} \\ -\dfrac{i-1}{(i+r)r} & \dfrac{2}{(i+r)r} \end{pmatrix}.$$

For $N_j = \left(\begin{pmatrix} j \\ j \\ 2 \end{pmatrix} \right)$ we get

$$e^{\mathrm{T}} B^{-1} N_j = j \frac{2i+r-j}{i(i+r)}. \tag{3.5}$$

Proposition 4.3 *The basis*

$$B = \left(\begin{pmatrix} i \\ i \\ 2 \end{pmatrix} \begin{pmatrix} i+r \\ i+r \\ 2 \end{pmatrix} \right)$$

satisfies the optimality condition

(a) for (3.3) if and only if $r = 1$ (i arbitrary);

(b) for (3.4) if and only if $i = 1$ and $i + r = n$.

Proof

(a) If $r \geq 2$, we get from (3.5) for $j = i + 1$

$$\begin{aligned} e^{\mathrm{T}} B^{-1} N_{i+1} &= j \frac{2i+r-j}{i(i+r)} \\ &= \frac{i(i+r)+r-1}{i(i+r)} \\ &> 1, \end{aligned}$$

so that the optimality condition for (3.3) is not satisfied for $r > 1$, showing that $r = 1$ is necessary.

Now let $r = 1$. Then for $j < i$ we have, according to (3.5),

$$\begin{aligned} e^{\mathrm{T}} B^{-1} N_j &= j \frac{2i+1-j}{i(i+1)} \\ &= \frac{j + i^2 - (j-i)^2}{i(i+1)} \\ &< \frac{i(i+1) - (j-i)^2}{i(i+1)} \\ &< 1, \end{aligned}$$

[1] $BB^{-1} = I$, the identity matrix!

whereas for $j > i+1$ we get

$$\begin{aligned} e^{\mathrm{T}}B^{-1}N_j &= j\frac{2i+1-j}{i(i+1)} \\ &= \frac{j(i+1)+j(i-j)}{i(i+1)} \\ &< 1, \end{aligned}$$

the last inequality resulting from the fact that subtracting the denominator from the numerator yields

$$j(i+1)+j(i-j)-i(i+1) = \underbrace{(j-i)}_{>1}\underbrace{[(i+1)-j]}_{<0} < 0.$$

Hence in both cases the optimality condition for (3.3) is *strictly* satisfied.

(b) If $i+r<n$ then we get from (3.5) for $j=n$

$$\begin{aligned} e^{\mathrm{T}}B^{-1}N_n &= \frac{n(i+r)+n(i-n)}{i(i+r)} \\ &< 1 \end{aligned}$$

since

$$\begin{aligned} \{\text{numerator}\} - \{\text{denominator}\} &= n(i+r)+n(i-n)-i(i+r) \\ &= (n-i)(i+r-n) \\ &< 0. \end{aligned}$$

Finally, if $i>1$ then, with (3.5), we have for $j=1$

$$\begin{aligned} e^{\mathrm{T}}B^{-1}N_n &= \frac{2i+r-1}{i(i+r)} \\ &= \frac{(i-1)+(i+r)}{i(i+r)} \\ &= \frac{i-1}{i(i+r)}+\frac{1}{i} \\ &< \tfrac{1}{3}+\tfrac{1}{2} \\ &< 1. \end{aligned}$$

Hence the only possible choice for a basis satisfying the optimality condition for problem (3.4) is $i=1, r=n-1$.

□

As can be seen from the simplex method, a basis that satisfies the optimality condition strictly does determine a unique optimal solution if it is feasible.

Hence we now have to find from the optimal bases

$$B = \begin{pmatrix} i & i+1 \\ \binom{i}{2} & \binom{i+1}{2} \end{pmatrix}$$

the one that is feasible for (3.3).

A basis $B = \begin{pmatrix} i & i+1 \\ \binom{i}{2} & \binom{i+1}{2} \end{pmatrix}$ is feasible for (3.3) if and only if

$$\begin{aligned} B^{-1}\begin{pmatrix} S_{1,n} \\ S_{2,n} \end{pmatrix} &= \begin{pmatrix} 1 & -\dfrac{2}{i} \\ -\dfrac{i-1}{i+1} & \dfrac{2}{i+1} \end{pmatrix} \begin{pmatrix} S_{1,n} \\ S_{2,n} \end{pmatrix} \\ &= \begin{pmatrix} S_{1,n} - \dfrac{2}{i} S_{2,n} \\ -\dfrac{i-1}{i+1} S_{1,n} + \dfrac{2}{i+1} S_{2,n} \end{pmatrix} \\ &\geq 0, \end{aligned}$$

or, equivalently, if

$$(i-1)S_{1,n} \leq 2S_{2,n} \leq iS_{1,n}.$$

Hence we have to choose i such that $i-1 = \lfloor 2S_{2,n}/S_{1,n} \rfloor$, where $\lfloor \alpha \rfloor$ is the integer part of α (i.e. the greatest integer less than or equal to α). With this particular i the optimal value of (3.3) amounts to

$$S_{1,n} - \frac{2}{i} S_{2,n} - \frac{i-1}{i+1} S_{1,n} + \frac{2}{i+1} S_{2,n} = \frac{2}{i+1} S_{1,n} - \frac{2}{i(i+1)} S_{2,n}.$$

Thus we have found a lower bound for $P(\tilde{\nu} \geq 1)$ as

$$P(\tilde{\nu} \geq 1) \geq \frac{2}{i+1} S_{1,n} - \frac{2}{i(i+1)} S_{2,n}, \quad \text{with } i-1 = \left\lfloor \frac{2S_{2,n}}{S_{1,n}} \right\rfloor. \tag{3.6}$$

For the optimal basis of (3.4)

$$B = \begin{pmatrix} 1 & n \\ 0 & \binom{n}{2} \end{pmatrix}$$

we have

$$B^{-1} = \begin{pmatrix} 1 & -\dfrac{2}{n-1} \\ 0 & \dfrac{2}{n(n-1)} \end{pmatrix}$$

and hence

$$B^{-1}\begin{pmatrix} S_{1,n} \\ S_{2,n} \end{pmatrix} = \begin{pmatrix} S_{1,n} - \dfrac{2}{n-1}S_{2,n} \\ \dfrac{2}{n(n-1)}S_{2,n} \end{pmatrix}.$$

The last vector is nonnegative since the definition of the binomial moments implies $(n-1)S_{1,n} - 2S_{2,n} \geq 0$ and $S_{2,n} \geq 0$. This yields for (3.4) the optimal value $S_{1,n} - (2/n)S_{2,n}$. Therefore we finally get an upper bound for $P(\tilde{\nu} \geq 1)$ as

$$P(\tilde{\nu} \geq 1) \leq S_{1,n} - \frac{2}{n}S_{2,n}. \tag{3.7}$$

In conclusion, recalling that

$$F_{\tilde{\xi}}(z) = 1 - P(\tilde{\nu} \geq 1),$$

we have shown the following.

Proposition 4.4 *The distribution function $F_{\tilde{\xi}}(z)$ is bounded according to*

$$F_{\tilde{\xi}}(z) \geq 1 - \left(S_{1,n} - \frac{2}{n}S_{2,n}\right)$$

and

$$F_{\tilde{\xi}}(z) \leq 1 - \left(\frac{2}{i+1}S_{1,n} - \frac{2}{i(i+1)}S_{2,n}\right), \quad \textit{with } i-1 = \left\lfloor \frac{2S_{2,n}}{S_{1,n}} \right\rfloor.$$

Example 4.1 We defined in (3.1) the binomial moments of $\tilde{\nu}$ as

$$S_{k,n} := E_{\tilde{\xi}}\left[\binom{\tilde{\nu}}{k}\right] = \sum_{i=0}^{n}\binom{i}{k} P(\{\xi \mid \tilde{\nu}(\xi) = i\}),\ k = 0, 1, \cdots, n.$$

Another way to introduce these moments is the following. With the same notation as at the beginning of this section, let us define new random variables $\tilde{\chi}_i : \Xi \longrightarrow \mathbb{R},\ i = 1, \cdots, n$, as the indicator functions

$$\tilde{\chi}_i(\xi) := \begin{cases} 1 & \text{if } \xi \in B_i, \\ 0 & \text{otherwise.} \end{cases}$$

Then clearly $\tilde{\nu} = \sum_{i=1}^{n}\tilde{\chi}_i$, and

$$\binom{\tilde{\nu}}{k} = \binom{\tilde{\chi}_1 + \cdots \tilde{\chi}_n}{k} = \sum_{1 \leq i_1 \leq \cdots \leq i_k \leq n} \tilde{\chi}_{i_1}\tilde{\chi}_{i_2}\cdots\tilde{\chi}_{i_k}.$$

Taking the expectation on both sides yields for the binomial moments $S_{k,n}$

$$\begin{aligned} E_{\tilde{\xi}}\left[\binom{\tilde{\nu}}{k}\right] &= \sum_{1 \le i_1 \le \cdots \le i_k \le n} E_{\tilde{\xi}}\left(\tilde{\chi}_{i_1}\tilde{\chi}_{i_2}\cdots\tilde{\chi}_{i_k}\right) \\ &= \sum_{1 \le i_1 \le \cdots \le i_k \le n} P\left(B_{i_1} \cap \cdots \cap B_{i_k}\right). \end{aligned}$$

This formulation indicates the possibility of estimating the binomial moments from large samples through the relation

$$S_{k,n} = \sum_{1 \le i_1 \le \cdots \le i_k \le n} E_{\tilde{\xi}}\left(\tilde{\chi}_{i_1}\tilde{\chi}_{i_2}\cdots\tilde{\chi}_{i_k}\right)$$

if they are difficult to compute directly.

Consider now the following example. Assume that we have a four-dimensional random vector $\tilde{\xi}$ with mutually independent components. Let $z \in \mathbb{R}^4$ be chosen such that with $p_i = P(A_i)$, $i = 1, 2, 3, 4$, we have

$$p^{\mathrm{T}} = (0.9, 0.95, 0.99, 0.92).$$

Consequently, for $q_i = P(B_i) = 1 - p_i$ we get

$$q^{\mathrm{T}} = (0.1, 0.05, 0.01, 0.08).$$

Obviously we get $F_{\tilde{\xi}}(z) = \prod_{i=1}^4 p_i = 0.778734$. From the above representation of the binomial moments, we have

$$\begin{aligned} S_{1,n} &= \sum_{i=1}^{4} q_i &= 0.24 \\ S_{2,n} &= \sum_{i=1}^{3}\sum_{j=i+1}^{4} q_i q_j &= 0.0193 \end{aligned}$$

such that we get from (3.7) for $P(\tilde{\nu} \ge 1)$ the upper bound

$$P_U = 0.24 - \frac{2}{4} \times 0.0193 = 0.23035.$$

According to (3.6), we find $i-1 = \left\lfloor \frac{2\times 0.0193}{0.24} \right\rfloor = 0$ and hence $i = 1$, so that (3.6) yields the lower bound

$$P_L = \frac{2}{2} \times 0.24 - \frac{2}{2} \times 0.0193 = 0.1757.$$

In conclusion, we get for $F_{\tilde{\xi}}(z) = 0.778734$ the bounds $1 - P_U \le F_{\tilde{\xi}}(z) \le 1 - P_L$, and hence

$$0.76965 \le F_{\tilde{\xi}}(z) \le 0.8243.$$

Observe that these bounds could be derived without any specific information about the type of the underlying probability distribution (except the assumption of independent components made only for the sake of a simple presentation). □

Further bounds have been derived for $P(\tilde{\nu} \geq 1)$ using binomial moments up to the order m, $2 < m < n$, as well as for $P(\tilde{\nu} \geq r)$, $r > 1$. For some of them explicit formulae could also be derived, while others require the computational solution of optimization problems with algorithms especially designed for the particular problem structures.

4.4 Bibliographical Notes

One of the first attempts to state deterministic equivalent formulations for chance-constrained programs can be found in Charnes and Cooper [4].

The discussion of convexity of joint chance constraints with stochastic right-hand sides was initiated by Prékopa [7, 8, 11], investigating log-concave measures, and could be extended to quasi-concave measures through the results of Borell [1], Brascamp and Lieb [3] and Rinott [16]. Marti [5] derived convexity statements in particular for separate chance constraints, for various distribution functions and probability levels, including the one mentioned first by van de Panne and Popp [19] for the multivariate normal distribution and described in Section 4.2.

Prékopa [9] proposed an extension of Zoutendijk's method of feasible directions for the solution of (1.1), which was implemented under the name STABIL by Prékopa et al. [15]. For more general types of chance-contrained problems solution, approaches have also been considered by Prékopa [10, 12]. After all, the case of joint chance constraints with nondiscrete random matrix is considered to be a hard problem.

As described in Section 4.1, Mayer developed a special reduced gradient method for (1.1) and implemented it as PROCON [6]. For the evaluation of the probability function $G(x) = P(\{\xi \mid Tx \geq \xi\})$ and its gradient $\nabla G(x)$, an efficient Monte-Carlo technique due to Szántai [17] was used.

An alternative method following the lines of Veinott's supporting hyperplane algorithm was implemented by Szántai [18].

There has been for some time great interest in getting (sharp) bounds for distribution functions and, more generally, for probabilities of certain events in complex systems (e.g. reliabilities of special technical installations). In Section 4.3 we only sketch the direction of thoughts in this field. Among the wide range of literature on the subject, we just refer to the more recent papers of Prékopa [13, 14] and of Boros and Prékopa [2], from which the interested reader may trace back to earlier original work.

Exercises

1. Given a random vector $\tilde{\xi}$ with support Ξ in $\mathbb{R}^k$, assume that for $A \subset \Xi$ and $B \subset \Xi$ we have $P(A) = P(B) = 1$. Show that then also $P(A \cap B) = 1$.
2. Under the assumptions of Proposition 4.2, the support of the distribution is $\Xi = \{\xi^1, \cdots, \xi^r\}$, with $P(\xi = \xi^j) = p_j > 0 \; \forall j$. Show that for $\alpha > 1 - \min_{j\in\{1,\cdots,r\}} p_j$ the only event $A \subset \Xi$ satisfying $P(A) \geq \alpha$ is $A = \Xi$.
3. Show that for the random variable $\tilde{\zeta}(x)$ introduced in Section 4.2 with $\sigma^2_{\tilde{\zeta}}(x)$, $\sigma_{\tilde{\zeta}}(x)$ is also a convex function in x.
4. In Section 4.3 we saw that the binomial moments $S_{0,n}, S_{1,n}, \cdots, S_{n,n}$ determine uniquely the probabilities $v_i = P(\tilde{\nu} = i)$, $i = 0, 1, \cdots, n$, as the solution of (3.2). From the first equation, it follows, owing to $S_{0,n} = 1$, that $\sum_{i=0}^{n} v_i = 1$. To get lower and upper bounds for $P(\tilde{\nu} \geq 1)$, we derived the linear programs (3.3) and (3.4) by omitting, among others, the first equation.

   (a) Show that in any case (provided that $S_{1,n}$ and $S_{2,n}$ are binomial moments) for the optimal solution $\hat{v}$ of (3.3), $\sum_{i=1}^{n} \hat{v}_i \leq 1$.

   (b) If for the optimal solution $\hat{v}$ of (3.3) $\sum_{i=1}^{n} \hat{v}_i < 1$ then we have $v_0 = 1 - \sum_{i=1}^{n} \hat{v}_i > 0$. What does this mean with respect to $F_{\tilde{\xi}}(z)$?

   (c) Solving (3.4) can result in $\sum_{i=1}^{n} \hat{v}_i > 1$. To what extent does this result improve your knowledge about $F_{\tilde{\xi}}(z)$?

References

[1] Borell C. (1975) Convex set functions in d-space. *Period. Math. Hungar.* 6: 111–136.

[2] Boros E. and Prékopa A. (1989) Closed-form two-sided bounds for probabilities that at least r and exactly r out of n events occur. *Math. Oper. Res.* 14: 317–342.

[3] Brascamp H. J. and Lieb E. H. (1976) On extensions of the Brunn–Minkowski and Prekopa–Leindler theorems, including inequalities for log concave functions, and with an application to the diffusion euation. *J. Funct. Anal.* 22: 366–389.

[4] Charnes A. and Cooper W. W. (1959) Chance-constrained programming. *Management Sci.* 5: 73–79.

[5] Marti K. (1971) Konvexitätsaussagen zum linearen stochastischen

Optimierungsproblem. *Z. Wahrsch. theorie u. verw. Geb.* 18: 159–166.
[6] Mayer J. (1988) Probabilistic constrained programming: A reduced gradient algorithm implemented on pc. Working Paper WP-88-39, IIASA, Laxenburg.
[7] Prékopa A. (1970) On probabilistic constrained programming. In Kuhn H. W. (ed) *Proc. of the Princeton Symposioum on Math. Programming*, pages 113–138. Princeton University Press, Princeton, New Jersey.
[8] Prékopa A. (1971) Logarithmic concave measures with applications to stochastic programming. *Acta Sci. Math. (Szeged)* 32: 301–316.
[9] Prékopa A. (1974) Eine Erweiterung der sogenannten Methode der zulässigen Richtungen der nichtlinearen Optimierung auf den Fall quasikonkaver Restriktionen. *Math. Operationsforsch. Statist., Ser. Opt.* 5: 281–293.
[10] Prékopa A. (1974) Programming under probabilistic constraints with a random technology matrix. *Math. Operationsforsch. Statist., Ser. Opt.* 5: 109–116.
[11] Prékopa A. (1980) Logarithmic concave measures and related topics. In Dempster M. A. H. (ed) *Stochastic Programming*, pages 63–82. Academic Press, London.
[12] Prékopa A. (1988) Numerical solution of probabilistic constrained programming problems. In Ermoliev Y. and Wets R. J.-B. (eds) *Numerical Techniques for Stochastic Optimization*, pages 123–139. Springer-Verlag, Berlin.
[13] Prékopa A. (1988) Boole-bonferroni inequalities and linear programming. *Oper. Res.* 36: 145–162.
[14] Prékopa A. (1990) Sharp bounds on probabilities using linear programming. *Oper. Res.* 38: 227–239.
[15] Prékopa A., Ganczer S., Deák I., and Patyi K. (1980) The STABIL stochastic programming model and its experimental application to the electricity production in Hungary. In Dempster M. A. H. (ed) *Stochastic Programming*, pages 369–385. Academic Press, London.
[16] Rinott Y. (1976) On convexity of measures. *Ann. Prob.* 4: 1020–1026.
[17] Szántai T. (1987) Calculation of the multivariate probability distribution function values and their gradient vectors. Working Paper WP-87-82, IIASA, Laxenburg.
[18] Szántai T. (1988) A computer code for solution of probabilistic-constrained stochastic programming problems. In Ermoliev Y. M. and Wets R. J.-B. (eds) *Numerical Techniques for Stochastic Optimization*, pages 229–235. Springer-Verlag, Berlin.
[19] van de Panne C. and Popp W. (1963) Minimum cost cattle feed under probabilistic problem constraint. *Management Sci.* 9: 405–430.

5

Preprocessing

The purpose of this chapter is to discuss different aspects of preprocessing the data associated with a stochastic program. The term "preprocessing" is rather vague, but whatever it could possibly mean, our intention here is to discuss anything that will enhance the model understanding and/or simplify the solution procedures. Thus "preprocessing" refers to any analysis of a problem that takes place before the final solution of the problem. Some tools will focus on the issue of model understanding, while others will focus on issues related to choice of solution procedures. For example, if it can be shown that a problem has (relatively) complete recourse, we can apply solution procedures where that is required. At the same time, the fact that a problem has complete recourse is of value to the modeller, since it says something about the underlying problem (or at least the model of the underlying problem).

5.1 Problem Reduction

Reducing the problem size can be of importance in a setting of stochastic programming. Of course, it is always useful to remove unnecessary rows and columns. In the setting of a single deterministic linear programming problem it may not pay off to remove rows and columns. That is, it may cost more to figure out which columns and rows are not needed than it costs to solve the overall problem with the extra data in it. In the stochastic setting, the same coefficient matrix is used again and again, so it definitely pays to reduce the problem size. The problem itself becomes smaller, and even, more importantly, the number of possible bases can be substantially reduced (especially if we are able to remove rows). This can be perticularly important when using the stochastic decomposition method (where we build up a collection of dual feasible bases) and in trickling down within the setting of the L-shaped decomposition method. Let us start by defining a *frame* and showing how to compute it.

```
procedure framebylp(W:(m × n) matrix);
begin
   n1 := n;
   q := 0;
   for i := n1 downto 1 do begin
      LP(W \ W_i, W_i, q, y,feasible);
      if feasible then begin
         W_i := W_n;
         n := n − 1;
      end;
   end;
end;
```

Figure 1 Finding a frame.

5.1.1 Finding a Frame

Let us repeat the definition of pos W:

$$\text{pos } W = \{t \mid t = Wy, \ y \geq 0\}.$$

In words, pos W is the set of all positive (nonnegative) linear combinations of columns of the matrix W. A subset of the columns, determining a matrix W', is called a frame if pos W = pos W', and equality is not preserved if any one column is removed from W'. So, by finding the frame of a given matrix, we remove all columns that are not needed to describe the pointed cone pos W. As an example, if we use a two-phase simplex method to solve a linear programming problem, only the columns of W' are needed in phase 1.

If W is a matrix, and j is an index, let $W \setminus W_j$ be the matrix W with column j removed.

A simple approach for finding a frame is outlined in Figure 1. To do that, we need a procedure that solves LPs. It can be found in Figure 7. The matrix W in **procedure** framebylp in Figure 1 is both input and output. On entry, it contains the matrix for which we seek the frame; on exit, it contains those columns that were in the frame. The number of columns, n, is changed accordingly.

To summarize, the effect of the frame algorithm is that as many columns as possible are removed from a matrix W without changing the pointed cone spanned by the columns. We have earlier discussed generators of cones. In this case we may say that the columns in W, after the application of **procedure** framebylp, are generators of pos W. Let us now turn to the use of this algorithm.

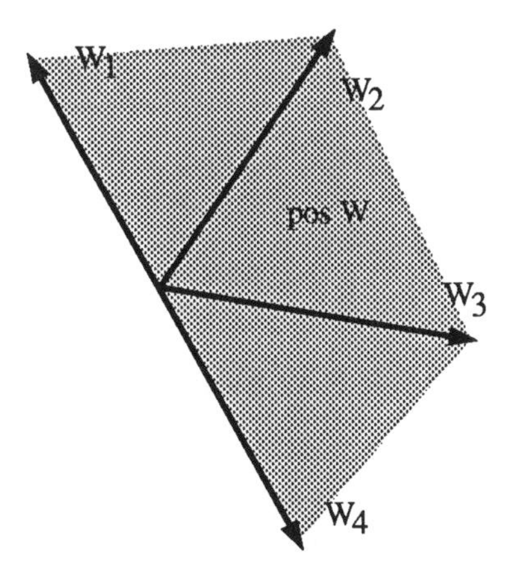

Figure 2 Illustration of the frame algorithm.

5.1.2 Removing Unnecessary Columns

This can be useful in a couple of different settings. Let us first see what happens if we simply apply the frame algorithm to the recourse matrix W. We shall then remove columns that are not needed to describe feasibility. This is illustrated in Figure 2. Given the matrix $W = (W_1, W_2, W_3, W_4)$, we find that the shaded region represents pos W and the output of a frame algorithm is either $W = (W_1, W_2, W_4)$ or $W = (W_1, W_3, W_4)$. The **procedure** framebylp will produce the first of these two cases.

Removing columns not needed for feasibility can be of use when verifying feasibility in the L-shaped decomposition method (see page 159). We are there to solve a given LP for all $\xi \in \mathcal{A}$. If we apply frame to W before checking feasibility, we get a simpler problem to look at, without losing information, since the removed columns add nothing in terms of feasibility. If we are willing to live with two version of the recourse matrix, we can therefore reduce work while computing.

From the modelling perspective, note that columns thrown out are only needed if the cost of the corresponding linear combination is higher than that of the column itself. The variable represented by the column does not add to our production possibilities—only, possibly, to lower our costs. In what follows in this subsection let us assume that we have only right-hand side randomness, and let us, for simplicity, denote the cost vector by q. To see if a column can reduce our costs, we define

$$\overline{W} := \begin{pmatrix} q & 1 \\ W & 0 \end{pmatrix},$$

that is, a matrix containing the coefficient matrix, the cost vector and an extra column. To see the importance of the extra column, consider the following interpretation of pos $\overline{W}$ (remember that pos W equals the set of all positive linear combinations of columns from W):

$$\text{pos}\begin{pmatrix} q_1 & \cdots & q_n & 1 \\ W_1 & \cdots & W_n & 0 \end{pmatrix} = \left\{ \begin{pmatrix} q \\ W \end{pmatrix} \middle| W = \sum_{\lambda_k \geq 0} \lambda_k W_k,\ q \geq \sum_{\lambda_k \geq 0} \lambda_k q_k \right\}.$$

In other words, finding a frame of $\overline{W}$ means removing all columns

$$\begin{pmatrix} q_j \\ W_j \end{pmatrix} \text{ with } W_j = \sum_{\lambda_k \geq 0} \lambda_k W_k, \text{ and } q_j \geq \sum \lambda_k q_k$$

in a sequential manner until we are left with a minimal (but not necessarily unique) set of columns. A column thrown out in this process will *never* be part of an optimal solution, and is hence not needed. It can be dropped. From a modelling point of view, this means that the modeller has added an activity that is clearly inferior. Knowing that it is inferior should add to the modellers understanding of his model.

A column that is not a part of the frame of pos W, but is a part of the frame of pos $\overline{W}$, is one that does not add to our production possibilities, but its existence might add to our profit.

5.1.3 Removing Unnecessary Rows

There is a large amount of research on the topic of eliminating redundant constraints. In this section we shall focus on the use of frames in removing unnecessary rows. Not very surprisingly, this problem has a dual relationship to that of removing columns. Let us first look at it from a general point of view, and then see how we can apply the results in stochastic programming.

Assume we have the system

$$Wy \leq h,\ y \geq 0.$$

Let W^j be the jth row of W, such that the jth inequality is given by $W^j y \leq h_j$. A row j is not needed if there exists a vector $\alpha \geq 0$ such that

$$\sum_{i \neq j} \alpha_i W^i = W^j$$

and

$$\sum_{i \neq j} \alpha_i h_i \leq h_j.$$

Finding which rows satisfy this is equivalent to finding the frame of

$$\text{pos}\begin{pmatrix} h^{\mathrm{T}} & 1 \\ W^{\mathrm{T}} & 0 \end{pmatrix}$$

where T indicates the transpose. Of course, if we have $\geq$ or $=$ in the original setting, we can easily transform that into a setting with only $\leq$.

The next question is where we can use this in our setting. The first, and obvious, answer is to apply it to the first-stage (deterministic) set of constraints $Ax = b$. (On the other hand, note that we may not apply frame to the first-stage coefficient matrix in order to remove unnecessary columns; these columns may be necessary after feasibility and optimality cuts have been added.)

It is more difficult to apply these results to the recourse problem. In principle, we have to check if a given row is unnecessary with *all* possible combinations of x and $\tilde{\xi}$. This may happen with inequality constraints, but it is not very likely with equality constraints. With inequalities, we should have to check if an inequality $W^j y \leq h_j$ was implied by the others, even when the jth inequality was at its tightest and the others as loose as possible. This is possible, but not within the frame setting.

We have now discussed how the problem can be reduced in size. Let us now assume that all possible reductions have been performed, and let us start discussing feasibility. This will clearly be related to topics we have seen in earlier chapters, but our focus will now be more specifically directed towards preprocessing.

5.2 Feasibility in Linear Programs

The tool for understanding feasibility in linear programs is the cone pol pos W. We have discussed it before, and it is illustrated in Figure 3. The important aspect of Figure 3 is that a right-hand side h represents a feasible recourse problem if and only if $h \in \text{pos } W$. But this is equivalent to requiring that $h^{\mathrm{T}} y \leq 0$ for all $y \in \text{pol pos } W$. In particular, it is equivalent to requiring that $h^{\mathrm{T}} y \leq 0$ for all y that are generators of pol pos W. In the figure there are two generators. You should convince yourself that a vector is in pos W if and only if it has a nonpositive inner product with the two generators of pol pos W.

Therefore what we shall need to find is a matrix W^*, to be referred to as the *polar matrix* of W, whose columns are the generators of pol pos W, so that we get

$$\text{pos } W^* = \text{pol pos } W.$$

Assume that we knew a column w^* from W^*. For h to represent a feasible recourse problem, it must satisfy $h^{\mathrm{T}} w^* \leq 0$.

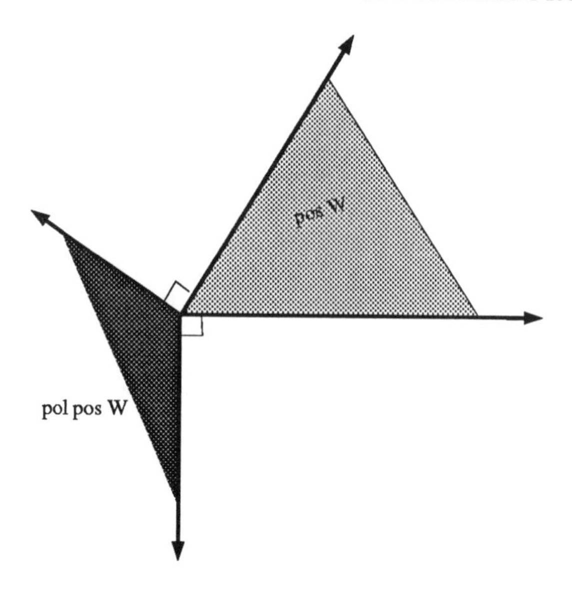

Figure 3 Finding the generators of pol pos W.

There is another important aspect of the *polar cone* pos W^* that we have not yet discussed. It is indicated in Figure 3 by showing that the generators are pairwise normals. However, that is slightly misleading, so we have to turn to a three-dimensional figure to understand it better. We shall also need the term *facet*. Let a cone pos W have dimension k. Then every cone K positively spanned by $k-1$ generators from pos W, such that K belongs to the boundary of pos W, is called a facet. Consider Figure 4.

What we note in Figure 4 is that the generators are not pairwise normals, but that the *facets* of one cone have generators of the other as normals. This goes in both directions. Therefore, when we state that $h \in$ pos W if and only if $h^{\mathrm{T}}y \leq 0$ for all generators of pol pos W, we are in fact saying that either h represents a feasible problem because it is a linear combination of columns in W *or* because it satisfies the inequality implied by the facets of pos W. In still other words, the point of finding W^* is not so much to describe a new cone, *but to replace the description of* pos W *in terms of generators with another in terms of inequalities.*

This is useful if the number of facets is not too large. Generally speaking, performing an inner product of the form $b^{\mathrm{T}}y$ is very cheap. For those who know anything about parallel processing, it is worth noting that it can be pipelined on a vector processor and the different inner products can be done in parallel. And, of course, as soon as we find one positive inner product, we can stop—the given recourse problem is infeasible.

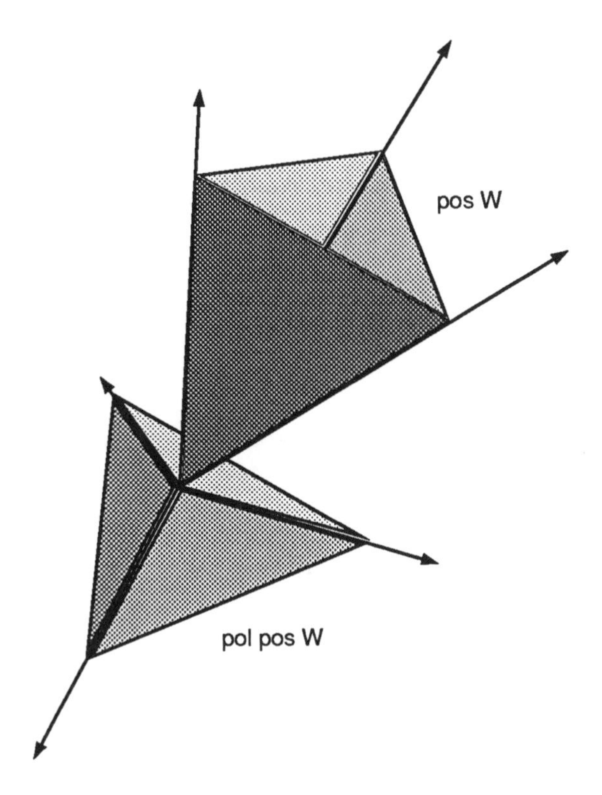

Figure 4 Three-dimensional picture of pos W and pol pos W = pos W^*.

Readers familiar with extreme point enumeration will see that going from a generator to a facet representation of pos W is indeed extreme point enumeration. As such, it is a problem with exponential complexity. Therefore we cannot in general expect to find W^* in reasonable time. However, taking a practical view of the matter, it is our suggestion that an attempt we made. The results are generally only interesting if there are relatively few facets, and those cases are the easiest. Figure 5 presents a procedure for finding the facets. It is called **procedure** support because it finds a minimal selection of supporting hyperplanes (not necessarily unique) of pos W, such that pos W is fully described. In practice, it has been shown to possess the desired property that it solves quickly if there are few facets. An example is presented shortly to help in understanding this **procedure** support.

The **procedure** support finds the polar matrix W^*, and thereby the support of pos W. The matrix W is reduced by the application of **procedure** framebylp, but is otherwise unchanged on exit. The process is initialized with

procedure support(W, W^*:matrices);
begin
  framebylp(W);
  $done := false$;
  for $i := 1$ **to** n **do if not** $done$ **then begin**
    $\alpha := W_i^T W^*$;
    $I_+ := \{k|\alpha[k] > 0\}$;
    $I_- := \{k|\alpha[k] < 0\}$;
    $I_0 := \{k|\alpha[k] = 0\}$;
    $done := (I_- \cup I_0 = \emptyset)$;
    if $done$ **then** $W^* := 0$;
    if $I_+ \neq \emptyset$ **and not** $done$ **then begin**
      if $I_- = \emptyset$ **then** $W^* := W^*_{I_0}$;
      else begin
        for all $k \in I_+$ **do**
          for all $j \in I_-$ **do**
            $C_{kj} := W^*_k - (\alpha[k]/\alpha[j])W^*_j$;
        $W^* := W^*_{I_0} \cup W^*_{I_-} \cup_{kj} C_{kj}$;
        framebylp(W^*);
      end; (* else *)
    end; (* if *)
  end; (* for *)
end;

Figure 5 Finding the support.

a matrix W^* that spans the entire column (range) space. We do this by letting

$$W^* := \begin{pmatrix} 1 & 0 & \cdots & 0 & -1 \\ 0 & 1 & \cdots & 0 & -1 \\ \vdots & \vdots & \ddots & \vdots & \vdots \\ 0 & 0 & \cdots & 1 & -1 \end{pmatrix}$$

or

$$W^* := \begin{pmatrix} 1 & 0 & \cdots & 0 & -1 & 0 & \cdots & 0 \\ 0 & 1 & \cdots & 0 & 0 & -1 & \cdots & 0 \\ \vdots & \vdots & \ddots & \vdots & \vdots & \vdots & \ddots & \vdots \\ 0 & 0 & \cdots & 1 & 0 & 0 & \cdots & -1 \end{pmatrix}$$

On exit W^* is the polar matrix. We initiate support by a call to framebylp in order to remove all columns from W that are not needed to describe pos W.

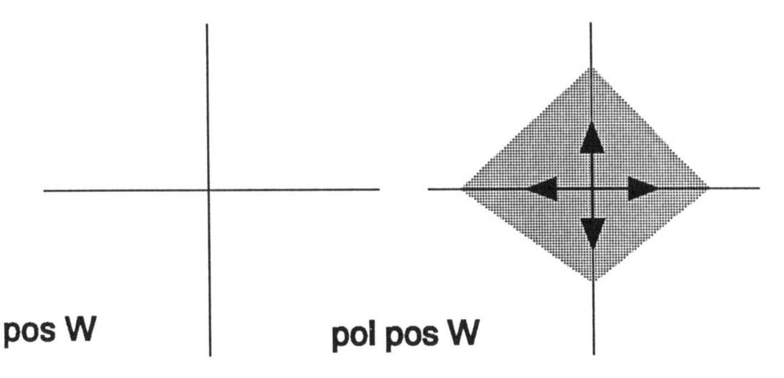

Figure 6 The cones pos W and pol pos W before any column has been added to W.

Example 5.1 Let us turn to a small example to see how **procedure** support progresses. Since pos W and pol pos W live in the same dimension, we can draw them side by side.

Let us initially assume that

$$W = \begin{pmatrix} 3 & 1 & -1 & -2 \\ 1 & 1 & 2 & 1 \end{pmatrix}.$$

The first thing to do, according to **procedure** support, is to subject W to a frame finding algorithm, to see if some columns are not needed. If we do that (check it to see that you understand frames) we end up with

$$W = \begin{pmatrix} 3 & -2 \\ 1 & 1 \end{pmatrix}.$$

Having reduced W, we then initialize W^* to span the whole space. Consult Figure 6 for details. We see there that

$$W^* = \begin{pmatrix} 1 & 0 & -1 & 0 \\ 0 & 1 & 0 & -1 \end{pmatrix}.$$

Consult **procedure** support. From there, it can be seen that the approach is to take one column from W at a time, and with it perform some calculations. Figure 6 shows the situation before we consider the first column of W. Calling it pos W is therefore a bit imprecise. The main point, however, is that the left and right parts correspond. If W has no columns then pol pos W spans the whole space.

Now, let us take the first column from W. It is given by $W_1 = (3,1)^{\mathrm{T}}$. We next find the inner products between W_1 and all four columns of W^*. We get

$$\alpha = (3, 1, -3, -1)^{\mathrm{T}}.$$

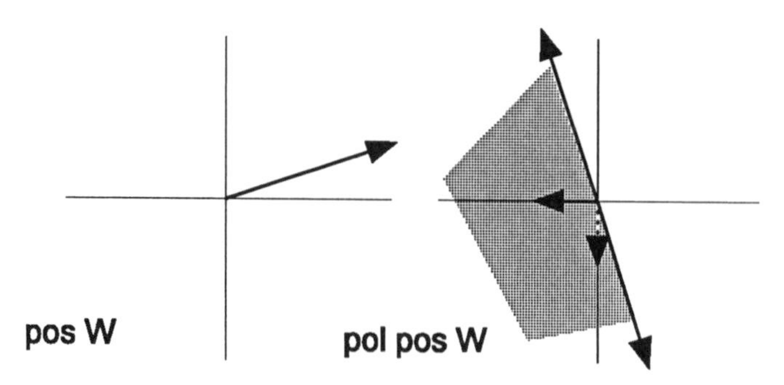

Figure 7 The cones pos W and pol pos W after one column has been added to W.

In other words, the sets $I_+ = \{1,2\}$ and $I_- = \{3,4\}$ have two members each, while $I_0 = \emptyset$. What this means is that two of the columns must be removed, namely those in I^+, and two kept, namely those in I^-. But to avoid losing parts of the space, we now calculate four columns C_{kj}. First, we get $C_{13} = C_{24} = 0$. They are not interesting. But the other two are useful:

$$C_{14} = \begin{pmatrix} 1 \\ 0 \end{pmatrix} + 3\begin{pmatrix} 0 \\ -1 \end{pmatrix} = \begin{pmatrix} 1 \\ -3 \end{pmatrix}, \quad C_{23} = \begin{pmatrix} 0 \\ 1 \end{pmatrix} + \tfrac{1}{3}\begin{pmatrix} -1 \\ 0 \end{pmatrix} = \begin{pmatrix} -\frac{1}{3} \\ 1 \end{pmatrix}.$$

Since our only interests are directions, we scale the latter to $(-1,3)^{\mathrm{T}}$. This brings us into Figure 7. Note that one of the columns in pos W^* is drawn with dots. This is done to indicate that if **procedure** framebylp is applied to W^*, that column will disappear. (However, that is not a unique choice.)

Note that if W had had only this one column then W^*, as it appears in Figure 7, is the polar matrix of that one-column W. This is a general property of **procedure** support. At any iteration, the present W^* is the polar matrix of the matrix containing those columns we have so far looked at.

Now let us turn to the second column of W. We find

$$\alpha^{\mathrm{T}} = (-2,1)W^* = (-2,1)\begin{pmatrix} -1 & 1 & -1 \\ 3 & -3 & 0 \end{pmatrix} = (5,-5,2)$$

We must now calculate two extra columns, namely C_{12} and C_{32}. The first gives 0, so it is not of interest. For the latter we get

$$C_{32} = \begin{pmatrix} -1 \\ 0 \end{pmatrix} + \tfrac{2}{5}\begin{pmatrix} 1 \\ -3 \end{pmatrix} = \begin{pmatrix} -\frac{3}{5} \\ -\frac{6}{5} \end{pmatrix},$$

which we scale to $(-1,-2)^{\mathrm{T}}$. This gives us Figure 8. To the left we have pos W, with W being the matrix we started out with, and to the right its polar cone.

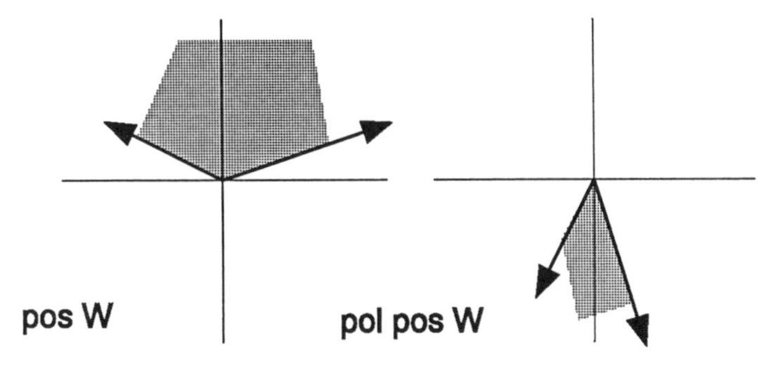

Figure 8 The cones pos W and pol pos W after two columns have been added to W.

A column represents a feasible problem if it is inside pos W, or equivalently, if it has a nonpositive inner product with all generators of pos $W^* =$ pol pos W. □

Assume we could indeed find W^* using **procedure** support. Let w^* be some column in W^*. For feasibility, we must have

$$(w^*)^{\mathrm{T}}[h_0 + H\xi - T(\xi)x] \leq 0 \quad \text{for all } \xi.$$

Hence

$$(w^*)^{\mathrm{T}}T(\xi)x \geq (w^*)^{\mathrm{T}}(h_0 + H\xi) \quad \text{for all } \xi.$$

If randomness affects both h and T, as indicated above, we must, at least in principle, create one inequality per ξ for each column from W^*. However, if $T(\xi) \equiv T_0$, we get a much easier set-up by calculating

$$(w^*)^{\mathrm{T}}T_0x \geq h_0 + \max_{t\in\Xi}\left[(w^*)^{\mathrm{T}}H\right]t,$$

where Ξ is the support of $\tilde{\xi}$. If we do this for all columns of W^* and add the resulting inequalities in terms of x to $Ax = b$, we achieve relatively complete recourse. Hence we see that relatively complete recourse can be *generated.* This is why the term is useful. It is very hard to *test* for relatively complete recourse. With relatively complete recourse we should never have to worry about feasibility.

Since the inequalities resulting from the columns of W^* can be dominated by others (in particular, if $T(\xi)$ is truly random), the new rows, together with those in $Ax = b$, should be subjected to row removal, as outlined earlier in this chapter.

5.2.1 A Small Example

Let us return to the example we discussed in Section 1.2. We have now named the right-hand side elements b_1, b_2 and b_3, since they are the focus of the discussion here (in the numerical example they had the values 100, 180 and 162):

$$\begin{array}{lrcl}\min\{2x_{raw1} + 3x_{raw2}\} & & & \\ \text{s. t.} & x_{raw1} + \;\; x_{raw2} & \leq & b_1, \\ & 2x_{raw1} + 6x_{raw2} & \geq & b_2, \\ & 3x_{raw1} + 3x_{raw2} & \geq & b_3, \\ & x_{raw1} \qquad\qquad & \geq & 0, \\ & x_{raw2} & \geq & 0. \end{array}$$

The interpretation is that b_1 is the production limit of a refinery, which refines crude oil from two countries. The variable x_{raw1} represents the amount of crude oil from Country 1 and x_{raw2} the amount from Country 2. The quality of the crudes is different, so one unit of crudes from Country 1 gives two units of Product 1 and three units of Product 2, whereas the crudes from the second country gives 6 and 3 units of the same products. Company 1 wants at least b_2 units of Product 1 and Company 2 at least b_3 units of Product 2.

If we now calculate the inequalities describing pos W, or alternatively the generators of pol pos W, we find that there are three of them:

$$\begin{array}{lcl} b_1 & & \geq 0 \\ 6b_1 - b_2 & & \geq 0 \\ 3b_1 & - b_3 & \geq 0. \end{array}$$

The first should be easy to interpret, and it says something that is not very surprising: the production capacity must not be negative. That we already knew. The second one is more informative. *Given appropriate units on crudes and products*, it says that the demand of Company 1 must not exceed six times the production capacity of the refinery. Similarly, the third inequality says that the demand of Company 2 must not exceed three times the production capacity of the refinery. (The inequalities are not as meaningless as they might appear at first sight: remember that the units for refinery capacity and finished products are not the same.) These three inequalities, one of which was obvious, are examples of constraints that are not explicitly written down by the modeller, but still are implied by him or her. And they should give the modeller extra information about the problem.

In case you wonder where the feasibility constraints are, what we have just discussed was a one-stage deterministic model, and what we obtained was three inequalities that can be used to check feasibility of certain instances of that model. For example, the numbers used in Section 1.2 satisfy all three constraints, and hence that problem was feasible. (In the example $b_1 = 100$, $b_2 = 180$ and $b_3 = 162$.)

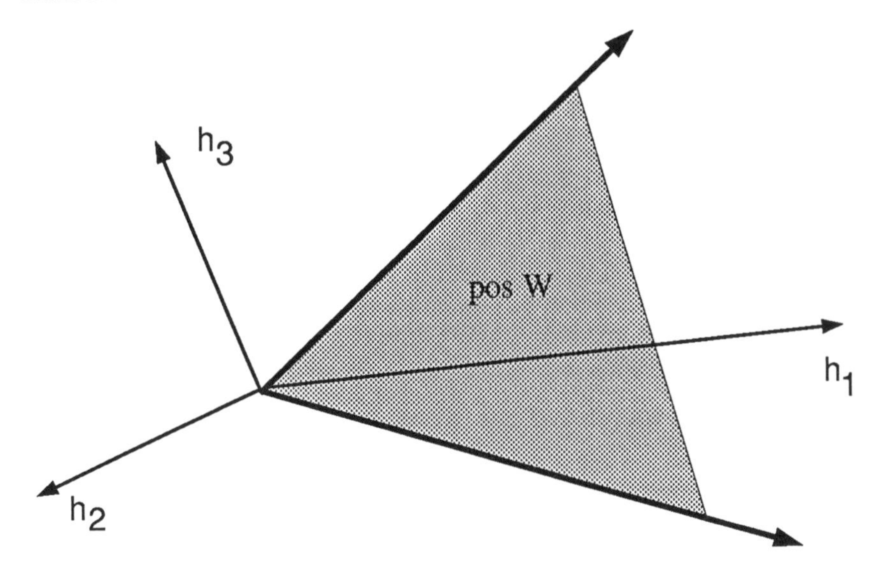

Figure 9 Illustration of feasibility.

5.3 Reducing the Complexity of Feasibility Tests

In Chapter 3, (page 150), we discussed the set $\mathcal{A}$ that is a set of ξ values such that if $h_0 + H\xi - T(\xi)x$ produces a feasible second-stage problem for all $\xi \in \mathcal{A}$ then the problem will be feasible for all possible values of $\tilde{\xi}$. We pointed out that in the worst case $\mathcal{A}$ had to contain all extreme points in the support of $\tilde{\xi}$.

Assume that the second stage is given by

$$Q(x,\xi) = \min\{q(\xi)^{\mathrm{T}}y \mid Wy = h_0 + H\xi - T_0x,\ y \geq 0\},$$

where W is fixed and $T(\xi) \equiv T_0$. This covers many situations. In R^2 consider the example in Figure 9, where $\tilde{\xi} = (\tilde{\xi}_1, \tilde{\xi}_2, \tilde{\xi}_3)$.

Since $h_1 \in \text{pos } W$, we can safely fix $\tilde{\xi}_1$ at its lowest possible value, since if things are going to go wrong, then they must go wrong for $\xi_1^{\min}$. Or, in other words, if $h_0 + H\hat{\xi} - T_0x \in \text{pos } W$ for $\hat{\xi} = (\xi_1^{\min}, \hat{\xi}_2, \hat{\xi}_3)$ then so is any other vector with $\tilde{\xi}_2 = \hat{\xi}_2$ and $\tilde{\xi}_3 = \hat{\xi}_3$, regardless of the value of $\tilde{\xi}_1$. Similarly, since $-h_2 \in \text{pos } W$, we can fix $\tilde{\xi}_2$ at its largest possible value. Neither h_3 nor $-h_3$ are in pos W, so there is nothing to do with $\tilde{\xi}_3$.

Hence to check if x yields a feasible solution, we must check if

$$h_0 + H\xi - T_0x \in \text{pos } W \text{ for } \xi = (\xi_1^{\min}, \xi_2^{\max}, \xi_3^{\min})^{\mathrm{T}} \text{ and } \xi = (\xi_1^{\min}, \xi_2^{\max}, \xi_3^{\max})^{\mathrm{T}}$$

Hence in this case $\mathcal{A}$ will contain only two points instead of $2^3 = 8$. In general, we see that whenever a column from H, in either ts positive or negative direction, is found to be in pos W, we can halve the number of points in $\mathcal{A}$. In some cases we may therefore reduce the testing to one single problem.

It is of importance to understand that the reduction in the size of $\mathcal{A}$ has two positive aspects. First, if we do not have (or do not know that we have) relatively complete recourse, the test for feasibility, and therefore generation of feasibility cuts, becomes much easier. But equally important is the fact that it tells us something about our problem. If a column from H is in pos W, we have found a direction in which we can move as far as we want without running into feasibility problems. This will, in a real setting, say something about the random effect we have modelled using that column.

5.4 Bibliographical Notes

Preprocessing and similar procedures have been used in contexts totally different from ours. This is natural, since questions of model formulations and infeasibilities are equally important in all areas of mathematical programming. For further reading, consult e.g. Roodman [7], Greenberg [3, 4, 5] or Chinneck and Dravnieks [1].

An advanced algorithm for finding frames can be found in Wets and Witzgall [12]. Later developments include the work of Rosen et al. [8] and Dulá et al. [2]. The algorithm for finding a support was described by Tschernikov[9], and later also by Wets [11]. For computational tests using the procedure see Wallace and Wets [10]. Similar procedures for networks will be discussed in Chapter 6.

For an overview of methods for extreme point enumeration see e.g. Mattheiss and Rubin [6].

Exercises

1. Let W be the coefficient matrix for the following set of linear equations:

$$\begin{array}{rcl} x + \frac{1}{2}y - z + s_1 & & = 0, \\ 2x + z & + s_2 & = 0, \\ x, \quad y, \quad z, \quad s_1, & s_2 & \geq 0. \end{array}$$

   (a) Find a frame of pos W.
   (b) Draw a picture of pos W, and find the generators of pol pos W by simple geometric arguments.
   (c) Find the generators of pol pos W by using **procedure** support in Figure 5. Make sure you draw the cones pos W and pol pos W after each iteration of the algorithm, so that you see how it proceeds.

2. Let the following set of equations be given:

$$\begin{array}{rrrl} x + & y & + z & \leq 4, \\ 2x & & + z & \leq 5, \\ & y & + z & \leq 8, \\ x, & y, & z & \geq 0. \end{array}$$

   (a) Are there any columns that are not needed for feasibility? (Remember the slack variables!)
   (b) Let W contain the columns that were needed from question (a), including the slacks. Try to find the generators of pol pos W by geometric arguments, i.e. draw a picture.

3. Consider the following recourse problem constraints:

$$\begin{pmatrix} 1 & 3 \\ 3 & 1 \end{pmatrix} y = \begin{pmatrix} 2 \\ 7 \end{pmatrix} + \begin{pmatrix} 2 & -1 & 0 & -1 & -4 \\ 2 & -2 & -1 & 1 & -1 \end{pmatrix} \xi + \begin{pmatrix} 5 & 1 \\ 3 & 2 \end{pmatrix} x$$

with $y \geq 0$. Assume that all random variables are independent, with support $[0, 1]$. Look back at Section 5.3, where we discussed how we could simplify the feasibility test if we were not aware of relatively complete recourse. We there defined a set $\mathcal{A}$ that was such that if the recourse problem was feasible for all $\xi \in \mathcal{A}$ then it was feasible for all ξ. In the worst case $\mathcal{A}$ has, in our case, $2^5 = 32$ elements. By whatever method you find useful (what about a picture?), reduce this number to six, and list the six elements.

References

[1] Chinneck J. W. and Dravnieks E. W. (1991) Locating minimal infeasible constraint sets in linear programs. *ORSA J.Comp.* 3: 157–168.

[2] Dulá J. H., Helgason R. V., and Hickman B. L. (1992) Preprocessing schemes and a solution method for the convex hull problem in multidimensional space. In Balci O. (ed) *Computer Science and Operations Research: New Developments in their Interfaces*, pages 59–70. Pergamon Press, Oxford.

[3] Greenberg H. J. (1982) A tutorial on computer-assisted analysis. In Greenberg H. J., Murphy F. H., and Shaw S. H. (eds) *Advanced Techniques in the Practice of Operations Research*. Elsevier, New York.

[4] Greenberg H. J. (1983) A functional description of ANALYZE: A computer-assisted analysis. *ACM Trans. Math. Software* 9: 18–56.

[5] Greenberg H. J. (1987) Computer-assisted analysis for diagnosing infeasible or unbounded linear programs. *Math. Prog. Study* 31: 79–97.

[6] Mattheiss T. H. and Rubin D. S. (1980) A survey and comparison of methods for finding all vertices of convex polyhedral sets. *Math. Oper. Res.* 5: 167–185.

[7] Roodman G. M. (1979) Post-infeasibility analysis in linear programming. *Management Sci.* 9: 916–922.

[8] Rosen J. B., Xue G. L., and Phillips A. T. (1992) Efficient computation of extreme points of convex hulls in $\mathbb{R}^d$. In Pardalos P. M. (ed) *Advances in Optimization and Parallel Computing*, pages 267–292. North-Holland, Amsterdam.

[9] Tschernikow S. N. (1971) *Lineare Ungleichungen*. VEB Deutscher Verlag der Wissenschaften, Berlin. (Translated from Russian).

[10] Wallace S. W. and Wets R. J.-B. (1992) Preprocessing in stochastic programming: The case of linear programs. *ORSA Journal on Computing* 4: 45–59.

[11] Wets R. J.-B. (1990) Elementary, constructive proofs of the theorems of Farkas, Minkowski and Weyl. In Gabszewicz J., Richard J.-F., and Wolsey L. (eds) *Economic Decision Making: Games, Econometrics and Optimization: Contributions in Honour of Jacques Dreze*, pages 427–432. North-Holland, Amsterdam.

[12] Wets R. J.-B. and Witzgall C. (1967) Algorithms for frames and lineality spaces of cones. *J. Res. Nat. Bur. Stand.* 71B: 1–7.

6

Network Problems

The purpose of this chapter is to look more specifically at networks. There are several reasons for doing this. First, networks are often easier to understand. Some of the results we have outlined earlier will be repeated here in a network setting, and that might add to understanding of the results. Secondly, some results that are stronger than the corresponding LP results can be obtained by utilizing the network structure. Finally, some results can be obtained that do not have corresponding LP results to go with them. For example, we shall spend a section on PERT problems, since they provide us with the possibility of discussing many important issues.

The overall setting will be as before. We shall be interested in two- or multistage problems, and the overall solution procedures will be the same. Since network flow problems are nothing but specially structured LPs, everything we have said before about LPs still hold. The bounds we have outlined can be used, and the L-shaped decomposition method, with and without bounds, can be applied as before. We should like to point out, though, that there exists one special case where scenario aggregation looks more promising for networks than for general LPs: that is the situation where the overall problem is a network. This may require some more explanation.

When we discuss networks in this chapter, we refer to a situation in which the second stage (or the last stage in a multistage setting) is a network. We shall mostly allow the first stage to be a general linear program. This rather limited view of a network problem is caused by properties of the L-shaped decomposition method (see page 159). The computational burden in that algorithm is the calculation of $\mathcal{Q}(\hat{x})$, the expected recourse cost, and to some extent the check of feasibility. Both those calculations concern only the recourse problem. Therefore, if that problem is a network, network algorithms can be used to speed up the L-shaped algorithm.

What if the first-stage problem is also a network? Example 2.2 (page 112) was such an example. If we apply the L-shaped decomposition method to that problem, the network structure of the master problem is lost as soon as feasibility and optimality cuts are added. This is where scenario aggregation, outlined in Section 2.6, can be of some use. The reason is that, throughout the calculations, individual scenarios remain unchanged in terms of constraints, so that structure is not lost. A nonlinear term is added to the objective function, however, so if the original problem was linear, we are now in a setting of quadratic objectives and linear (network) constraints. If the original problem was a nonlinear network, the added terms will not increase complexity at all.

6.1 Terminology

Consider a network with arcs $\mathcal{E} = \{1, \ldots, m\}$ and nodes $\mathcal{N} = \{1, \ldots, n\}$. An arc $k \in \mathcal{E}$ will be denoted by $k \sim (i, j)$, indicating that it starts at i and ends at j. The capacity of k will be denoted by $\gamma(k)$ and the cost by $q(k)$. For each node $i \in \mathcal{N}$, let $\beta(i)$ be the external flow. We let $\beta(i) > 0$ denote supply and $\beta(i) < 0$ demand.

We say that a network flow problem is *capacitated* if all arcs k have $\gamma(k) < \infty$. Otherwise,we say that the network is *uncapacitated.* Most networks are a mixture of these two cases, and their properties will then be mixtures of what we discuss for the two cases in this chapter.

By $G(Y)$,we understand a network consisting of the nodes in $Y \subseteq \mathcal{N}$ and all arcs in $\mathcal{E}$ connecting nodes in Y. Of course, $G(\mathcal{N})$ is the original network.

For two arbitrary sets $Y, Y' \subset \mathcal{N}$, let $\{k \sim (i, j) \mid i \in Y, j \in Y'\} \subseteq \mathcal{E}$ be denoted by $[Y, Y']^+$ and let $\{k \sim (i, j) \mid j \in Y, i \in Y'\} \subseteq \mathcal{E}$ be denoted by $[Y, Y']^-$.

For $Y \subset \mathcal{N}$ define $Q^+ = [Y, \mathcal{N} \setminus Y]^+$ and $Q^- = [Y, \mathcal{N} \setminus Y]^-$. We call $Q = Q^+ \cup Q^- = [Y, \mathcal{N} \setminus Y]$ a *cut.* Whenever we refer to Y and Q, without stating their relationship, we are assuming that $Q = [Y, \mathcal{N} \setminus Y]$. For each $Y \subseteq \mathcal{N}$, let $b(Y) \in \{0, 1\}^n$ be an index vector for the set Y, i.e. $b(Y, i) = 1$ if $i \in Y$, and 0 otherwise. Similarly, for each $Q \subseteq \mathcal{E}$, let $a(Q^+) \in \{0, 1\}^m$ be an index vector for the set Q^+, i.e. $a(Q^+, k) = 1$ if $k \in Q^+$ and 0 otherwise.

The *node–arc incidence matrix* for a network will be denoted by W', and is defined by

$$W'(i, k) = \begin{cases} 1 & \text{if } k \sim (i, j) \text{ for some } j, \\ -1 & \text{if } k \sim (j, i) \text{ for some } j, \\ 0 & \text{otherwise.} \end{cases}$$

The rows in the node–arc incidence matrix are linearly dependent. For the system $W'y = b$ to have a solution, we know from Chapter 1 that $\text{rk } W' = \text{rk } (W' \mid b)$. In a network this requirement means that there must be one node where the external flow equals exactly the negative sum of the external flows in the other nodes. This node is called the *slack node.* It is customary not to include a row for that node in W'. Hence W' has only $n - 1$ rows, and it has full rank provided the network is connected. A network is *connected* if for all $Y \subset \mathcal{N}$ we have $Q = [Y, \mathcal{N} \setminus Y] \neq \emptyset$.

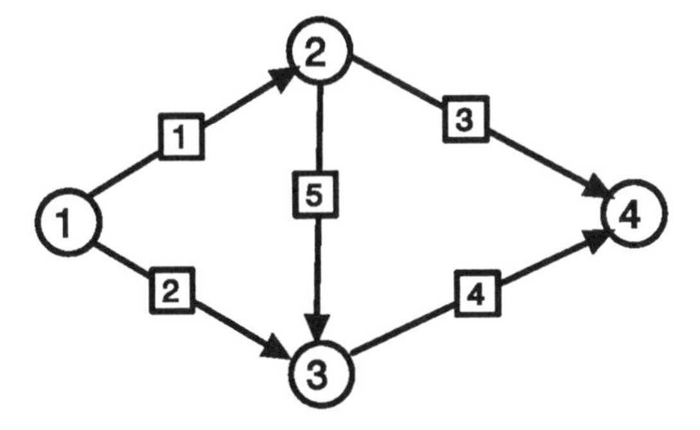

Figure 1 Network used to demonstrate definitions.

We shall also need the following sets:

$$F^+(Y) = \{\text{nodes } j \mid k \sim (i,j) \text{ for } i \in Y\} \cup Y,$$

$$B^+(Y) = \{\text{nodes } j \mid k \sim (j,i) \text{ for } i \in Y\} \cup Y.$$

The set $F^+(Y)$ contains Y itself plus all nodes that can be reached directly (i.e. in one step) from a node in Y. Similarly $B^+(Y)$ contains Y and all nodes from which Y can be reached directly.

Two other sets that are very similar to $F^+(Y)$ and $B^+(Y)$ are

$$F^*(Y) = \{\text{nodes } j \mid \exists \text{ a directed path from some node } i \in Y \text{ to node } j\} \cup Y,$$

$$B^*(Y) = \{\text{nodes } j \mid \exists \text{ a directed path from node } j \text{ to some node } i \in Y\} \cup Y.$$

Thus the sets F^+ and B^+ pick up immediate successors and predecessors, whereas F^* and B^* pick up *all* successors and predecessors.

Example 6.1 Let us consider Figure 1 to briefly illustrate most of the concepts we have introduced.

The node set $\mathcal{N} = \{1,2,3,4\}$, and the arc set $\mathcal{E} = \{1,2,3,4,5\}$. An example of an arc is $5 \sim (2,3)$, since arc 5 starts at node 2 and ends at node 3. Let $Y = \{1,3\}$ and $Y' = \{2\}$. The network $G(Y)$ consists of nodes 1 and 3, and arc 2, since that is the only arc connecting nodes in Y. Furthermore, for the same Y and Y', we have $[Y,Y']^+ = \{1\}$, since arc 1 is the only arc going *from* nodes 1 or 3 to node 2. Similarly $[Y,Y']^- = \{5\}$. If we define $Q = [Y, \mathcal{N} \setminus Y]$ then $Q^+ = \{1,4\}$ and $Q^- = \{5\}$. Therefore $Q = \{1,4,5\}$ is a cut.

Again, with the same definition of Y, we have

$$b(Y) = (1,0,1,0)^{\mathrm{T}}, \quad a(Q^+) = (1,0,0,1,0)^{\mathrm{T}}.$$

Furthermore, we have

$$F^+(\{1\}) = \{1,2,3\}, \quad F^*(\{1\}) = \{1,2,3,4\},$$

since we can reach nodes 2 and 3 in one step, but we need two steps to reach node 4. Node 1 itself is in both sets by definition.

Two examples of predecessors of a node are

$$B^+(\{1\}) = \{1\}, \quad B^*(\{2,3\}) = \{1,2,3\},$$

since node 1 has no predecessors, and nodes 2 and 3 can be reached from node 1.

A common problem in network flows is the *min cost network flow problem.* It is given as follows.

$$\begin{array}{lrcrcrcrcrcl}
\min & q(1)y(1) & + & q(2)y(2) & + & q(3)y(3) & + & q(4)y(4) & + & q(5)y(5) & & \\
\text{s.t.} & y(1) & + & y(2) & & & & & & & = & \beta(1), \\
& -y(1) & & & + & y(3) & & & + & y(5) & = & \beta(2), \\
& & - & y(2) & & & + & y(4) & - & y(5) & = & \beta(3), \\
& & & & - & y(3) & - & y(4) & & & = & \beta(4), \\
& & & & & & & & & y(k) & \le & \gamma(k), \quad k = 1, \ldots, 5, \\
& & & & & & & & & y(k) & \ge & 0, \quad k = 1, \ldots, 5.
\end{array}$$

The coefficient matrix for this problem has rank 3. Therefore the node–arc incidence matrix has three rows, and is given by

$$W' = \begin{pmatrix} 1 & 1 & 0 & 0 & 0 \\ -1 & 0 & 1 & 0 & 1 \\ 0 & -1 & 0 & 1 & -1 \end{pmatrix}.$$

□

6.2 Feasibility in Networks

In Section 3.2 and Chapter 5 we discussed feasibility in linear programs. Refer back to those results. As will become apparent shortly, it is easier to obtain feasibility results for networks than for LPs. Let us first run through the development, and then later see how this fits in with the LP results.

A well-known result concerning feasibility in networks states that if the net flow across every cut in a network is less than or equal to the capacity of that cut, then the problem is feasible. More formally, this can be stated as follows, using $\beta^{\mathrm{T}} = (\beta(1), \ldots, \beta(n))$ and $\gamma^{\mathrm{T}} = (\gamma(1), \ldots, \gamma(m))$.

Proposition 6.1 *A capacitated network flow problem with total supply equal to total demand is feasible iff for every cut* $Q = [Y, \mathcal{N} \setminus Y]$*, and* $b(Y)^{\mathrm{T}}\beta \le a(Q^+)^{\mathrm{T}}\gamma$.

If the arcs in the network are uncapacitated, the result is somewhat simpler, namely the following.

Proposition 6.2 *An uncapacitated network flow problem with total supply equal to total demand is feasible iff for every cut* $Q = [Y, \mathcal{N} \setminus Y]$, *with* $Q^+ = \emptyset$ *and* $b(Y)^{\mathrm{T}}\beta \leq 0$.

The latter proposition is of course just a special case of the first where we take into account that whenever a cut contains an uncapacitated arc, the capacity of that cut is $+\infty$, and hence the inequality from Proposition 6.1 is always satisfied.

The above two propositions are very simple in nature. However, from a computational point of view, they are not very useful. Both require that we look at *all* subsets Y of $\mathcal{N}$, in other words 2^n subsets. For reasonably large n it is not computationally feasible to try to enumerate subsets this way. Another problem that might not be that obvious when reading the two propositions is that they are not "if and only if" statements in a very useful sense. There is no guarantee that inequalities arising from the propositions are indeed needed. We might—and most probably will—end up with inequalities that are implied by other inequalities. A key issue in this respect is the connectedness of a network. We defined earlier that a network was connected if for all $Y \subset \mathcal{N}$ we have that $Q = [Y, \mathcal{N} \setminus Y] \neq \emptyset$. It is reasonably easy to check connectedness of a network. Details are given in **function** Connected in Figure 2. Note that we use F^* and B^*. If they are not available, we can also use F^+ and B^+, or calculate F^* and B^*, which is quite simple.

Using the property of connectedness, it is possible to prove the following stronger result.

Proposition 6.3 *Let* $Q = [Y, \mathcal{N} \setminus Y]$. *For capacitated networks the inequalities*

$$b(Y)^{\mathrm{T}}\beta \leq a(Q^+)^{\mathrm{T}}\gamma, \quad b(\mathcal{N} \setminus Y)^{\mathrm{T}}\beta \leq a(Q^-)^{\mathrm{T}}\gamma$$

are both needed if and only if $G(Y)$ *and* $G(\mathcal{N} \setminus Y)$ *are both connected.*

For uncapacitated networks we are, as shown in Proposition 6.2, interested only in cuts Q with $Q^+ = \emptyset$. Given this restriction, we can show the following result.

Proposition 6.4 *For an uncapacitated network a cut* $Q = [Y, \mathcal{N} \setminus Y]$ *with* $Q^+ = \emptyset$ *is needed if and only if* $G(Y)$ *and* $G(\mathcal{N} \setminus Y)$ *are both connected.*

As we can see, these results are very similar. They both state that an inequality, satisfying the requirements of Proposition 6.1 or 6.2 must be kept

```
function Connected(W : set of nodes) : boolean;
begin
   PickNode(i, W);
   Qlist := {i};
   Visited := {i};
   while Qlist ≠ ∅ do begin
      PickNode(i, Qlist);
      Qlist := Qlist \ {i};
      s := (B*(i) ∪ F*(i)) ∩ (W \ Visited);
      Qlist := Qlist ∪ s;
      Visited := Visited ∪ s;
   end;
   Connected := (Visited = W);
end;
```

Figure 2 Function checking network connectedness.

if Y and $\mathcal{N} \setminus Y$ both generate connected networks, otherwise the inequality can be dropped.

Example 6.2 Let us look at the small example network in Figure 3 to at least partially see the relevance of the last proposition. The following three inequalities are examples of inequalities describing feasibility for the example network:

$$\begin{aligned} \beta(2) \qquad\qquad &\leq \gamma(d) \qquad\qquad\quad + \gamma(f), \\ \beta(3) &\leq \qquad\quad \gamma(e), \\ \beta(2) + \beta(3) &\leq \gamma(d) + \gamma(e) + \gamma(f). \end{aligned}$$

Proposition 6.3 states that the latter inequality is not needed, because $G(\{2,3\})$ is not connected. From the inequalities themselves, we easily see that if the first two are satisfied, then the third is automatically true. It is perhaps slightly less obvious that, for the very same reason, the inequality

$$\beta(1) + \beta(4) + \beta(5) \leq \gamma(a) + \gamma(c)$$

is also not needed. It is implied by the requirement that total supply must equal total demand plus the companions of the first two inequalities above. (Remember that each node set gives rise to two inequalities). More specifically, the inequality can be obtained by adding the following two inequalities and

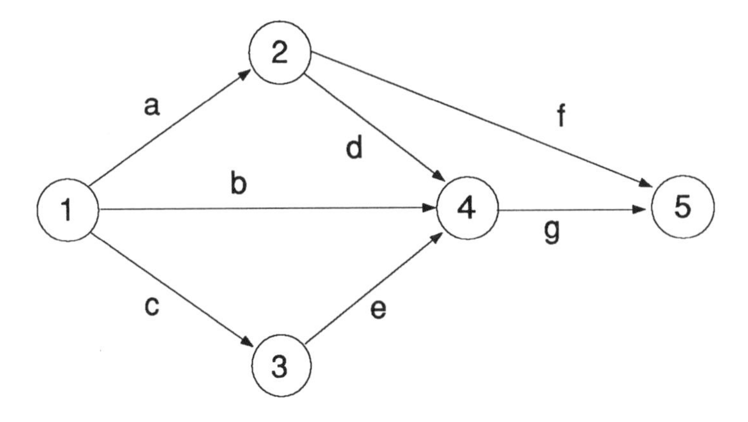

Figure 3 Example network 1.

one equality (representing supply equals demand):

$$\begin{aligned} \beta(1) + \beta(2) \qquad\qquad + \beta(4) + \beta(5) &\leq \gamma(c), \\ \beta(1) \qquad\qquad + \beta(3) + \beta(4) + \beta(5) &\leq \gamma(a), \\ -\beta(1) - \beta(2) - \beta(3) - \beta(4) - \beta(5) &= \ 0. \end{aligned}$$

□

Once you have looked at this for a while, you will probably realize that the part of Proposition 6.3 that says that if $G(Y)$ or $G(\mathcal{N} \setminus Y)$ is disconnected then we do not need any of the inequalities is fairly obvious. The other part of the proposition is much harder to prove, namely that if $G(Y)$ and $G(\mathcal{N} \setminus Y)$ are both connected then the inequalities corresponding to Y and $\mathcal{N} \setminus Y$ are both needed. We shall not try to outline the proof here.

Propositions 6.3 and 6.4 might not seem very useful. A straightforward use of them could still require the enumeration of all subsets of $\mathcal{N}$, and for each such subset check if $G(Y)$ and $G(\mathcal{N} \setminus Y)$ are both connected. However, we can obtain more than that.

The first important observation is that the results refer to the connectedness of two networks—both the one generated by Y and the one generated by $\mathcal{N} \setminus Y$. Consider the capacitated case. Let $Y_1 = \mathcal{N} \setminus Y$. If both networks are connected, we have two inequalities that we need, namely

$$b(Y)^{\mathrm{T}}\beta \leq a(Q^{+})^{\mathrm{T}}\gamma$$

and

$$b(Y_1)^{\mathrm{T}}\beta = b(\mathcal{N} \setminus Y)^{\mathrm{T}}\beta \leq a(Q^{-})^{\mathrm{T}}\gamma.$$

On the other hand, if at least one of the networks is disconnected, neither inequality will be needed. Therefore checking *each* subset of $\mathcal{N}$ means doing twice as much work as needed. If we are considering Y and discover that both $G(Y)$ and $G(Y_1 = \mathcal{N} \setminus Y)$ are connected, we write down both inequalities at the same time. An easy way to achieve this is to disregard some node (say node n) from consideration in a full enumeration. This way, we will achieve $n \in \mathcal{N} \setminus Y$ for all Y we investigate. Then for each cut where the connectedness requirement is satisfied we write down two inequalities. This will halve the number of subsets to be checked.

For uncapacitated networks there is no corresponding result. If we have a Y with $Q^+ = \emptyset$ then $Q^- \neq \emptyset$, since otherwise we should have a disconnected network to start out with. But, of course, the uncapacitated problem is easier in other respects.

In some cases it is possible to reduce the complexity of a calculation by *collapsing nodes*. By these, we understand the process of replacing a set of nodes by one new node. Any other node that had an arc to or from one of the collapsed nodes will afterwards have an arc to or from the new node: one for each original arc. Of course, in the uncapacitated case it is enough with one arc from one of the original nodes to the new node. A simple but important use of node collapsing is given by the following proposition.

Proposition 6.5 *For an uncapacitated network, if $Q^+ = [Y, \mathcal{N} \setminus Y]^+ = \emptyset$ then $F^*(i) \subseteq Y$ if $i \in Y$.*

From this the following easily follows.

Proposition 6.6 *If $j_1, j_2, \ldots, j_K$ is a set of arcs in an uncapacitated network such that $j_k \sim (i_k, i_{k+1})$ and $i_1 = i_{K+1}$ then the nodes $i_1, \ldots, i_K$ will always be on the same side of a cut Q if $Q^+ = \emptyset$.*

We utilize this by collapsing all directed circuits in the network. As an example, consider Figure 4, which is almost like Figure 3, except that arc b has been turned around.

Since arcs a, d and b, as well as arcs b, c and e, constitute directed circuits, we can collapse these circuits and arrive at the network in Figure 5. Of course, it is now much easier to investigate all possible subsets of $\mathcal{N}$.

Let us briefly summarize where we are with respect to describing feasibility. We first formulated Propositions 6.1 and 6.2 to arrive at sufficient inequalities for both capacitated and uncapacitated networks. We found that in both cases we had to look at all subsets Y of $\mathcal{N}$. However, for the uncapacitated case we only had to consider those cuts Q for which $Q^+ = \emptyset$. We next turned to Propositions 6.3 and 6.4, showing that, among the inequalities coming from Propositions 6.1 and 6.2, only those having both $G(Y)$ and $G(\mathcal{N} \setminus Y)$ connected were needed. Finally, we realized that any directed circuit in an uncapacitated

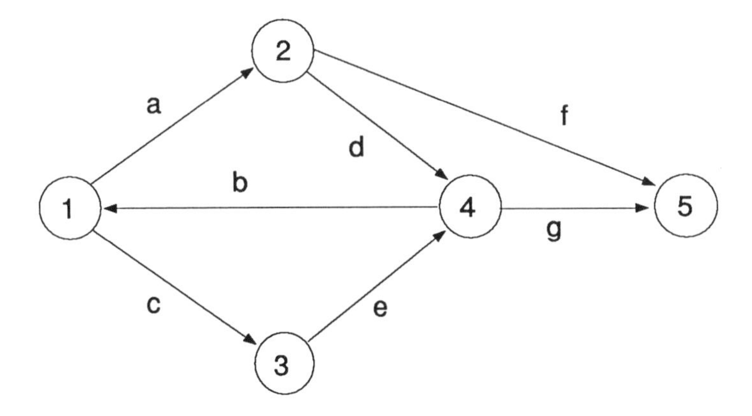

Figure 4 Example network 2, assumed to be uncapacitated.

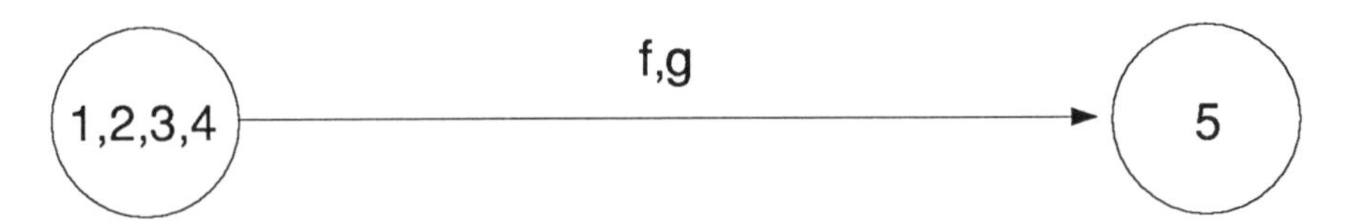

Figure 5 Example network 2 after collapsing the directed circuit.

network could be collapsed, since those nodes will always appear on the same side of a cut with $Q^+ = \emptyset$.

To simplify statements later on we shall also need a way to simply state which inequalities we want to write down. We present two algorithms for this: one for the capacitated and one for the uncapacitated case. They are given in Figures 6 and 7.

Note that we allow the procedures to be called with $Y = \emptyset$. This is a technical devise to ensure consistent results, but you should not let that confuse you at the present time. Based on Propositions 6.3 and 6.4, it is possible to develop procedures that in some cases circumvent the exponential complexity arising from checking all subsets of $\mathcal{N}$. We shall use Figure 8 to illustrate some of our points.

Proposition 6.7 *If $B^+(i) \cup F^+(i) = \{i, j\}$ then nodes i and j can be collapsed after the inequalities generated by CreateIneq($\{i\}$) have been created.*

This result holds for both the capacitated and uncapacitated case, as long as we use the relevant version of CreateIneq. The only set Y where $i \in Y$ but $j \notin Y$, at the same time as both $G(Y)$ and $G(\mathcal{N} \setminus Y)$ are connected, is the set where $Y = \{i\}$. The reason is that node j blocks node i's connection to all

procedure CreateIneq(Y : set of nodes);
begin
  if $A(Y) \neq \emptyset$ **then begin**
    create the inequality $b(A(Y))^{\mathrm{T}}\beta \leq a(Q^{+})^{\mathrm{T}}\gamma$;
    create the inequality $b(A(\mathcal{N} \setminus Y))^{\mathrm{T}}\beta \leq a(Q^{-})^{\mathrm{T}}\gamma$;
  end
  else begin
    create the inequality $b(A(\mathcal{N}))^{\mathrm{T}}\beta \leq 0$;
    create the inequality $-b(A(\mathcal{N}))^{\mathrm{T}}\beta \leq 0$;
  end;
end;

Figure 6 Algorithm for generating inequalities—capacitated case.

procedure CreateIneq(Y : set of nodes);
begin
  if $A(Y) \neq \emptyset$ **then**
    create the inequality $b(A(Y))^{\mathrm{T}}\beta \leq 0$;
  else
    create the inequality $-b(A(\mathcal{N}))^{\mathrm{T}}\beta \leq 0$;
end;

Figure 7 Algorithm for generating inequalities—uncapacitated case.

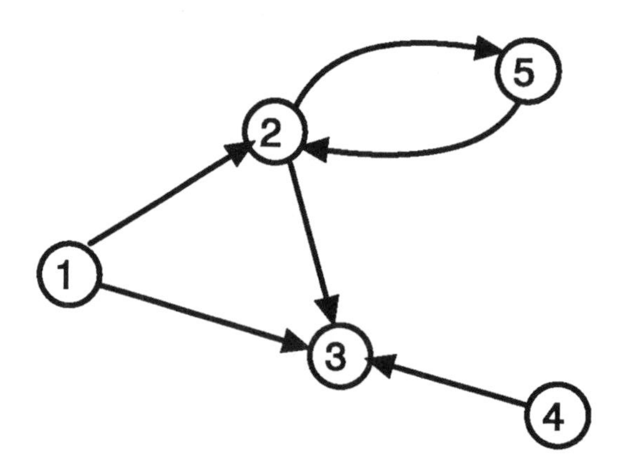

Figure 8 Example network used to illustrate Proposition 6.7.

```
procedure AllFacets;
begin
   TreeRemoval;
   CreateIneq(∅);
   Y := ∅;
   if network capacitated then W := 𝒩 \ {n};
   if network uncapacitated then W := 𝒩;
   Facets(Y, W);
end;
```

Figure 9 Main program for full enumeration of inequalities satisfying Propositions 6.3 and 6.4.

other nodes. Therefore, after calling CreateIneq($\{i\}$), we can safely collapse node i into node j. Examples of this can be found in Figure 8, (see e.g. nodes 4 and 5). This result is easy to implement, since all we have to do is run through all nodes, one at a time, and look for nodes satisfying $B^+(i) \cup F^+(i) = \{i, j\}$. Whenever collapses take place, F^+ and B^+ (or, alternatively, F^* and B^*) must be updated for the remaining nodes.

By repeatedly using this proposition, we can remove from the network all trees (and trees include "double arcs" like those between nodes 2 and 5). We are then left with circuits and paths connecting circuits. For capacitated networks, the circuits can be both directed and undirected. For uncapacitated networks they will all be undirected, since directed circuits are collapsed according to Proposition 6.6. In the example in Figure 8 we are left with nodes 1, 2 and 3. We shall assume that there is a procedure TreeRemoval that takes care of this reduction.

There is one final remark to be made based on Propositions 6.3 and 6.4. For each set Y we must check the connectedness of both $G(Y)$ and $G(\mathcal{N} \setminus Y)$. We can skip the first if we simply make sure that $G(Y)$ is always connected. This can easily be achieved by building up Y (in the enumeration) such that it is always connected. We shall assume that we have available a procedure PickNode(Y, W, i) that picks a node i from W provided that node is reachable from Y in one step. Otherwise, it returns $i := 0$.

We now present a main program and a main procedure for the full enumeration. They are listed in Figures 9 and 10.

```
procedure Facets(Y, W: set of nodes);
begin
   PickNode(Y, W, i);
   if i ≠ 0 then begin
      W := W \ {i};
      Facets(Y, W);
      if network capacitated then Y := Y ∪ {i};
      if network uncapacitated then Y := Y ∪ F*(i);
      if network uncapacitated then W := W \ F*(i);
      Facets(Y, W);
      if Connected(𝒩 \ Y) then CreateIneq(Y);
   end;
end;
```

Figure 10 Recursive algorithm for generating facets.

6.2.1 Comparing the LP and Network Cases

We used Section 5.2 to discuss feasibility in linear programs. Since network flow problems are just special cases of linear programs, those results apply here as well, of course. On the other hand, we have just discussed feasibility in networks more specifically, and apparently the setting was very different. The purpose of this section is to show in some detail how these results relate to each other.

Let us first repeat the major discussions from Section 5.2. Using the cone pos $W = \{t \mid t = Wy,\ y \geq 0\}$, we defined the polar cone

$$\text{pos } W^* = \text{pol pos } W = \{t \mid t^{\mathrm{T}} y \leq 0 \text{ for all } y \in \text{pos } W\}.$$

The interesting property of the cone pos W^* is that the recourse problem is feasible if and only if a given right-hand side has a non-positive inner product with all generators of the cone. And if there are not too many generators, it is much easier to perform inner products than to check if a linear program is feasible. Refer to Figure 4 for an illustration in three dimensions.[1] To find the polar cone, we used **procedure** support in Figure 5. The major computational burden in that procedure is the call to **procedure** framebylp, outlined in Figure 1. In principle, to determine if a column is part of the frame, we must remove the column from the matrix, put it as a right-hand side, and see if the corresponding system of linear equations has a solution or not. If it has a solution, the column is not part of the frame, and can be removed. An

[1] Figures and **procedures** referred to in this Subsection are contained in Chapter 5

important property of this procedure is that to determine if a column can be discarded, we have to use *all* other columns in the test. This is a major reason why **procedure** framebylp is so slow when the number of columns gets very large.

So, a generator w^* of the cone pos W^* has the property that a right-hand side h must satisfy $h^T w^* \leq 0$ to be feasible. In the uncapacitated network case we saw that a right-hand side β had to satisfy $b(Y)^T\beta \leq 0$ to represent a feasible problem. Therefore the index vector $b(Y)$ corresponds exactly to the column w^*. And calling **procedure** framebylp to remove those columns that are not in the frame of the cone pos W^* corresponds to using Proposition 6.4. Therefore the index vector of a node set from Proposition 6.4 corresponds to the columns in W^*.

Computationally there are major differences, though. First, to find a candidate for W^*, we had to start out with W, and use **procedure** support, which is an iterative procedure. The network inequalities, on the other hand, are produced more directly by looking at all subsets of nodes. But the most important difference is that, while the use of **procedure** framebylp, as just explained, requires all columns to be available in order to determine if one should be discarded, Proposition 6.4 is *totally local*. We can pick up an inequality and determine if it is needed without looking at any other inequalities. With possibly millions of candidates, this difference is crucial.

We did not develop the LP case for explicit bounds on variables. If such bounds exist, they can, however, be put in as explicit constraints. If so, a column w^* from W^* corresponds to the index vector $\begin{pmatrix} b(Y) \\ -a(Q^+) \end{pmatrix}$.

6.3 Generating Relatively Complete Recourse

Let us now discuss how the results obtained in the previous section can help us, and how they can be used in a setting that deserves the term preprocessing. Let us first repeat some of our terminology, in order to see how this fits in with our discussions in the LP setting.

A two-stage stochastic linear programming problem where the second-stage problem is a directed capacitated network flow problem can be formulated as follows:

$$\begin{aligned} &\min_x \left[c^T x + \mathcal{Q}(x)\right] \\ &\text{s.t. } \; Ax = b, \; x \geq 0, \end{aligned}$$

where

$$\mathcal{Q}(x) = \sum Q(x, \xi^j) p^j$$

and

$$Q(x,\xi) = \min_{y^1}\{(q^1)^{\mathrm{T}}y^1 \mid W'y^1 = h_0^1 + H^1\xi - T^1(\xi)x,\ 0 \le y^1 \le h_0^2 + H^2\xi - T^2(\xi)x\},$$

where W' is the node–arc incidence matrix for the network. To fit into a more general setting, let

$$W = \begin{pmatrix} W' & 0 \\ I & I \end{pmatrix}$$

so that $Q(x,\xi)$ can also be written as

$$Q(x,\xi) = \min_y\{q^{\mathrm{T}}y \mid Wy = h_0 + H\xi - T(\xi)x,\ y \ge 0\}$$

where $y = \begin{pmatrix} y^1 \\ y^2 \end{pmatrix}$, y^2 is the slack of y^1, $q = \begin{pmatrix} q^1 \\ 0 \end{pmatrix}$, $h_0 = \begin{pmatrix} h_0^1 \\ h_0^2 \end{pmatrix}$, $T(\xi) = \begin{pmatrix} T^1(\xi) \\ T^2(\xi) \end{pmatrix}$ and $H = \begin{pmatrix} H^1 \\ H^2 \end{pmatrix}$. Given our definition of β and γ, we have, for a given $\hat{x}$,

$$\begin{pmatrix} \beta \\ \gamma \end{pmatrix} = h_0 + H\xi - T(\xi)\hat{x} = h_0 + \sum_i h_i\xi_i - T(\xi)\hat{x}.$$

Using the inequalities derived in the previous section, we can proceed to transform these inequalities into inequalities in terms of x. By adding these inequalities to the first-stage constraints $Ax = b$, we get relatively complete recourse, i.e. we guarantee that any x satisfying the (expanded) first-stage constraints will yield a feasible second-stage problem for any value of $\tilde{\xi}$. An inequality has the form

$$b[A(Y)]^{\mathrm{T}}\beta = \sum_{i\in A(Y)} \beta(i) \le \sum_{k\in Q+} \gamma(k) = a(Q^+)^{\mathrm{T}}\gamma.$$

Let us replace β and γ with their expressions in terms of x and $\tilde{\xi}$. An inequality then says that the following must be true for all values of x and all realizations ξ of $\tilde{\xi}$:

$$b[A(Y)]^{\mathrm{T}}\left[h_0^1 + \sum_i h_i^1\xi_i - T^1(\xi)x\right] \le a(Q^+)^{\mathrm{T}}\left[h_0^2 + \sum_i h_i^2\xi_i - T^2(\xi)x\right].$$

Collecting all x terms on the left-hand side and all other terms on the right-hand side we get the following expression:

$$\begin{aligned}\Big[-b(A(Y))^{\mathrm{T}}\Big(T_0^1 + \sum_j T_j^1\xi_j\Big) + a(Q^+)^{\mathrm{T}}\Big(T_0^2 + \sum_j T_j^2\xi_j\Big)\Big]x \\ \le \sum_i\{-b[A(Y)]^{\mathrm{T}}h_i^1 + a(Q^+)^{\mathrm{T}}h_i^2\}\xi_i - b[A(Y)]^{\mathrm{T}}h_0^1 + a(Q^+)^{\mathrm{T}}h_0^2.\end{aligned}$$

Since this must be true for all possible values of $\tilde{\xi}$, we get one such inequality for each ξ. If $T(\xi) \equiv T_0$, we can make this more efficient by calculating only one cut, given by the following inequality:

$$\begin{aligned}&\{-b[A(Y)]^{\mathrm{T}}T_0^1 + a(Q^+)^{\mathrm{T}}T_0^2\}x \\ &\quad \le \min_{\xi\in\Xi}\sum_i\{-b[A(Y)]^{\mathrm{T}}h_i^1 + a(Q^+)^{\mathrm{T}}h_i^2\}\xi_i - b[A(Y)]^{\mathrm{T}}h_0^1 + a(Q^+)^{\mathrm{T}}h_0^2.\end{aligned}$$

The minimization is of course very simple in the independent case, since the minimization can be moved inside the sum. When facets have been transformed into inequalities in terms of x, we might find that they are linearly dependent. We should therefore subject them, together with the constraints $Ax = b$, to a procedure that removes redundant constraints. We have discussed this subject in Chapter 5.

The above results have two applications. Both are related to preprocessing. Let us first repeat the one we briefly mentioned above, namely that, after the inequalities have been added to $Ax = b$, we have relatively complete recourse, i.e. any x satisfying the (expanded) first-stage constraints will automatically produce a feasible recourse problem for all values of $\tilde{\xi}$. This opens up the avenue to methods that require this property, and it can help in others where this is really not needed. For example, we can use the L-shaped decomposition method (page 159) without concern about feasibility cuts, or apply the stochastic decomposition method as outlined in Section 3.8.

Another—and in our view more important—use of these inequalities is in model understanding. As expressions in x, they represent *implicit* assumptions made by the modeller in terms of the first-stage decisions. They are implicit because they were never written down, but they are there because otherwise the recourse problem can become infeasible. And, as part of the model, the modeller has made the requirements expressed in these implicit constraints. If there are not too many implicit assumptions, the modeller can relate to them, and either learn about his or her own model, or might decide that he or she did not want to make these assumption. If so, there is need for a revision of the model.

It is worth noting that the inequalities in terms of β and γ are also interesting in their own right. They show the modeller how the external flow and arc capacities must combine in order to produce a feasible recourse problem. Also, this can lead to understanding and/or model reformulation.

6.4 An Investment Example

Consider the simple network in Figure 11. It represents the flow of sewage (or some other waste) from three cities, represented by nodes 1, 2 and 3.

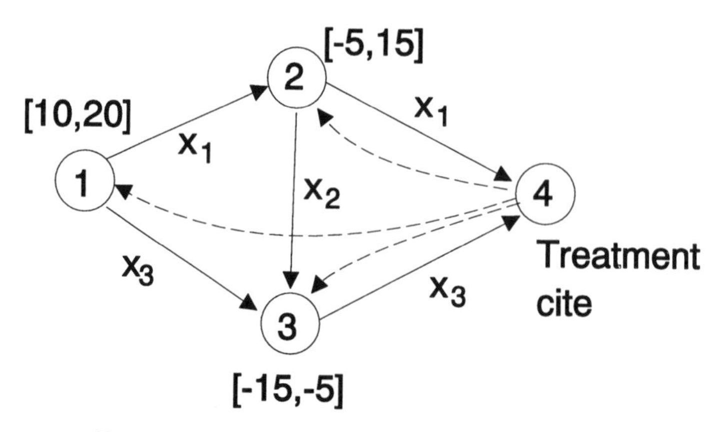

Figure 11 Transportation network for sewage, used for the example in Section 6.4.

All three cities produce sewage, and they have local treatment plants to take care of some of it. Both the amount of sewage from a city and its treatment capacity vary, and the net variation from a city is given next to the node representing the city. For example, City 1 always produces more than it can treat, and the surplus varies between 10 and 20 units per unit time. City 2, on the other hand, sometimes can treat up to 5 units of sewage from other cities, but at other times has as much as 15 units it cannot itself treat. City 3 always has extra capacity, and that varies between 5 and 15 units per unit time.

The *solid* lines in Figure 11 represent pipes through which sewage can be pumped (at a cost). Assume all pipes have a capacity of up to 5 units per unit time. Node 4 is a common treatment site for the whole area, and its capacity is so large that for practical purposes we can view it as being infinite. Until now, whenever a city had sewage that it could not treat itself, it first tried to send it to other cities, or site 4, but if that was not possible, the sewage was simply dumped in the ocean. (It is easy to see that that can happen. When City 1 has more than 10 units of untreated sewage, it must dump some of it.)

New rules are being introduced, and within a short period of time dumping sewage will not be allowed. Four projects have been suggested.

- Increase the capacity of the pipe from City 1 (via City 2) to site 4 with x_1 units (per unit time).
- Increase the capacity of the pipe from City 2 to City 3 with x_2 units (per unit time).
- Increase the capacity of the pipe from City 1 (via City 3) to site 4 with x_3 units (per unit time).
- Build a new treatment plant in City 1 with a capacity of x_4 units (per unit

time).

It is not quite clear if capacity increases can take on any values, or just some predefined ones. Also, the cost structure of the possible investments are not yet clear. Even so, we are asked to analyse the problem, and create a better basis for decisions.

The first thing we must do, to use the procedures of this chapter, is to make sure that, technically speaking, we have a network (as defined at the start of the chapter). A close look will reveal that a network must have equality constraints at the node, i.e. flow in must equal flow out. That is not the case in our little network. If City 3 has spare capacity, we do not *have to* send extra sewage to the city, we simply leave the capacity unused if we do not need it. The simplest way to take care of this is to introduce some new arcs in the network. They are shown with dotted lines in Figure 11. Finally, to have supply equal to demand in the network (remember from Proposition 6.1 that this is needed for feasibility), we let the external flow in node 4 be the negative of the sum of external flows in the other three nodes.

You may wonder if this rewriting makes sense. What does it mean when "sewage" is sent along a dotted line in the figure? The simple answer is that the amount exactly equals the unused capacity in the city to which the arc goes. (Of course, with the given numbers, we realize that no arc will be needed from node 4 to node 1, but we have chosen to add it for completeness.)

Now, to learn something about our problem, let us apply Proposition 6.3 to arrive at a number of inequalities. You may find it useful to try to write them down. We shall write down only some of them. The reason for leaving out some is the following observation: any node set Y that is such that Q^+ contains a dotted arc from Figure 11 will be uninteresting, because

$$a(Q^+)^{\mathrm{T}}\gamma = \infty,$$

so that the inequality says nothing interesting. The remaining inequalities are as follows (where we have used that all existing pipes have a capacity of 5 per unit time).

$$\left.\begin{array}{lll}
\beta_1 & \leq 10 + x_1 & + x_3 + x_4, \\
\quad\beta_2 & \leq 10 + x_1 + x_2, & \\
\qquad\beta_3 & \leq 5 & + x_3, \\
\beta_1 + \beta_2 + \beta_3 & \leq 10 + x_1 & + x_3 + x_4, \\
\beta_1 + \beta_2 & \leq 15 + x_1 + x_2 & + x_3 + x_4, \\
\beta_1 + \beta_3 & \leq 10 + x_1 & + x_3 + x_4, \\
\beta_2 + \beta_3 & \leq 10 + x_1 & + x_3.
\end{array}\right\} \qquad (4.1)$$

Let us first note that if we set all $x_i = 0$ in (4.1), we end up with a number of constraints that are not satisfied for all possible values of β. Hence, as we already know, there is presently a chance that sewage will be dumped.

However, our interest is mainly to find out about which investments to make. Let us therefore rewrite (4.1) in terms of x_i rather than β_i:

$$\left.\begin{array}{llllllllll}
x_1 & & +x_3 & +x_4 & \geq \beta_1 & & & -10 \geq & 10, \\
x_1 & +x_2 & & & \geq & \beta_2 & & -10 \geq & 5, \\
 & & +x_3 & & \geq & & \beta_3 & -\ 5 \geq & -10, \\
x_1 & & +x_3 & +x_4 & \geq \beta_1 & +\beta_2 & +\beta_3 & -10 \geq & 20, \\
x_1 & +x_2 & +x_3 & +x_4 & \geq \beta_1 & +\beta_2 & & -15 \geq & 20, \\
x_1 & & +x_3 & +x_4 & \geq \beta_1 & & +\beta_3 & -10 \geq & 5, \\
x_1 & & +x_3 & & \geq & \beta_2 & +\beta_3 & -10 \geq & 0.
\end{array}\right\} \quad (4.2)$$

The last inequality in each constraint of (4.2) is obtained by simply maximizing over the possible values of β, since what is written down must be true for *all* values of β. We can now start to remove some of the constraints because they do not say anything, or because they are implied by others. When this cannot be done manually, we can use the methods outlined in Section 5.1.3. In the arguments that follow, remember that $x_i \geq 0$.

First, we can remove inequalities 1 and 6, because they are weaker than inequality 4. But inequality 4, having the same right-hand side as number 5, but fewer variables on the left-hand side, implies number 5, and the latter can therefore be dropped. Inequality number 3 is uninteresting, and so is number 7 (since we clearly do not plan to make negative investments). This leaves us with only two inequalities, which we shall repeat:

$$\left.\begin{array}{ll}
x_1 + x_2 & \geq 5, \\
x_1 \qquad + x_3 + x_4 & \geq 20.
\end{array}\right\} \quad (4.3)$$

Even though we know nothing so far about investment costs and pumping costs through the pipes, we know a lot about what limits the options. Investments of at least five units must be made on a combination of x_1 and x_2. What this seems to say is that the capacity out of City 2 *must* be increased by at least 5 units. It is slightly more difficult to interpret the second inequality. If we see both building pipes and a new plant in City 1 as increases in treatment capacity (although they are of different types), the second inequality seems to say that a total of 15 units must be built to facilitate City 1. However, a closer look at which cut generated the inequality reveals that a more appropriate interpretation is to say that the three cities, when they are seen as a whole, must obtain extra capacity of 15 units. It was the node set $Y = \{1, 2, 3\}$ that generated the cut.

The two constraints (4.3) are all we need to pass on to the planners. If these two, very simple, constraints are taken care of, sewage will never have to be dumped. Of course, if the investment problem is later formulated as a linear program, the two constraints can be added, thereby guaranteeing feasibility, and, from a technical point of view, relatively complete recourse.

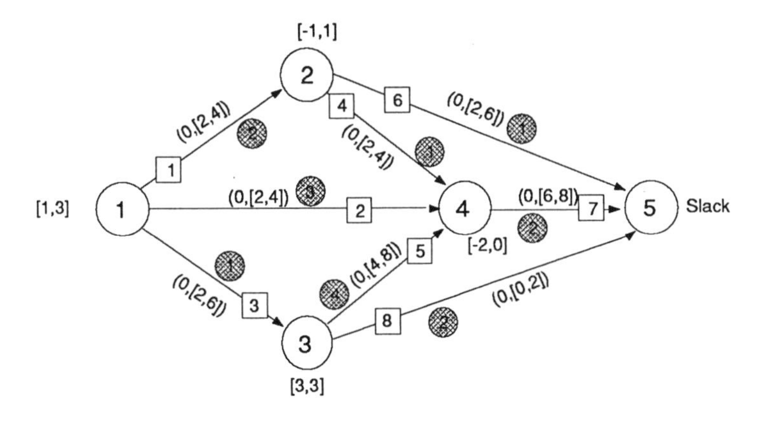

Figure 12 Network illustrating the different bounds.

6.5 Bounds

We discussed some bounds for general LPs in Chapter 3. These of course also apply to networks, since networks are nothing but special cases of linear programs. The Jensen lower bound can be found by replacing each random variable (external flow or arc capacity) by its mean and solving the resulting deterministic network flow problem. The Edmundson–Madansky upper bound is found by evaluating the network flow problem at all extreme points of the support. (If the randomness sits in the objective function, the methods give opposite bounds, just as we discussed for the LP case.)

Figure 12 shows an example that will be used in this section to illustrate bounds. The terminology is as follows. Square brackets, for example $[a, b]$, are used to denote supports of random variables. Placed next to a node, they show the size of the random external flow. Placed in a setting like $(c, [a, b])$, the square bracket shows the support of the upper bound on the arc flow for the arc next to which it is placed. In this setting, c is the lower bound on the flow. It can become negative in some of the methods. The circled number next to an arc is the unit arc cost, and the number in a square *on* the arc is the arc number. For simplicity, we shall assume that all random variables are independent and uniformly distributed.

Figure 13 shows the set-up for the Jensen lower bound for the example from Figure 12. We have now replaced each random variable by its mean, assuming that the distributions are symmetric. The optimal flow is

$$f = (2, 0, 0, 0, 2, 2, 1, 1)^{\mathrm{T}},$$

with a cost of 18.

Although the Edmundson–Madansky distribution is very useful, it still has

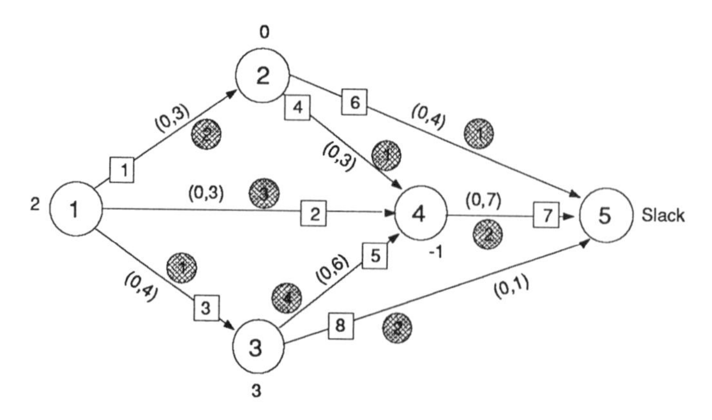

Figure 13 Example network with arc capacities and external flows corresponding to the Jensen lower bound.

the problem that the objective function must be evaluated in an exponential number of points. If there are k random variables, we must work with 2^k points. This means that with more than about 10 random variables we are not in business. Thus, since there are 11 random variable in the example, we must solve 2^{11} problems to find the upper bound. We have not done that here. In what follows, we shall demonstrate how to obtain a piecewise linear upper bound that does not exhibit this exponential characterization. A weakness of this bound is that it may be $+\infty$ even if the problem is feasible. That may not happen to the Edmundson–Madansky upper bound. We shall continue to use the network in Figure 12 to illustrate the ideas.

6.5.1 Piecewise Linear Upper Bounds

Let us illustrate the method in a simplified setting. Define $\phi(\xi, \eta)$ by

$$\phi(\xi, \eta) = \min_y \{q^{\mathrm{T}} y \mid W' y = b + \xi,\ 0 \leq y \leq c + \eta\},$$

where all elements of the random vectors $\tilde{\xi} = (\tilde{\xi}_1^{\mathrm{T}}, \tilde{\xi}_2^{\mathrm{T}}, \ldots)^{\mathrm{T}}$ and $\tilde{\eta} = (\tilde{\eta}_1^{\mathrm{T}}, \tilde{\eta}_2^{\mathrm{T}}, \ldots)^{\mathrm{T}}$ are mutually independent. Furthermore, let the supports be given by $\Xi(\tilde{\xi}) = [A, B]$ and $\Xi(\tilde{\eta}) = [0, C]$. The matrix W' is the node–arc incidence matrix for a network, with one row removed. That row represents the slack node. The external flow in the slack node equals the negative sum of the external flows in the other nodes. The goal is to create an upper bounding function $U(\xi, \eta)$ that is piecewise linear, separable and convex in ξ, as well as easily integrable in η:

$$U(\xi,\eta) = \phi(E\tilde{\xi},0) + H(\eta) + \sum_i \begin{cases} d_i^+(\xi_i - E\tilde{\xi}_i) & \text{if } \xi_i \geq E\tilde{\xi}_i, \\ d_i^-(E\tilde{\xi}_i - \xi_i) & \text{if } \xi_i < E\tilde{\xi}_i, \end{cases}$$

for some parameters $d_i^{\pm}$. The principles of the ξ part of this bound were outlined in Section 3.4.4 and will not be repeated in all details here. We shall use the developments from that section here, simply by letting $\eta = 0$ while developing the ξ part. Because this is a restriction (constraint) on the original problem, it produces an upper bound. Then, afterwards, we shall develop $H(\eta)$. In Section 3.4.4 we assumed that $E\tilde{\xi} = 0$. We shall now drop that assumption, just to illustrate that it was not needed, and to show how many parameters can be varied in this method.

Let us first see how we can find the ξ part of the function, leaving $\eta = 0$. First, let us calculate

$$\phi(E\tilde{\xi},0) = \min_y \{q^{\mathrm{T}}y \mid W'y = b + E\tilde{\xi},\ 0 \leq y \leq c\} = q^{\mathrm{T}}y^0.$$

This is our basic setting, and all other values of ξ will be seen as deviations from $E\tilde{\xi}$. Note that since y^0 is "always" there, we shall update the arc capacities to become $-y^0 \leq y \leq c - y^0$. For this purpose, we define $\alpha^1 = -y^0$ and $\beta^1 = c - y^0$. Let e_i be a unit vector of appropriate dimension with a $+1$ in position i.

Next, define a counter r and let $r := 1$. Now, check out the case when $\xi_1 > E\tilde{\xi}_1$ by solving

$$\min_y \{q^{\mathrm{T}}y \mid W'y = e_r(B_r - E\tilde{\xi}_r),\ \alpha^r \leq y \leq \beta^r\} = q^{\mathrm{T}}y^{r+} = d_r^+(B_r - E\tilde{\xi}_r). \tag{5.1}$$

Similarly, check out the case with $\xi_1 < E\tilde{\xi}_1$ by solving

$$\min_y \{q^{\mathrm{T}}y \mid W'y = e_r(A_r - E\tilde{\xi}_r),\ \alpha^r \leq y \leq \beta^r\} = q^{\mathrm{T}}y^{r-} = d_r^-(A_r - E\tilde{\xi}_r). \tag{5.2}$$

Now, based on $y^{r\pm}$, we shall assign portions of the arc capacities to the random variable $\tilde{\xi}_r$. These portions will be given to $\tilde{\xi}_r$ and left unused by other random variables, even when $\tilde{\xi}_r$ does not need them. The portions will correspond to paths in the network connecting node r to the slack node (node 5 in the example). That is done by means of the following problem, where we calculate what is left for the next random variable:

$$\alpha_i^{r+1} = \alpha_i^r - \min\{y_i^{r+}, y_i^{r-}, 0\}. \tag{5.3}$$

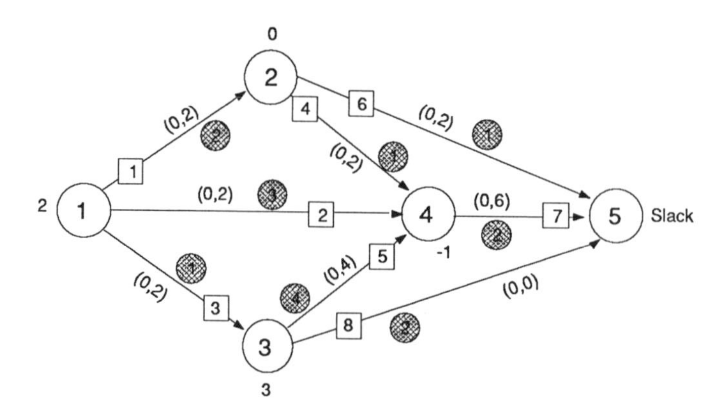

Figure 14 Network needed to calculate $\phi(E\tilde{\xi}, 0)$ for the network in Figure 12.

What we are doing here is to find, for each variable, how much $\tilde{\xi}_r$, in the worst case, uses of arc i in the negative direction. That is then subtracted from what we had before. There are three possibilities. We may have both (5.1) and (5.2) yielding nonnegative values for the variable i. Then nothing is used of the available "negative capacity" α_i^r. Then $\alpha_i^{r+1} = \alpha_i^r$. Alternatively, when (5.1) has $y_i^{r+} < 0$, it will in the worst case use y_i^{r+} of the available "negative capacity". Finally, when (5.2) has $y_i^{r-} < 0$, in the worst case we use y_i^{r-} of the capacity. Therefore, α_i^{r+1} is what is left for the next random variable. Similarly, we find

$$\beta_i^{r+1} = \beta_i^r - \max\{y_i^{r+}, y_i^{r-}, 0\}, \tag{5.4}$$

where β_i^{r+1} shows how much is still available of the capacity on arc i in the forward (positive) direction.

We next increase the counter r by one and repeat (5.1)–(5.4). This takes care of the piecewise linear functions in ξ.

Let us now look at our example in Figure 12. To calculate the ξ part of the bound, we put all arc capacities at their lowest possible value and external flows at their means. This is shown in Figure 14.

The optimal solution in Figure 14 is given by

$$y^0 = (2, 0, 0, 0, 3, 2, 2, 0)^{\mathrm{T}},$$

with a cost of 22. The next step is update the arc capacities in Figure 14 to account for this solution. The result is shown in Figure 15.

Since the external flow in node 1 varies between 1 and 3, and we have so far solved the problem for a supply of 2, we must now find the cost associated with a supply of 1 and a demand of 1 in node 1. For a supply of 1 we get the

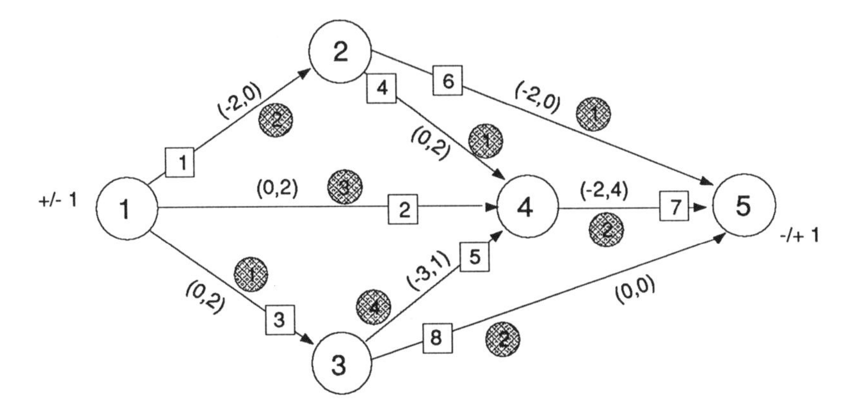

Figure 15 Arc capacities after the update based on $\phi(E\tilde{\xi}, 0)$.

solution

$$y^{1+} = (0, 1, 0, 0, 0, 0, 1, 0)^{\mathrm{T}},$$

with a cost of 5. Hence $d_1^+ = 5$. For a demand of 1 we get

$$y^{1-} = (-1, 0, 0, 0, 0, -1, 0, 0)^{\mathrm{T}},$$

with a cost of -3, so that $d_1^- = 3$. Hence we have used one unit of the forward capacity of arcs 2 and 7, and one unit of the reverse capacity of arcs 1 and 6. Note that both solutions correspond to paths between node 1 and node 5 (the slack node). We update to get Figure 16.

For node 2 the external flow varies between -1 and 1, so we shall now check the supply of 1 and demand of 1 based on the arc capacities of Figure 16. For supply we get

$$y^{2+} = (0, 0, 0, 1, 0, 0, 1, 0)^{\mathrm{T}},$$

with a cost of 3. For the demand of 1 we obtain

$$y^{2-} = (0, 0, 0, 0, 0, -1, 0, 0)^{\mathrm{T}},$$

with a cost of -1. Hence $d_2^+ = 3$ and $d_2^- = 1$. Node 3 had deterministic external flow, so we turn to node 4. Node 4 had a demand between 0 and 2 units, and we have so far solved for a demand of 1. Therefore we must now look at a demand of 1 and a supply of 1 in node 4, based on the arc capacities in Figure 17. In that figure we have updated the capacities from Figure 16 based on the solutions for node 2.

A supply in node 4 gives us the solution

$$y^{4+} = (0, 0, 0, 0, 0, 0, 1, 0)^{\mathrm{T}},$$

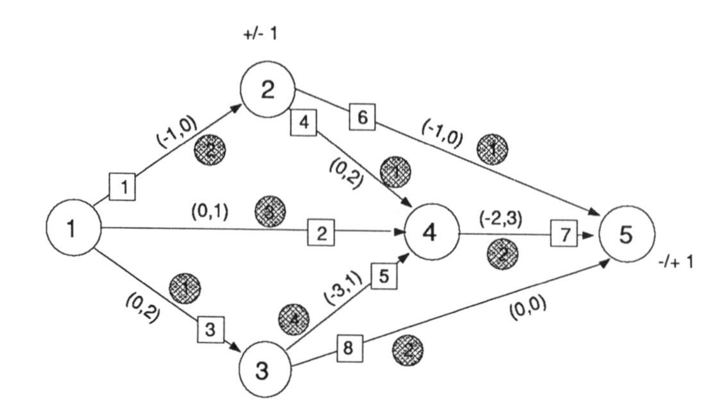

Figure 16 Arc capacities after the update based on $\phi(E\tilde{\xi}, 0)$ and node 1.

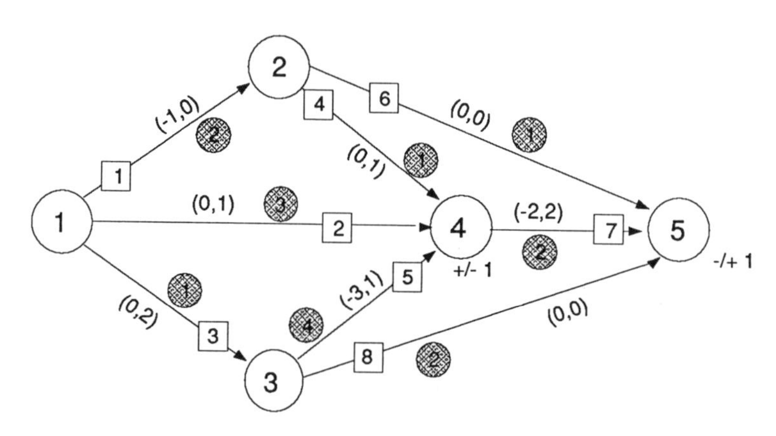

Figure 17 Arc capacities after the update based on $\phi(E\tilde{\xi}, 0)$ and nodes 1 and 2.

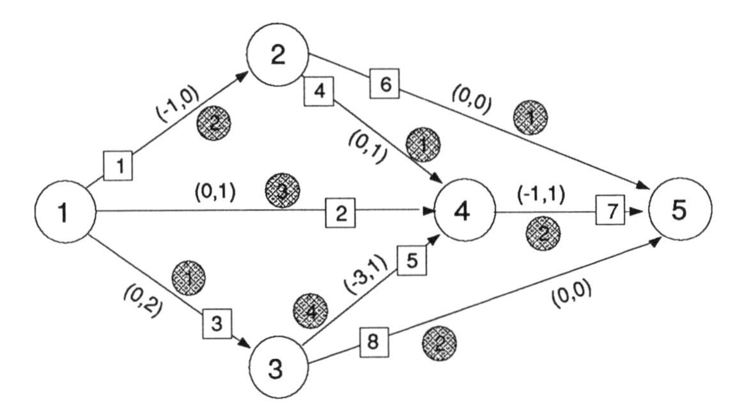

Figure 18 Arc capacities after the update based on $\phi(E\tilde{\xi}, 0)$ and external flow in all nodes.

with a cost of 2. One unit demand, on the other hand, gives us

$$y^{4-} = (0,0,0,0,0,-1,0)^{\mathrm{T}},$$

with a cost of -2. The parameters are therefore $d_4^+ = 2 = d_4^-$. This leaves the arc capacities in Figure 18.

What we have found so far is as follows:

$$\begin{aligned}
\phi(\xi,\eta) = {} & 22 + H(\eta) \\
& + \begin{cases} 5(\xi_1 - 2) & \text{if } \xi_1 \geq 2, \\ 3(\xi_1 - 2) & \text{if } \xi_1 < 2, \end{cases} \\
& + \begin{cases} 3\xi_2 & \text{if } \xi_2 \geq 0, \\ \xi_2 & \text{if } \xi_2 < 0, \end{cases} \\
& + \begin{cases} 2(\xi_4 + 1) & \text{if } \xi_4 \geq -1, \\ 2(\xi_4 + 1) & \text{if } \xi_4 < -1. \end{cases}
\end{aligned}$$

If, for simplicity, we assume that all distributions are uniform, we easily integrate the upper-bounding function $U(\xi,\eta)$ to obtain

$$\begin{aligned}
U = {} & 22 + H(\eta) \\
& + \int_1^2 3(\xi_1 - 2)\tfrac{1}{2}d\xi_1 + \int_2^3 5(\xi_1 - 2)\tfrac{1}{2}d\xi_1 \\
& + \int_{-1}^0 \xi_2 \tfrac{1}{2}d\xi_2 + \int_0^1 3\xi_2\tfrac{1}{2}d\xi_2 \\
& + \int_{-2}^{-1} 2(\xi_4 + 1)\tfrac{1}{2}d\xi_4 + \int_{-1}^0 2(\xi_4 + 1)\tfrac{1}{2}d\xi_4 \\
= {} & 22 + H(\eta) - 3 \times \tfrac{1}{4} + 5 \times \tfrac{1}{4} - 1 \times \tfrac{1}{4} + 3 \times \tfrac{1}{4} - 2 \times \tfrac{1}{4} + 2 \times \tfrac{1}{4} \\
= {} & 23 + H(\eta).
\end{aligned}$$

Note that there is no contribution from ξ_4 to the upper bound. The reason is that the recourse function $\phi(\xi,\eta)$ is linear in ξ_4. This property of discovering

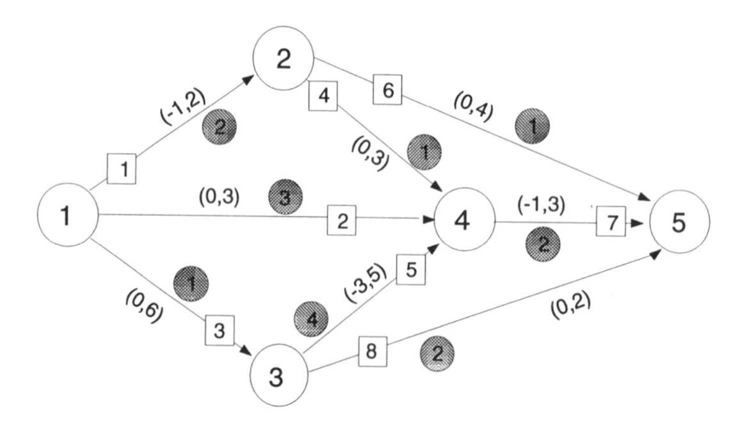

Figure 19 Arc capacities used to calculate $H(\eta)$ for the example in Figure 12.

that the recourse function is linear in some random variable is shared with the Jensen and Edmundson–Madansky bounds.

We then turn to the η part of the bound. Note that if (5.3) and (5.4) were calculated after the final $y^{r\pm}$ had been found, the α and β show what is left of the deterministic arc capacities after all random variable $\tilde{\xi}_i$ have received their shares. Let us call these α^* and β^*. If we add to each upper bound in Figure 18 the value C (remember that the support of the upper arc capacities was $\Xi = [0, C]$), we get the arc capacities of Figure 19. Now we solve the problem

$$\min_y \{q^{\mathrm{T}} y \mid W' y = 0,\ \alpha^* \leq y \leq \beta^* + C\} = q^{\mathrm{T}} y^*. \tag{5.5}$$

With zero external flow in Figure 19, we get the optimal solution

$$y^* = (0, 0, 0, 0, -1, 0, -1, 1)^{\mathrm{T}},$$

with a cost of -4. This represents cycle flow with negative costs. The cycle became available as a result of making arc 8 having a positive arc capacity. If, again for simplicity, we assume that η_8 is uniformly distributed over $[0, 2]$, we find that the capacity of that cycle has a probability of being 1 equal to 0.5. The remaining probability mass is uniformly distributed over $[0, 1]$. We therefore get

$$EH(\eta) = -4 \times 1 \times \tfrac{1}{2} - 4 \int_0^1 \tfrac{1}{2} x \, dx = -2 - 1 = -3.$$

The total upper bound for this example is thus $23 - 3 = 20$, compared with the Jensen lower bound of 18.

In this example the solution y^* of (5.5) contained only one cycle. In general, y^* may consist of several cycles, possibly sharing arcs. It is then necessary to pick y^* apart into individual cycles. This can be done in such a way that all cycles have nonpositive costs (those with zero costs can then be discarded), and such that all cycles that use a common arc use it in the same direction. We shall not go into details of that here.

6.6 Project Scheduling

We shall spend a whole section on the subject of project scheduling, and we shall do so in a setting of PERT (project evaluation and review technique) networks. There are several reasons for looking specifically at this class of problems. First, project scheduling is widely used, and therefore known to many people. Even though it seems that the setting of CPM (critical path method) is more popular among industrial users, the difference is not important from a principle point of view. Secondly, PERT provides us with a genuine opportunity to discuss some modelling issues related to the relationship between time periods and stages. We shall see that PERT has sometimes been cast in a two-stage setting, but that it can be hard to interpret that in a useful way. Thirdly, the more structure a problem has, the better bounds can often be found. PERT networks provide us with a tool for showing how much structure provide tight bounds.

Before we continue, we should like to point out a possible confusion in terms. When PERT was introduced in 1959, it was seen as a method for analysing projects with stochastic activity durations. However, the way in which randomness was treated was quite primitive (in fact, it is closely related to the Jensen lower bound that we discussed in Section 3.4.1). Therefore, despite the historical setting, many people today view PERT as a deterministic approach, simply disregarding what the original authors said about randomness. When we use the term PERT in the following, we shall refer to the mathematical formulation with its corresponding deterministic solution procedure, and not to its original random setting.

There are many ways to formulate the PERT problem. For our purpose, the following will do. A PERT network is a network where arcs correspond to *activities*, and nodes to *events*. If arc $k \sim (i,j)$ then activity k can start when event i has taken place, and event j can take place when all activities k', with $k' \sim (i',j)$, have finished. A PERT network must be acyclic, otherwise an activity must finish before it can start—a meaningless situation. Because of acyclicity, we can number nodes, such that if $k \sim (i,j)$ then $i < j$. As a consequence, node 1 represents the event "We are ready to start" and n the event "The project is finished". Let π_i be the time event i takes place, and let us define $\pi_1 := 0$. Furthermore, let q_k be the duration of activity k. Since an event can take place only after all activities preceding it have finished, we must have

$$\pi_j \geq \pi_i + q_k \text{ for all } k \sim (i, j).$$

Since π_n is the time at which the project finishes, we can calculate the minimal project completion time by solving

$$\left.\begin{array}{ll} \min \pi_n & \\ \text{s.t. } \pi_j - \pi_i \geq q_k & \text{for all } k \sim (i, j), \\ \qquad \pi_1 = 0. & \end{array}\right\} \tag{6.1}$$

It is worth noting that (6.1) is not really a decision problem. There are namely no decisions. We are only calculating consequences of an existing setting of relations and durations.

6.6.1 PERT as a Decision Problem

As pointed out, (6.1) is not a decision problem, since there are no decisions to be made. Very often, activity durations are not given by nature, but can be affected by how much resources we put into them. For example, it takes longer to build a house with one carpenter than with two. Assume we have available a budget of B units of resources, and that if we spend one unit on activity k, its duration will decrease by a_k time units. A possible decision problem is then to spend the budget in such a way that the project duration is minimized. This can be achieved by solving the following problem:

$$\left.\begin{array}{ll} \min \pi_n & \\ \text{s.t. } \pi_j - \pi_i \geq q_k - a_k x_k & \text{for all } k \sim (i, j), \\ \qquad \sum_k x_k \leq B, & \\ \qquad \pi_1 = 0, & \\ \qquad x_k \geq 0. & \end{array}\right\} \tag{6.2}$$

Of course, there might be other constraints, such as $x_k \leq c_k$, but they can be added to (6.2) as needed.

6.6.2 Introduction of Randomness

It seems natural to assume that activity durations are random. If so, the project duration is also random, and we can no longer talk about finding the minimal project duration time. However, a natural alternative seems to be to look for the *expected* (minimal) project duration time. In (6.1) and (6.2) the goal would then be to minimize $E\pi_n$. However, we must now be careful about how we interpret the problems. Problem (6.1) is simple enough. There are still no decisions, so we are only trying to calculate when, on expectation, the project will finish, if all activities start as soon as they can. But when we turn to (6.2) we must be careful. In what order do things happen? Do we first decide on x, and then simply sit back (as we did with (6.1)) and observe what happens? Or do we first observe what happens, and then make decisions on x? These are substantially different situations. It is of importance that you understand the modelling aspects of this difference. (There are solution differences as well, but they are less interesting now.)

If we interpret (6.2) as a problem where x is determined before the activity durations are known, we have in fact a standard two-stage stochastic program. The first-stage decision is to find x, and the second-stage "decision" to find the project duration given x and a realization of $q(\tilde{\xi})$. (We put $q(\tilde{\xi})$ to show that q is indeed a random variable.) But—and this is perhaps the most important question to ask in this section—is this a good model? What does it mean? First, it is implicit in the model that, while the original activity durations are random, the changes $a_k x_k$ are not. In terms of probability distributions, therefore what we have done is to reduce the means of the distributions describing activity durations, but without altering the variances. This might or might not be a reasonable model. Clearly, if we find this unreasonable, we could perhaps let a_k be a random variable as well, thereby making also the effect of the investment x_k uncertain.

The above discussion is more than anything a warning that whenever we introduce randomness in a model, we must make sure you know what the randomness means. But there is a much more serious model interpretation if we see (6.2) as a two-stage problem. It means that we think we are facing a project where, before it is started, we can make investments, but where afterwards, however badly things go, we shall never interfere in order to fix shortcomings. Also, even if we are far ahead of schedule, we shall not cut back on investments to save money. We may ask whether such projects exist—projects where we are free to invest initially, but where afterwards we just sit back and watch, *whatever happens.*

From this discussion you may realize (as you have before—we hope) that the definition of stages is important when making models with stochasticity. In our view, project scheduling with uncertainty is a multistage problem, where decisions are made each time new information becomes available. This makes the problem extremely hard to solve (and even formulate—just try!) But this complexity cannot prevent us from pointing out the difficulties facing anyone trying to formulate PERT problems with only two stages.

We said earlier that there were two ways of interpreting (6.2) in a setting of uncertainty. We have just discussed one. The other is different, but has similar problems. We could interpret (6.2) with uncertainties as if we first observed the values of q and then made investments. This is the "wait-and-see solution". It represents a situation where we presently face uncertainty, but where all uncertainty will be resolved before decisions have to be made. What does that mean in our context? It means that before the project starts, all uncertainty related to activities disappears, everything becomes known, and we are faced with investments of the type (6.2). If the previous interpretation of our problem was odd, this one is probably even worse. In what sort of project will we have initial uncertainty, but before the first activity starts, everything, up to the finish of the project, becomes known? This seems almost as unrealistic as having a deterministic model of the project in the first place.

6.6.3 Bounds on the Expected Project Duration

Despite our own warnings in the previous subsection, we shall now show how the extra structure of PERT problems allows us to find bounds on the expected project duration time if activity durations are random. Technically speaking, we are looking for the expected value of the objective function in (6.1) with respect to the random variables $q(\tilde{\xi})$. There is a very large collection of different methods for bounding PERT problems. Some papers are listed at the end of this chapter. However, most, if not all, of them can be categorized as belonging to one or more of the following groups.

6.6.3.1 Series reductions

If there is a node with only one incoming and one outgoing arc, the node is removed, and the arcs replaced by one arc with a duration equal to the sum of the two arc durations. This is an exact reformulation.

6.6.3.2 Parallel reductions

If two arcs run in parallel with durations $\tilde{\xi}_1$ and $\tilde{\xi}_2$ then they are replaced with one arc having duration $\max\{\tilde{\xi}_1, \tilde{\xi}_2\}$. This is also an exact reformulation.

6.6.3.3 Disregarding path dependences

Let $\tilde{\pi}_i$ be a random variable describing when event i takes place. Then we can calculate

$$\tilde{\pi}_j = \max_{i \in B^+(i) \setminus \{i\}} \{\tilde{\pi}_i + q_k(\tilde{\xi})\}, \quad \text{with } k \sim (i,j),$$

as if all these random variables were independent. However, in a PERT network, the $\tilde{\pi}$s will normally be dependent (even if the qs are independent), since the paths leading up to the nodes usually share some arcs. Not only will they be dependent, but the correlation will always be positive, never negative.

Table 1 Joint distribution for $\tilde{\xi}_1$ and $\tilde{\xi}_2$, plus the calculation of $\max\{\tilde{\xi}_1, \tilde{\xi}_2\}$.

$\tilde{\xi}_1$	$\tilde{\xi}_2$	Prob.	max
1	1	0.3	1
1	2	0.2	2
2	1	0.2	2
2	2	0.3	2

Hence viewing the random variables as independent will result in an upper bound on the project duration. The reason is that $E\max\{\tilde{\xi}_1, \tilde{\xi}_2\}$ is smaller if the nonnegative $\tilde{\xi}_1$ and $\tilde{\xi}_2$ are (positively) correlated than if they are not correlated. A small example illustrates this

Example 6.3 Assume we have two random variables $\tilde{\xi}_1$ and $\tilde{\xi}_2$, with joint distribution as in Table 1. Note that both random variables have the same marginal distributions; namely, each of them can take on the values 1 or 2, each with a probability 0.5. Therefore $E\max\{\tilde{\xi}_1, \tilde{\xi}_2\} = 1.7$ from Table 1, but $0.25(1+2+2+2) = 1.75$ if we use the marginal distributions as independent distributions. Therefore, if $\tilde{\xi}_1$ and $\tilde{\xi}_2$ represent two paths with some joint arc, disregarding the dependences will create an upper bound.

□

6.6.3.4 Arc duplications

If there is a node i' with $B^+(i') = \{i, i'\}$, so that the node has only one incoming arc $k'' \sim (i, i')$, remove node i', and for each $j \in F^+(i') \setminus \{i'\}$ replace $k' \sim (i', j)$ by $k \sim (i, j)$. The new arc has associated with it the random duration $q_k(\tilde{\xi}) := q_{k'}(\tilde{\xi}) + q_{k''}(\tilde{\xi})$. If arc k'' had a deterministic duration, this is an exact reformulation. If not, we get an upper bound based on the previous principle of disregarding path dependences. (This method is called arc duplication because we duplicate arc k'' and use one copy for each arc k'.) An exactly equal result applies if there is only one outgoing arc. This result is illustrated in Figure 20, where $F^+(i') = \{i', 1, 2, 3\}$.

If there are several incoming and several outgoing arcs, we may pair up all incoming arcs with all outgoing arcs. This always produces an upper bound based on the principle of disregarding path dependences.

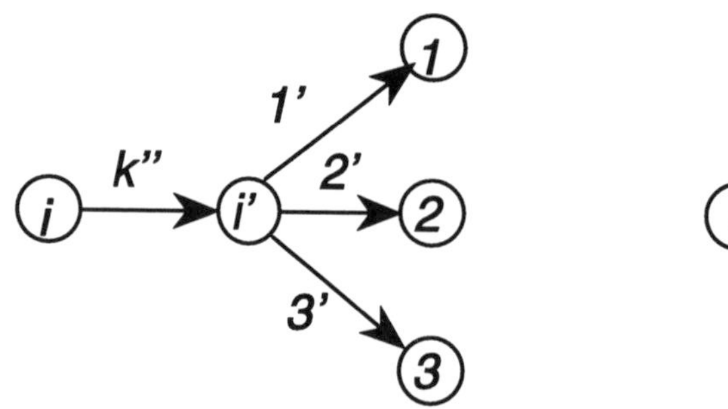

Figure 20 Arc duplication.

6.6.3.5 Using Jensen's inequality

Since our problem is convex in ξ, we get a lower bound whenever a $q_k(\tilde{\xi})$ or a $\tilde{\pi}_i$ (as defined above) is replaced by its mean.

Note that if we have a node and choose to apply arc duplication, we get an exact reformulation if all incoming arcs and all outgoing arcs have deterministic durations, an upper bound if they do not, and a lower bound if we first replace the random variables on the incoming and outgoing arcs by their means and then apply arc duplication. If there is only one arc in or one arc out, we take the expectation for that arc, and then apply arc duplication, observing an overall lower bound.

6.7 Bibliographical Notes

The vocabulary in this chapter is mostly taken from Rockafellar [25], which also contains an extremely good overview of deterministic network problems. A detailed look at network recourse problem is found in Wallace [28].

The original feasibility results for networks were developed by Gale [10] and Hoffman [13]. The stronger versions using connectedness were developed by Wallace and Wets. The uncapacitated case is given in [31], while the capacitated case is outlined in [33] (with a proof in [32]). More details of the algorithms in Figures 9 and 10 can also be found in these papers. Similar results were developed by Prékopa and Boros [23]. See also Kall and Prékopa [14].

As for the LP case, model formulations and infeasibility tests have of course been performed in many contexts apart from ours. In addition to the references given in Chapter 5, we refer to Greenberg [11, 12] and Chinneck [3].

The piecewise linear upper bound is taken from Wallace [30]. At the very end of our discussion of the piecewise linear upper bound, we pointed out that the solution y^* to (5.5) could consist of several cycles sharing arcs. A detailed discussion of how to pick y^* apart, to obtain a *conformal realization* can be found in Rockafellar [25], page 476. How to use it in the bound is detailed in [30]. The bound has been strengthened for pure arc capacity uncertainty by Frantzeskakis and Powell [8].

Special algorithms for stochastic network problems have also been developed; see e.g. Qi [24] and Sun et al. [27].

We pointed out at the beginning of this chapter that scenario aggregation (Section 2.6) could be particularly well suited to problems that have network structure in all periods. This has been utilized by Mulvey and Vladimirou for financial problems, which can be formulated in a setting of generalized networks. For details see [19, 20]. For a selection of papers on financial problems (not all utilizing network structures), consult Zenios [36, 37], and, for a specific application, see Dempster and Ireland [5].

The above methods are well suited for parallel processing. This has been done in Mulvey and Vladimirou [18] and Nielsen and Zenios [21].

Another use of network structure to achieve efficient methods is described in Powell [22] for the vehicle routing problem.

The PERT formulation was introduced by Malcolm et al. [17]. An overview of project scheduling methods can be found in Elmaghraby [7]. A selection of bounding procedures based on the different ideas listed above can be found in the following: Fulkerson [9], Kleindorfer [16], Shogan [26], Kamburowski [15] and Dodin [6]. The PERT problem as an investment problem is discussed in Wollmer [34].

The max flow problem is another special network flow problem that is much studied in terms of randomness. We refer to the following papers, which discuss both bounds and a two-stage setting: Cleef and Gaul [4], Wollmer [35], Aneja and Nair [1], Carey and Hendrickson [2] and Wallace [29].

Exercises

1. Consider the network in Figure 21. The interpretation is as for Figure 12 regarding parameters, except that we for the arc capacities simply have written a number next to the arc. All lower bounds on flow are zero. Calculate the Jensen lower bound, the Edmundson-Madansky upper bound, and the piecewise linear upper bound for the expected minimal cost in the network.

2. When outlining the piecewise linear upper bound, we found a function that was linear both above and below the expected value of the random variable. Show how (5.1) and (5.2) can be replaced by a *parametric* linear program to get not just one linear piece above the expectation and one below, but

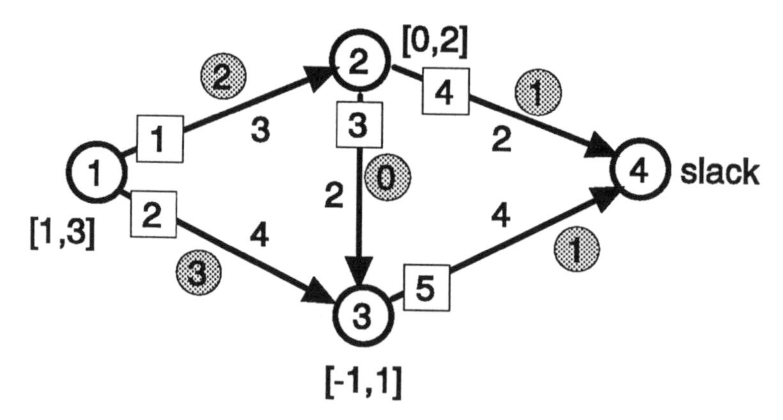

Figure 21 Example network for calculating bounds.

rather piecewise linearity on both sides. Also, show how (5.3) and (5.4) must then be updated to account for the change.

3. The *max flow problem* is the problem of finding the maximal amount of flow that can be sent from node 1 to node n in a capacitated network. This problem is very similar to the PERT problem, in that paths in the latter correspond to cuts in the max flow problem. Use the bounding ideas listed in Section 6.6.3 to find bounds on the expected max flow in a network with random arc capacities.

4. In our example about sewage treatment in Section 6.4 we introduced four investment options.

   (a) Assume that a fifth investment is suggested, namely to build a pipe with capacity x_5 directly from City 1 to site 4. What are the constraints on x_i for $i = 1, \ldots, 5$ that must now be satisfied for the problem to be feasible?

   (b) Disregard the suggestion in question (a). Instead, it is suggested to see the earlier investment 1, i.e. increasing the pipe capacity from City 1 to cite 4 via City 2 as two different investment. Now let x_1 be the increased capacity from City 1 to City 2, and x_5 the increased capacity from City 2 to cite 4 (the dump). What are now the constraints on x_i for $i = 1, \ldots, 5$ that must be satisfied for the problem to be feasible? Make sure you interpret the constraints.

References

[1] Aneja Y. P. and Nair K. P. K. (1980) Maximal expected flow in a network subject to arc failures. *Networks* 10: 45–57.

[2] Carey M. and Hendrickson C. (1984) Bounds on expected performance of networks with links subject to failure. *Networks* 14: 439–456.

[3] Chinneck J. W. (1990) Localizing and diagnosing infeasibilities in networks. Working paper, Systems and Computer Engineering, Carleton University, Ottawa, Ontario.

[4] Cleef H. J. and Gaul W. (1980) A stochastic flow problem. *J. Inf. Opt. Sci.* 1: 229–270.

[5] Dempster M. A. H. and Ireland A. M. (1988) A financial expert decision support system. In Mitra G. (ed) *Mathematical Methods for Decision Support*, pages 415–440. Springer-Verlag, Berlin.

[6] Dodin B. (1985) Reducability of stochastic networks. *OMEGA Int. J. Management* 13: 223–232.

[7] Elmaghraby S. (1977) *Activity Networks: Project Planning and Control by Network Models.* John Wiley & Sons, New York.

[8] Frantzeskakis L. F. and Powell W. B. (1989) An improved polynomial bound for the expected network recourse function. Technical report, Statistics and Operations Research Series, SOR-89-23, Princeton University, Princeton, New Jersey.

[9] Fulkerson D. R. (1962) Expected critical path lengths in PERT networks. *Oper. Res.* 10(6): 808–817.

[10] Gale D. (1957) A theorem of flows in networks. *Pac. J. Math.* 7: 1073–1082.

[11] Greenberg H. J. (1987) Diagnosing infeasibility in min-cost network flow problems. part I: Dual infeasibility. *IMA J. Math. in Management* 1: 99–109.

[12] Greenberg H. J. (1988/9) Diagnosing infeasibility in min-cost network flow problems. part II: Primal infeasibility. *IMA J. Math. in Management* 2: 39–50.

[13] Hoffman A. J. (1960) Some recent applications of the theory of a multivariate random variable. *Proc. Symp. Appl. Math.* 10: 113–128.

[14] Kall P. and Prékopa A. (eds) (1980) *Recent Results in Stochastic Programming*, volume 179 of *Lecture Notes in Econ. Math. Syst.* Springer-Verlag, Berlin.

[15] Kamburowski J. (1985) Bounds in temporal analysis of stochastic networks. *Found. Contr. Eng.* 10: 177–185.

[16] Kleindorfer G. B. (1971) Bounding distributions for a stochastic acyclic network. *Oper. Res.* 19: 1586–1601.

[17] Malcolm D. G., Roseboom J. H., Clark C. E., and Fazar W. (1959) Applications of a technique for R&D program evaluation. *Oper. Res.*

7: 646–696.

[18] Mulvey J. M. and Vladimirou H. (1989) Evaluation of a parallel hedging algorithm for stochastic network programming. In Sharda R., Golden B. L., Wasil E., Balci O., and Stewart W. (eds) *Impact of Recent Computer Advances on Operations Research*, pages 106–119. North-Holland, New York.

[19] Mulvey J. M. and Vladimirou H. (1989) Stochastic network optimization models for investment planning. *Ann. Oper. Res.* 20: 187–217.

[20] Mulvey J. M. and Vladimirou H. (1991) Applying the progressive hedging algorithm to stochastic generalized networks. *Ann. Oper. Res.* 31: 399–424.

[21] Nielsen S. and Zenios S. A. (1993) A massively parallel algorithm for nonlinear stochastic network problems. *Oper. Res.* 41: 319–337.

[22] Powell W. B. (1988) A comparative review of alternative algorithms for the dynamic vehicle allocation problem. In Golden B. and Assad A. (eds) *Vehicle Routing: Methods and Studies*, pages 249–291. North-Holland, Amsterdam.

[23] Prékopa A. and Boros E. (1991) On the existence of a feasible flow in a stochastic transportation network. *Oper. Res.* 39: 119–129.

[24] Qi L. (1985) Forest iteration method for stochastic transportation problem. *Math. Prog. Study* 25: 142–163.

[25] Rockafellar R. T. (1984) *Network Flows and Monotropic Optimization.* John Wiley & Sons, New York.

[26] Shogan A. W. (1977) Bounding distributions for a stochastic PERT network. *Networks* 7: 359–381.

[27] Sun J., Tsai K. H., and Qi L. (1993) A simplex method for network programs with convex separable piecewise linear costs and its application to stochastic transshipment problems. In Du D. Z. and Pardalos P. M. (eds) *Network Optimization Problems: Algorithms, Applications and Complexity*, pages 283–300. World Scientific, Singapore.

[28] Wallace S. W. (1986) Solving stochastic programs with network recourse. *Networks* 16: 295–317.

[29] Wallace S. W. (1987) Investing in arcs in a network to maximize the expected max flow. *Networks* 17: 87–103.

[30] Wallace S. W. (1987) A piecewise linear upper bound on the network recourse function. *Math. Prog.* 38: 133–146.

[31] Wallace S. W. and Wets R. J.-B. (1989) Preprocessing in stochastic programming: The case of uncapacitated networks. *ORSA J.Comp.* 1: 252–270.

[32] Wallace S. W. and Wets R. J.-B. (1993) The facets of the polyhedral set determined by the Gale–Hoffman inequalities. *Math. Prog.* 62: 215–222.

[33] Wallace S. W. and Wets R. J.-B. (1995) Preprocessing in stochastic programming: The case of capacitated networks. *ORSA J.Comp.* 7: 44–

64.
[34] Wollmer R. D. (1985) Critical path planning under uncertainty. *Math. Prog. Study* 25: 164–171.
[35] Wollmer R. D. (1991) Investments in stochastic maximum flow problems. *Ann. Oper. Res.* 31: 459–467.
[36] Zenios S. A. (ed) (1993) *Financial Optimization*. Cambridge University Press, Cambridge, UK.
[37] Zenios S. A. (1993) A model for portfolio management with mortgage-backed securities. *Ann. Oper. Res.* 43: 337–356.

Index

[1] Italic page numbers (e.g. *531*) indicate to literature.